Thin Films, Atomic Layer Deposition, and 3D Printing

Thin Films, Atomic Layer Deposition, and 3D Printing explains the concept of thin films, atomic layers deposition, and the Fourth Industrial Revolution (4IR) with an aim to illustrate existing resources and give a broader perspective of the involved processes as well as provide a selection of different types of 3D printing, materials used for 3D printing, emerging trends and applications, and current top-performing 3D printers using different technologies. It covers the concept of the 4IR and its role in current and future human endeavors for both experts/nonexperts. The book includes figures, diagrams, and their applications in real-life situations.

Features:

- Provides comprehensive material on conventional and emerging thin film, atomic layer, and additive technologies.
- Discusses the concept of Industry 4.0 in thin films technology.
- Details the preparation and properties of hybrid and scalable (ultra) thin materials for advanced applications.
- Explores detailed bibliometric analyses on pertinent applications.
- Interconnects atomic layer deposition and additive manufacturing.

This book is aimed at researchers and graduate students in mechanical, materials, and metallurgical engineering.

Thin Films, Atomic Layer Deposition, and 3D Printing

Demystifying the Concepts and Their Relevance in Industry 4.0

Kingsley Ukoba and Tien-Chien Jen

CRC Press
Taylor & Francis Group
Boca Raton London New York

CRC Press is an imprint of the
Taylor & Francis Group, an **informa** business

First edition published 2024
by CRC Press
2385 NW Executive Center Drive, Suite 320, Boca Raton FL 33431

and by CRC Press
4 Park Square, Milton Park, Abingdon, Oxon, OX14 4RN

CRC Press is an imprint of Taylor & Francis Group, LLC

ISBN: 9781032416953 (hbk)
ISBN: 9781032428239 (pbk)
ISBN: 9781003364481 (ebk)

DOI: 10.1201/9781003364481

Typeset in Times
by Newgen Publishing UK

Contents

About Authors

Kingsley Ukoba is a lecturer and researcher in the Department of Mechanical Engineering Science of the University of Johannesburg, South Africa. He obtained a doctoral degree in Mechanical Engineering from the University of KwaZulu-Natal, Durban, in South Africa, graduating among the top 15 researchers. He coordinates the smart energy group for JENANO, headed by Professor Jen. He is among the authors for the African Integrated Assessment report by the United Nations Environment Programme, African Union Commission, CCAC, and Stockholm Environment Institute (SEI) joint publication. He is also a 2022 Engineering for Change (E4C) fellow sponsored by American Society of Mechanical Engineers (ASME) to support an Impact Project around Climate Action. He has authored book chapters and journal articles, has presented at conferences, and serves as a reviewer for high-impact journals and conferences.

Tien-Chien Jen is the head of Department of Mechanical Engineering Science at the University of Johannesburg, South Africa. He is the SARChI Chair for Green Hydrogen in South Africa. A director of the Atomic Layer Deposition Research Centre of the University of Johannesburg, South Africa. Prior to that, he was a faculty member at the University of Wisconsin, Milwaukee, United States of America. Prof Jen received his Ph.D. in Mechanical and Aerospace Engineering from the University of California, Los Angeles (UCLA), specializing in the thermal aspects of grinding. He is currently leading the drive for the application of atomic layer deposition equipment (first of its kind in Africa) for various applications (hydrogen, renewable energy, thin films, etc.). He is a fellow of American Society of Mechanical Engineers (ASME) and member of Academy of Science of South Africa (ASSAf), among others. Prof. Jen has written over 360 peer-reviewed articles, including 180 peer-reviewed journal papers, 16 book chapters, and 5 books to date.

1 Introduction

1.1 INTRODUCTION

The book aims to demystify the concept of thin films, atomic layer deposition, and the fourth industrial revolution (4IR). The concept of thin films was extensively discussed, emphasizing the classification, properties, and characterization. A detailed discussion of physical, chemical vapor, and solution-based thin films will be done. The application of thin films in emerging technologies and global challenges will also be discussed. A chapter discusses the concept of thin film and its application to corrosion. A chapter is devoted to atomic layer deposition. This is to demystify the deposition techniques. Atomic layer deposition is a top-notch deposition technique capable of depositing high-quality thin films for various applications. This will attract more readers who are struggling to understand the technique. A holistic discussion of the concept of atomic layer deposition will be made. This chapter is closely followed by a discussion on additive manufacturing using 3D printing. This technology is gaining interest due to the prospect and versatility of its application in solving various global human challenges. The chapter discusses the trend, classification, applications, and current top manufacturers of 3D printing. It also discusses 4IR and emerging technologies such as energy conversion and storage, biomedical, smart membrane, machine learning, and corrosion.

The book also discusses 3D printing technology from the basics to complex fundamentals to demystify the technology. The book provides a selection of different types of 3D printing, materials used for 3D printing, emerging trends and applications, and current top-performing 3D printers using different printing technology. The book is easy to comprehend, interesting, and informative to readers using appropriate figures, photographs, and diagrams and applying them in real-life situations.

In comparison to most existing publications on specific techniques or materials, the book will appeal to a broader audience. As a result, the book fills in gaps in existing resources and gives a broader spectrum of thin-film deposition information. The book is user-friendly and will be useful to academics, students, and nonprofessionals seeking to understand thin films, 3D printing, and the fourth industrial revolution. It can also guide product design, 3D printing, and process development.

The book describes the concept of thin film technology, atomic layer deposition, additive manufacturing, and emerging areas in the context of 4IR. It also discusses

DOI: 10.1201/9781003364481-1

the applications of thin film coatings in various current industries. The book explains in-depth the technology of 3D printing and its application in old and emerging fields. The following are some of the book's unique features:

i.   It is not limited to a single material or deposition method. As a result, it covers a wide range of topics for a larger audience.
ii. It highlights how thin film deposition and associated technology are used in solving global human challenges.
iii. It discusses in-depth 3D printing and application in solving global human challenges of housing, transport, and space exploration, among others.
iv. The book explains to both experts and nonexperts the concept of the fourth industrial revolution and its role in current and future human endeavors.

1.2   AIM

The aim of this book is to present detailed and user-friendly information that will be useful to academics, students, and nonprofessionals seeking to understand thin films, atomic layer deposition, 3D printing, and the fourth industrial revolution.

1.3   SCOPE

To achieve the aim of the project, the book is composed of 10 chapters, with the first four chapters focusing on theory of film formation and growth, whereas the other chapters present various applications of thin film coatings in modern society.

This book covers the following:

• Introduction to thin films and coating materials
• Methods of characterization and properties of thin films
• Applications of thin film coatings in modern society
• Emerging trends in thin film technologies and applications

1.4   CONCLUSION

This chapter introduced the book entitled *Thin Films, Atomic Layer Deposition, and 3D Printing: Demystifying the Concepts and Their Relevance in Industry 4.0*. The book discusses in detail the concept of thin films, atomic layer deposition, and 3D printing in an easy-to-understand manner. The ten-chapter monogram book is written to enlighten all classes of readers interested in the topics.

2 Demystifying Concept of Thin Films

Properties, Application, and Challenges in Era of 4IR

2.1 BACKGROUND OF THIN FILM

A thin film is a material layer with a thickness ranging from fractions of a nanometer (monolayer) to a couple of micrometers as shown in Figure 2.1 (Seshan, 2012). Thin film falls within 500 nm ultraviolet light wavelength. A thin film is a layer that can stretch indefinitely in two directions but not in the third. It is layers of material put on a bulk substrate to impart qualities that the base material cannot easily (or at all) achieve. Controlled synthesis of materials as thin films (a process known as deposition) is critical in many applications. There is no doubt that "thin" films can be thicker than "thick" films. For a certain application, the thickness of thin-film material might range from a few nanometers to many micrometers. The thickness could be as little as a couple of atoms on a "substrate" surface or another layer previously deposited. Multilayer comprises thin films aggregated and arranged in layers. Thin films are vital in the development and study of materials with new and distinctive properties, in addition to their application appeal. Multiferroic materials and superlattices, for example, enable the investigation of quantum processes. A thin film can be achieved from a material using two main methods: subtractive, or the Etch Back technique, and additive, or the Lift-Off process. Subtractive, also known as the Etch Back technique, entails covering the entire surface, followed by selective removal of parts to make the desired pattern.

The specific property of a thin film is the result of its unique manufacturing process, which involves the incremental addition of atoms or molecules (Myny, 2018). The essential feature of a thin film is thickness, which is inextricably related to other qualities that scale differently with thickness. As a result, thin films are not defined solely by their thickness. Thin films have a variety of qualities depending on their thickness (Kaiser, 2002).

Thin-film layers serve various functions in various applications, including optoelectronics, display, semiconductor, water desalination, and medical (Elsheikh et al., 2019). Thin films are used in optical coatings, tribological coatings, quantum well structures based on higher lattices, magnetic multilayers, and nanoscale coatings. Reflective coatings, antireflective (AR) coatings, solar cells, monitors, waveguides, and optical detector arrays are all made with optical thin films (Al-Assadi and Al-Assadi, 2021). Electrical or electronic thin films are used in the fabrication of

DOI: 10.1201/9781003364481-2

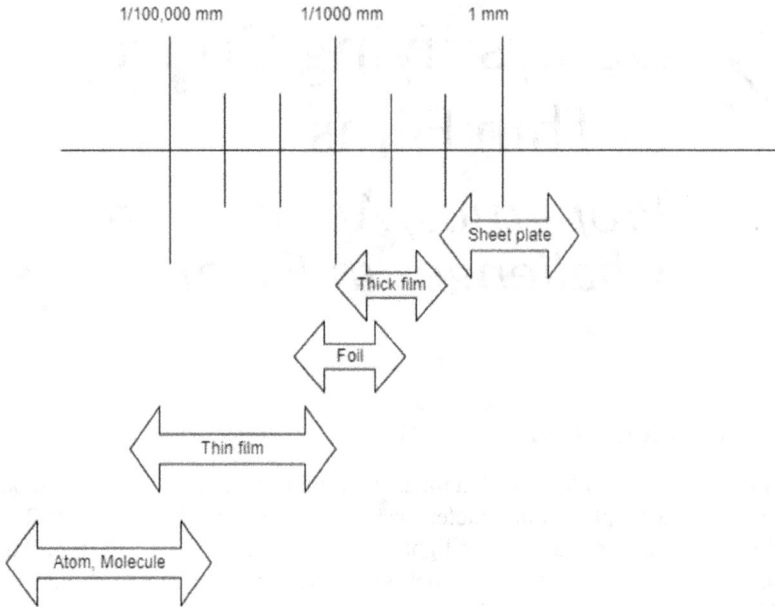

FIGURE 2.1   A schematic of dimensions of different types of films and sheets.

insulators, conductors, semiconductor devices, integrated circuits (ICs), and piezo-electric motors.

The history of thin film dates back to 5000 years ago in the application in optics by the Egyptians. Gold, an inorganic thin film, layers used in decorative and subsequent optical uses by the Egyptians during the middle Bronze Age over 5000 years ago were the oldest reported purposely manufactured inorganic thin films. During the 20th century, advances in thin-film deposition resulted in some amazing technological breakthroughs, such as magnetic recording media (floppy disks and magnetic strips on credit cards), electronic semiconductor devices, LEDs, optical coatings such as AR coatings, thin-film solar cells, and thin-film batteries. Their history, on the other hand, extends back thousands of years. Using mercury-based compositionally graded interfacial adhesion layers, the films were gilded on copper and bronze statues, jewelry, and religious objects. The first thin films formed from vapor phase were most likely metal layers that were unintentionally deposited on ceramic pots and rocks near hot charcoal fires used to refine metal ores. Johann Schroeder discovered a method for reducing arsenic oxide with charcoal in 1649, which was the first recorded planned development of metal films from the vapor phase.

2.1.1   PURPOSE AND APPLICATIONS OF THIN FILMS

Although the study of thin film phenomena dates back well over a century, it is really only over the last four decades that they have been used to a significant extent in practical situations. The requirement of microminiaturization made the use of thin and thick films virtually imperative. The development of computer technology led to a requirement for very high-density storage techniques, and it is this which has stimulated most of the research on the magnetic properties of thin films. Many thin film devices have been developed which have found themselves looking for an application or, perhaps more importantly, market. In general, these devices have resulted from research into the physical properties of thin films.

Secondly, as well as generating ideas for new devices, fundamental research has led to a dramatic improvement in understanding of thin films and surfaces. This in turn has resulted in a greater ability to fabricate devices with predictable, controllable, and reproducible properties. The cleanliness and nature of the substrate, the deposition conditions, postdeposition heat treatment, and passivation are vital process variables in thin film fabrication. Therefore, prior to this improvement in our understanding of thin films, it has not really been possible to apply them to real devices.

Thirdly, much of the finance for early thin film research originated from space and defense programs to which the device cost is less important than its lightweight and other advantages; the major applications of thin film technology are not now exclusively in these areas but rather often lie in the domestic sector in which low cost is essential (Ye et al., 2021; Yoshitake et al., 2003) [1,2].

Thin film materials have already been used in semiconductor devices, wireless communications, telecommunications, ICs, rectifiers, transistors, solar cells, light-emitting diodes, photoconductors, light crystal displays, magnetooptic memories, audio and video systems, compact discs, electrooptic coatings, memories, multilayer capacitors, flat-panel displays, smart windows, computer chips, magnetooptic discs, microelectromechanical systems (MEMS), and multifunctional emerging coatings, as well as other emerging cutting technologies such as:

2.1.1.1 Optical Coatings

A lens, mirror, or other optical component can have one or more thin layers of material applied to it to change how the optic transmits and reflects light. Antireflection coatings are one form of optical coating that is frequently used on camera and spectacle lenses to decrease undesired surface reflections. The high-reflector coating is another type that can be utilized to create mirrors that reflect more than 99.99% of the light that hits them. Dichroic thin film optical filters can be created using more complicated optical coatings that have high reflection over some wavelength ranges and antireflection over others.

2.1.1.2 Solar Cell

Nearly all of the commercial solar cells now on the market are based on thin-film solar cells. Thin film solar cells make use of semiconductor thin films that are often put onto a supporting substrate in an effort to lower the pricey silicon wafers. Due to the material's tremendous absorption, even though the active layers are only a few microns thick, they may nevertheless absorb a sizable amount of incident solar light (Bonnet, 2001).

Different processes can be used to deposit various layers (contact, buffer, absorber, antireflection, etc.) on a range of substrates (flexible or rigid, metal or nonmetal) [physical vapor deposition (PVD), chemical vapor deposition (CVD), spray pyrolysis, etc.]. Such adaptability enables layer engineering and customization to enhance device performance. The ability to further evolve the tandem-structure strategy into a more complex form known as the integrated tandem solar cell (ITSC) system is another benefit of thin film solar cells.

2.1.1.3   Semiconductor

The semiconductor industry has traditionally grown and etched the thin sheets of material that comprise electrical circuits for computers and other electronic devices on flat, two-dimensional chips. This thin coating, which is only a few hundred nanometers thick, may be applied to flexible materials like glass or plastic, creating a wide range of potential applications for flexible electronics. The semiconductor film can also be flipped while it is transferred to the new substrate, allowing for the inclusion of more components on the opposite side. This doubles the number of gadgets that could be used in the movie. Double-sided thin film semiconductor layers can be stacked on top of one another repeatedly to produce strong, low-power, three-dimensional electronic devices.

> It's crucial to remember that these are strained silicon or silicon germanium single-crystal films. We create the membrane in a way that introduces the strain. Strain causes a change in the crystal's atomic structure that allows for substantially quicker device speed while using less energy.

Flexible electronics are starting to have a big impact on noncomputer applications. The innovation could have a positive impact on active-matrix flat panel displays (FPDs), smart cards, RFID tags, medicinal applications, and solar cells. Using these approaches, flexible semiconductors might be incorporated into textiles to produce wearable electronics or computer monitors that can be raised like a window shade. This could lead to a paradigm change. Fast, low-power, multilayer electronics have a wide range of fascinating applications. Particularly intriguing are membranes made of silicon germanium. Compared to silicon, germanium has a far better capacity to absorb light. We can create devices with two to three orders of magnitude higher sensitivity by adding germanium without degrading the quality of the material. This improved sensitivity could be used to develop better low-light cameras or smaller, higher-resolution cameras.

2.1.1.4   Photoelectrochemical Cells (PEC)

In investigations involving photoelectrochemistry, an electrode is exposed to light that is absorbed by the electrode material, producing a current (a photocurrent).

The relationship between the photocurrent and wavelength, electrode voltage, and solution composition reveals details on the nature, energetics, and kinetics of the photoprocess. Photolytic reactions taking place in the solution close to the electrode surface can also create photocurrents at electrodes. To better understand the nature of the electrode-solution interface, photoelectrochemical investigations are commonly conducted. Chemiluminescence photocurrent can be thought of as the transformation of light energy into electrical and chemical energy; these processes are being looked into for possible practical uses. Since semiconductor electrodes are where the majority of the investigated photoelectrochemical processes take place, a brief overview of semiconductors and their interfaces with solutions is performed here. Gaining a microscopic understanding of electron-transfer mechanisms at solid-solution interfaces also benefits from taking into account semiconductor electrodes.

2.1.1.5 Optoelectronic

An optoelectronic thin-film chip that has a lens located downstream of the radiation-emitting region and at least one radiation-emitting region in the active zone of the thin-film layer. The lens is made of at least one partial region of the thin-film layer, and its lateral extent is greater than that of the radiation-emitting region. A layer sequence that is epitaxially produced on a growth substrate and from which the growth substrate is at least partially removed, for instance, can yield the thin-film layer. In other words, the substrate's thickness is decreased. The substrate is therefore thinned. Additionally, the complete growing substrate may be taken out of the thin-film layer. At least one active zone that can produce electromagnetic radiation exists in the thin-film layer. A layer or layer sequence with a p-n junction, a double heterostructure, a single quantum well structure, or a multiple quantum well structure, for instance, may supply the active zone.

The active zone should ideally have at least one radiation-emitting area. In this instance, a portion of the active zone can serve as the radiation-emitting region. As the optoelectronic thin-film chip operates, electromagnetic radiation is produced in the partial region of the active zone.

2.1.1.6 Flat Panel Displays

Entegris offers a wide range of liquid and gas contamination control solutions for the production of FPDs. These technologies were developed from the Mykrolis contamination control technologies. The world's most competitive and technologically advanced fabrication environment is that of the FPD industry. To meet the global consumer's hunger for larger displays, higher pixel resolution, and feature-rich performance at a lower price than the previous generation of technology, device designers and manufacturers continuously try to improve their products. Process engineers and designers are now placing a high priority on the requirement to control contamination in process streams that are made up of air, gas, and liquid. Entegris offers the solutions needed to prosper in these harsh circumstances.

2.1.1.7 Data Storage

The superparamagnetic effect complicates magnetic data storage media as the data storage density in cutting-edge microelectronic devices keeps rising. The application of thermomechanical data storage technology is one method for overcoming this challenge. In this method, data is written as an indentation on a surface by a nanoscale mechanical probe, read by a transducer integrated into the probe, and then deleted by the application of heat. The IBM millipede, which uses a polymer thin film as the data store medium, is an illustration of such a device. However, other types of media can also be used for thermomechanical data storage, and in the work that follows, we investigate the use of thin films of Ni-Ti shape memory alloy (SMA) as one such option.

Previous research has demonstrated that heating significantly recovers nanometer-scale indentations created in Ni-Ti SMA thin films in the martensite phase. The applicability of this approach is impacted by factors like indent closeness, repetitive thermomechanical cycling of indentations, and film thickness. SMA thin films are a suitable medium for thermomechanical data storage, yet there are still issues that need to be resolved, according to experimental evidence and theoretical expectations.

2.1.1.8 Super Capacitor

The concept of collecting charge in the electric dual layer that forms at the interface between a solid and an electrolyte has been patented since the 1950s, and after extensive development by the Standard Oil Company in Cleveland, Ohio (SOHIO), the company was forced to license the technology to Nippon Electric Company in the 1960s due to a lack of demand. The Pinnacle Research Institute created the first high-power double-layer capacitor for military use in 1982.

The capacitance of a traditional capacitor is $C = \epsilon A/d$, where ϵ is the dielectric constant, A is its surface area, and d is its thickness. A conventional capacitor stores energy electrostatically on two electrodes separated by a dielectric. Charges build up at the electrode/electrolyte border in a double-layer capacitor to generate two charge layers with a few Angstroms of spacing between them. Super capacitors are devices that store energy electrochemically. The electrical double-layer capacitance produced by pure electrostatic charge accumulation at the electrode interface and the pseudocapacitance created by a quick and reversible surface redox process at characteristic potential are the two types of capacitive behavior upon which super capacitors are based. In a hybrid car, a super capacitor is integrated as a power source, a power control, a multisource supply, a fuel-saving device, etc.

2.1.1.9 Gas Sensors

The creation of highly accurate and dependable sensors has been greatly aided by developments in microtechnology and the emergence of novel nonmaterial technologies. The scientific community has recently focused its efforts on macromolecules as the technology of sensors has developed significantly in recent years, especially for remote monitoring because of the growing need for environmental safety and providing new challenges and opportunities to the search for ever smaller devices capable of molecular level imaging and monitoring of pathological samples.

Gas sensors typically operate according to diverse principles, and over the years, a variety of gas-detecting components have been created, of which resistive metal oxide sensors make up a sizable portion. To achieve their best performance, these sensing elements must normally operate at a high temperature. As a result, there is increased power usage that is inappropriate for the inflation. The decaying carbon nanotube has received a lot of interest recently, and significant efforts have been made to take advantage of its peculiar electrical and mechanical features. Due to these characteristics, they are promising candidates for the active material building blocks of gas sensors, field emission devices, gas storage, and nanoelectronics. For many applications, the gas detection ability at ambient temperature is one of these.

2.2 THIN FILM TYPES

Thin films are used to produce properties that may be difficult to obtain in bulk materials. The invention of thin films has aided the advancement of many industries in the previous century. Semiconductors, magnetic recording media, ICs, LEDs, optical coatings (such as AR coatings), hard coatings for tool protection, pharmaceuticals, medicine, and a variety of other industries are among these.

They are used in a variety of sectors, and they also play an important part in the research and production of materials with unique and exceptional features, such as superlattices, which allow researchers to explore quantum processes. Furthermore, they are critical because they distinguish the properties and responses of the material surface from the bulk, and they have a diverse set of qualities that can be applied to a broad variety of applications. There are several different types of thin films, mainly classified based on the property of the end product, viz.:

i. Electrical or electronic thin films: Insulators, conductors, semiconductor devices, ICs, and piezoelectric motors are all made with this material.
ii. Optical thin films: Thin films with optical qualities are known as optical thin films. Backlight modules and polarizing films are the two most popular forms of optical films. OLED (organic light-emitting diode) panels and TFT (thin film transistor) liquid crystal panels (LCD) are two of the most common uses for these films. Reflective and antireflective coatings, solar cells, monitors, waveguides, and optical detector arrays are all made with this material. The Lawes' parotia's breast feathers are a natural example of optical thin films. Thin-film optics is responsible for the thin-film interference seen on many insect wings, and Ranunculus buttercups have lustrous flowers.
iii. Chemical thin films: These are utilized to build gas and liquid sensors, as well as to create resistance to alloying, diffusion, corrosion, and oxidation.
iv. Mechanical thin films: Tribological coatings for abrasion resistance, increased hardness and adhesion, and micromechanical characteristics.
v. Thermal thin films: These are utilized to make heat sinks and insulation coatings.
vi. Magnetic thin films: It is a 0.01–10 micron thick polycrystalline or monocrystal-line layer of a ferromagnetic metal, alloy, or magnetic oxide (such as ferrite). In computer technology, such films are utilized as memory elements, and in physics research, they are used as indicators. Sputtering is a process for depositing metal and oxide films by regulating the magnetic thin films' crystalline structure and surface roughness (Jilani et al., 2017).

2.3 THIN FILM DEPOSITION CLASSIFICATION

Thin film deposition has been classified using different approaches. It can be classified based on the state of the deposition process as vacuum based and solution based (Vyas, 2020) as shown in Figure 2.2a.

The fluid precursor undergoes a chemical change at a solid surface during vacuum deposition, resulting in a solid layer. Physical deposition, on the other hand, creates a thin solid film using mechanical, electromechanical, or thermodynamic processes. Also, it can further be classified based on the nature of deposition as solution based,

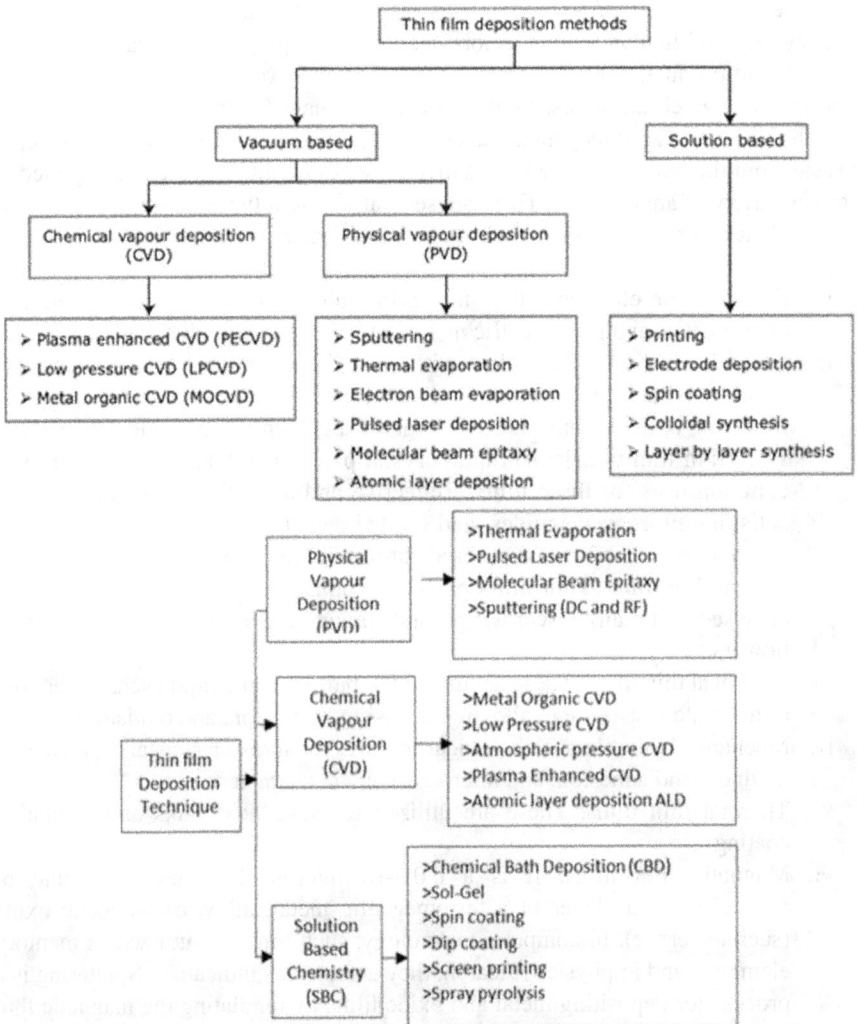

FIGURE 2.2 Classification of thin film deposition methods (a) based on state of deposition and (b) based on nature of deposition.

physical vapor, and chemical vapor as shown in Figure 2.2b (Ukoba et al., 2018a). Chemical deposition methods include gas-phase and solution deposition. CVD (Patra and Das, 2022), atomic layer epitaxy (Konh et al., 2022), and atomic layer deposition (ALD) (Oviroh et al., 2019) are examples of gas-phase techniques. Spray pyrolysis (Ukoba et al., 2018b), sol–gel (Lukong et al., 2022), spin (Shafi et al., 2022), and dip-coating (Zaremba et al., 2022) are examples of solution deposition processes. Pulsed laser deposition (Albu et al., 2022), PVD (Dan et al., 2022), molecular beam epitaxy (Bailey et al., 2022), and magnetron sputtering (Ren et al., 2022) are examples of physical processes. Chemical bath deposition (CBD) (Fazal et al., 2022), advanced reactive gas deposition (Fan et al., 2022), electron beam evaporation (Lorenz et al., 2022), vacuum evaporation (Mahana et al., 2022), and anodic oxidation (El-Shamy et al., 2022) are some of the other approaches.

Deposition Method: Some of the deposition methods are summarized in Table 2.1.

The schematic summary of selected deposition technique of thin films is shown in Figure 2.3.

2.3.1   SOLUTION-BASED DEPOSITION

The method is described as the solution-based deposition of an aqueous-based liquid phase onto the surface of a substrate. In most cases, the necessary material is dissolved in solutions and directly coated on the substrate surface, followed by the evaporation of the sedimentary (solvent) wet coating to obtain a dry layer (Chaudhary, 2021). The process allows for the deposit of thin films with great diversity and ease on a wide range of materials. The application of a liquid precursor to a substrate, which is then transformed into the appropriate coating material in a subsequent posttreatment step, is known as liquid film deposition. To create thin films with great efficiency and functionality, various nonvacuum solution-based deposition processes have been developed. This includes Spin coating, spray pyrolysis, CBD, dip coating, spray coating, and inkjet printing, among others. Because of its low cost, homogeneity, safety, and flexibility to scale up, spin coating is one of the most effective techniques for thin film production. A typical procedure includes putting a little amount of fluid in the center of a substrate and spinning it at a high speed. Another simple, cost-effective method that can be scaled up for commercial production is dip coating. Immersion, withdrawal, and evaporation are the three major technical stages of the dip coating process. Further heat treatment of the coating may be required to burn away remaining chemicals and induce crystallization of the functional oxides. Spray coating is a potential approach for preparing thin and thick films in research and industry. It is a simple method for producing thin films with uniform distribution on a tiny scale, ranging from a few nanometers to micrometers in thickness. Inkjet printing is a promising new technique for producing large-scale, flexible thin films. The inkjet printing technology makes it simple to customize a wide range of complex structures.

i) ***Chemical Bath Deposition***: It is also known as hemical solution deposition (Hodes, 2007). It is a thin-film deposition process that uses an aqueous precursor solution to produce solids from a solution or gas (Guire et al., 2013). CBD generates homogenous

TABLE 2.1
A Summary of Deposition Method

Deposition type	Description	Merit	Parameters	Demerit
Solution Based				
Chemical Bath Deposition	It is a thin-film deposition process that uses an aqueous precursor solution to produce solids from a solution or gas. Chemical Bath Deposition generates homogenous thin films of metal chalcogenides (primarily oxides, sulfides, and selenides).	CBD is generally cheap, convenient for large-area deposition, and has the capacity to tune thin film properties by modifying and controlling it.	Temperature of solution, stirring time, precursor material concentration, and substrate type.	There is waste of solution, and the substrate should be clean.
Dip-coating	It is the process of slowly dipping a substrate into a solution. After the liquid component of this solution has dried, the substrate is left with a solid thin coating.	It is affordable, and thickness of deposited layer can be finetuned easily.	Viscous drag, the forces of inertia, surface tension, and gravitational force.	The process is slow and screen blocking results in lower efficiency.
Electrodeposition	The method of controlled deposition of material on conducting surfaces using electric current from a solution containing ionic species is known as electrodeposition or electroplating. It has been widely utilized to create thin films with no binder. Electrodeposition can be divided into three types: electroplating, electrophoretic deposition, and underpotential deposition.	The capacity to wrap items with very thin metal layers to enhance their appearance.	Content, current density, and temperature. Step by step, the electrodeposition rate increases as the current density rises.	Lack of thick shell.

Screen Printing	Screen printing (also known as serigraphy and serigraph printing) is a printing process in which ink (or dye) is transferred onto a substrate using a mesh, except for areas made impermeable to the ink by a blocking stencil. A blade or squeegee is dragged across the screen to fill the open mesh apertures with ink, and then a reverse stroke causes the screen to touch the substrate along a line of contact for a moment. As the screen springs back after the blade has passed, the ink wets the substrate and is drawn out through the mesh holes. Because each color is printed separately, multiple screens can be utilized to create a multicolored image or design.	Process is simple, costs are inexpensive, and the deposition area is large.	The paste's viscosity, the amount of mesh screens, the snap-off distance between the screen and the substrate, and the pressure.	Low output rate, poor ink mileage, and lengthy drying time are all issues that must be addressed.
Sol–gel	The sol–gel method is a wet-chemical technique for making solid materials out of tiny molecules. The method entails converting monomers into a colloidal solution (sol), which serves as a precursor for forming an integrated network (or gel) of discrete particles or network polymers.	Simplest, most homogeneous, least expensive, most reliable, most reproducible.	Amount of water, temperature, pH, solvent, and precursor are all factors to consider.	Precursor costs are high, the process takes a long time, monolith synthesis is difficult, and the process chemistry is difficult to regulate and repeat.
Spray Pyrolysis	Spray pyrolysis is a method for coating large areas with films of very thin layers of uniform thickness. Spray coating is a process that involves forcing printing ink through a nozzle, resulting in a fine aerosol.	Cost-effectiveness, versatility, thickness range, and processing speed.		Nonuniformity of deposited films. Extra resources and cost are incurred to improve the deposited films.
Physical Vapor Deposition				

(continued)

TABLE 2.1 (Continued)
A Summary of Deposition Method

Deposition type	Description	Merit	Parameters	Demerit
Pulsed laser Deposition	A high-power pulsed laser beam is focused to impact a target of the desired composition in this physical vapor deposition technique. The material is then vaporized and deposited on a substrate facing the target as a thin film.	The main benefit is that it is conceptually simple and versatile.	Maximum pulse energy, wavelength.	Slow deposition rate, does not support large area deposition, target fragmentation in depositing some materials.
Molecular Beam Epitaxy	MBE is used to create thin films of diverse materials layer by layer.	Clean technique for obtaining films.		The ultra-high vacuum (UHV) environment is demanding.
Sputtering	Sputter deposition is a technique for depositing thin films that involves sputtering material from a "target" onto a "substrate." Sputtering is a physical process in which atoms in a solid-state (target) are bombarded by energetic ions, causing them to be freed and flow into the gas phase (mainly noble gas ions).	Deposit a wide range of metal and metal oxide nanoparticles (NPs) and nanoclusters (NCs), as well as insulators, alloys, composites, and organic molecules.		Sputtered films have an increased risk of contamination.
Thermal Evaporation	It's one of the most basic types of PVD, and it involves using a resistive heat source to evaporate a solid substance in a vacuum to make a thin film. In a high vacuum chamber, the material is heated until vapor pressure is achieved.	Good purity because of low pressure.	The main parameter is source material's vapor pressure at evaporation temperature.	First, pollution from the boat's outgassing or evaporation, as well as the heating circuits used during the high-temperature evaporation process. Second, obtaining a suitable boat material is difficult, especially when evaporating refractory materials. Also, poor step coverage, harder forming of alloys, and decreased throughput due to low vacuum.

| Sputtering | Sputtering is a process in which small particles of a solid material are ejected off its surface after being hit by intense plasma or gas particles. It can be an undesirable source of wear in precision components because it occurs naturally in space. It is, nevertheless, utilized in research and industry to execute accurate etching, carry out analytical methods, and deposit thin film layers in the creation of optical coatings, semiconductor devices, and nanotechnology products because of its ability to work on extremely small layers of material. It's a type of physical vapor deposition. | Uniform covering across a large area with little contaminants. | Operating pressure, direct current power, and substrate temperature are all factors to consider. | The bombardment's target region is too tiny, and the rate of deposition is often poor. |

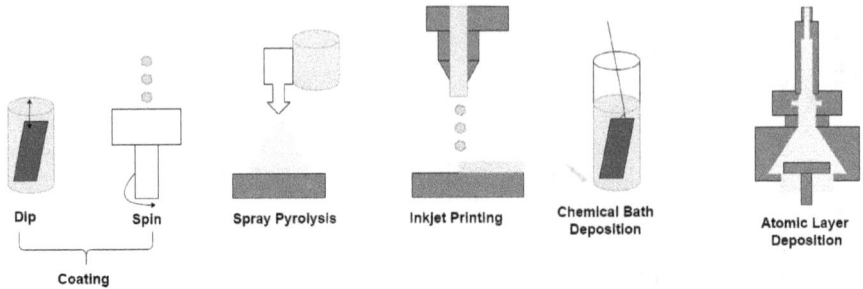

FIGURE 2.3 Schematic summary of various deposition methods.

thin films of metal chalcogenides (primarily oxides, sulfides, and selenides) and several less common ionic compounds utilizing heterogeneous nucleation (deposition or adsorption of water ions onto a solid substrate) (Nair et al., 1998). CBD dependably generates films with a simple technique that requires little equipment, at a low temperature (100°C), and at a reasonable cost (Cheng et al., 2006, Zou et al., 2015). CBD can also be used for batch processing or continuous deposition over a vast area (McPeak et al., 2011). CBD films are commonly employed in semiconductors, solar cells, and super capacitors, and there is growing interest in employing CBD to make nanomaterials. Figure 2.4 shows the schematic of the CBD method.

History of Chemical Bath Deposition (CBD): Justus Liebig wrote an article in 1865 explaining the use of CBD to make silver mirrors (to attach a reflecting layer of silver to the back of glass to produce a mirror) (Liebig, 1856). However, electroplating and vacuum deposition are more prevalent nowadays. Infrared detectors are thought to have employed lead sulfide (PbS) and lead selenide (PbSe) CBD films around World War II (WWII) (Hodes, 2007). When created by CBD, these films are photoconductive (Hodes, 2007). CBD has also been used to produce thin coatings in semiconductors for a long time. CBD is rarely utilized to create semiconductors nowadays due to the small size of deposited crystals, which is not optimal for semiconductors.

Process of Chemical Bath Deposition (CBD): Deposition is achieved by producing a solution in which deposition (the change from an aqueous to a solid substance) occurs only on the substrate: Metal salts and (typically) chalcogenide precursors are mixed with water to make an aqueous solution that contains the metal ions and chalcogenide ions that will create the deposited product. Temperature, pH, and salt concentration are varied until the solution reaches metastable supersaturation, which occurs when the ions are ready to deposit but cannot overcome the thermodynamic barrier to nucleation (forming solid crystals and precipitating out of the solution). One of the two procedures discussed below involves introducing a substrate that acts as a catalyst for nucleation, and the precursor ions cling to the substrate, forming a thin crystalline film. That is, the solution is in a state where the precursor ions or colloidal particles are "sticky," but not to each other. The precursor ions or particles cling to the substrate, and aqueous ions stick to solid ions, generating a solid compound that deposits to create

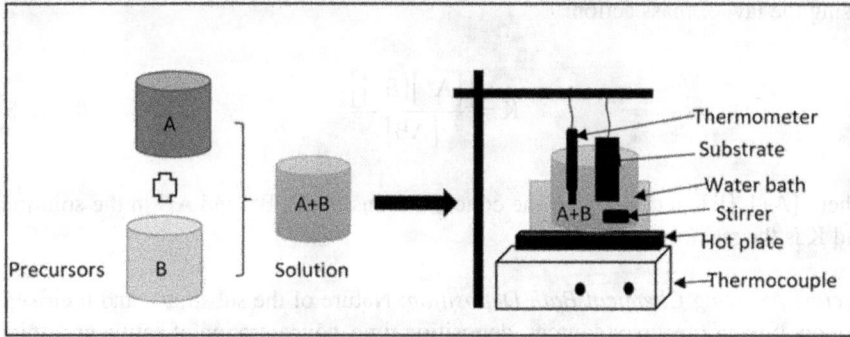

FIGURE 2.4 Schematic diagram of the chemical bath deposition.

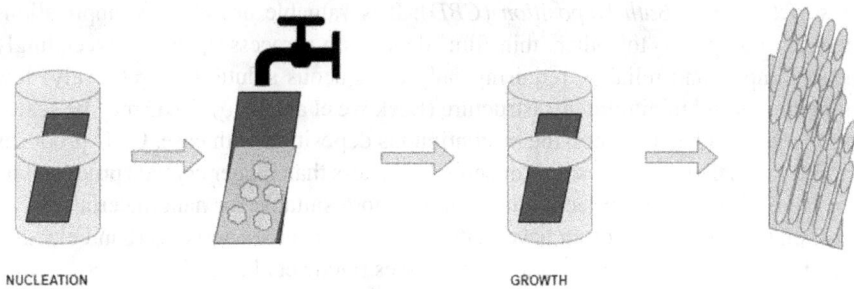

NUCLEATION   GROWTH

FIGURE 2.5 Stage of chemical bath deposition.

crystalline layers. Crystal size is affected by the pH, temperature, and content of the film, and these factors can be utilized to influence the pace of formation and structure of the film (Osherov et al., 2007). Figure 2.5 shows the stages of the CBD process.

CBD Operating Principle: The CBD method follows the same basic principles as all precipitation reactions, and it is based on the product's relative solubility (Lokhande et al., 2009, Moreno-Regino et al., 2019). Precipitation happens when the ionic product (IP) of reactants is more than the solubility product (K_{SP}) at a particular temperature. If the IP is less than the solubility product, the solid phase formed dissolves back into the solution, leaving no net precipitation (Aoba, 2004). The solubility product is an important notion to grasp to comprehend CBD's processes (K_{SP}). The solubility product calculates the solubility of a sparingly soluble ionic salt (including salts that are commonly referred to as "insoluble"). When a sparingly soluble salt, AB(s), is dissolved in water, a saturated solution comprising A^+ and B^- ions in contact with undissolved solid AB is formed, and equilibrium between the solid phase and ions in the solution is established in equations 2.1 and 2.2:

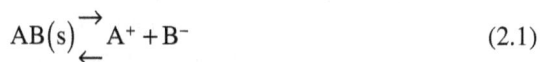

$$AB(s) \underset{\leftarrow}{\overset{\rightarrow}{}} A^+ + B^- \tag{2.1}$$

Using the law of mass action:

$$K = \frac{\left[\left[A^+\right]\left[B^-\right]\right]}{[AB]}$$

(2.2)

where [A+], [B], and [AB] are the concentrations of A^+, B^-, and AB in the solution, and K is the stability constant.

Factors Affecting Chemical Bath Deposition: Nature of the substrates and their separation, type of precursor sources, deposition time, concentration of cation and anion sources, bath temperature, effect of complexing agent, and chemical bath solution pH are parameters that affect the CBD method (Hone and Abza, 2019).

Uses of Chemical Bath Deposition (CBD): It is valuable in industrial applications because, compared to other thin-film deposition processes, it is exceedingly cheap, simple, and reliable, requiring only an aqueous solution at (relatively) low temperatures and minimum infrastructure (Ezekoye et al., 2013). CBD may be scaled up to large-area batch processing or continuous deposition with ease. CBD produces small crystals that are less useful for semiconductors than bigger crystals produced by other thin-film deposition techniques but are more suitable for nanomaterials. CBD films, on the other hand, often have better photovoltaic characteristics (band electron gap) than films made by conventional processes (Göde et al., 2007).

Applications of Chemical Bath Deposition (CBD): It has been used to deposit photovoltaics, optics, nanomaterials, and other process. Because many films have greater photovoltaic characteristics when deposited using CBD than when produced via other processes, photovoltaic cells are the most popular usage of films deposited by CBD (Abel et al., 2022). This is because CBD thin films have a higher size quantization, resulting in smaller crystals and a wider optical band gap than thin films made by other methods. Cadmium sulfide (CdS), a thin film prevalent in photovoltaic cells, is the substance most typically deposited by CBD and the substance most researched in CBD research articles because of its improved photovoltaic capabilities. Because CBD does not damage the substrate, CBD is also utilized to deposit buffer layers in solar cells. CBD films can be produced to absorb specific wavelengths while reflecting or transmitting others, depending on the application. This is because the electronic bandgap of films created using CBD may be accurately controlled. Antireflection and antidazzling coatings, solar thermal applications, optical filters, polarizers, total reflectors, and other applications can all benefit from this selective transmission (Shih, 2005). Antireflection, antidazzling, thermal control widow coatings, optical filters, complete reflectors, poultry protection and warming coatings, light emitting diodes, solar cell production, and varistors are all viable applications for CBD films. CBD has a lot of potential in the field of nanomaterials because the small crystal size allows for nanometer-scale formation, the properties and nanostructure of CBD films can be precisely controlled, and the uniform thickness, composition, and geometry of CBD films allows the film to retain the substrate's structure (Martínez-Benítez et al., 2022). CBD is unlike any other thin-film deposition technology because of its low cost and high dependability, even at the nanoscale scale. Polycrystalline and epitaxial films, porous networks, nanorods, superlattices, and composites can all be made using CBD (Switzer and Hodes, 2010).

ii) *Spin Coating*: A major deposition technique used for depositing thin films unto a substrate. It is a centrifugal force method for applying a uniform layer to a solid surface that requires a liquid–vapor interface.

History: Emslie et al. (1958) pioneered theoretical spin coating analysis, which has been expanded upon by other authors. Wilson et al. (2000) studied the rate of spreading in spin coating. Also, Danglad-Flores et al. (2018) discovered a method to predict the deposited film thickness.

Principle of Operation: A thin layer (a few nm to a few μm) is applied evenly across the surface of a substrate by coating a solution of the desired material in a solvent (an "ink") while the substrate is spinning. To put it another way, a liquid solution is deposited onto a spinning substrate to create a thin layer of solid material, such as a polymer. The centripetal force combined with the surface tension of the solution pushes the liquid coating into an even covering when the substrate is rotated at a high speed (typically >10 revolutions per second = 600 rpm). During this period, the solvent evaporates, leaving the desired substance in an even layer on the substrate. When a material and solvent solution is spun at high speeds, the centripetal force and the liquid's surface tension combine to form an equal covering. A little amount of coating material is usually put to the substrate's core, which is either rotating slowly or not at all. The coating substance is subsequently disseminated by centrifugal force while the substrate is rotated at speeds up to 10,000 rpm. A spin coater, or simply spinner, is a machine used for spin coating (Glaser, 2011). While the fluid spins off the borders of the substrate, the rotation continues until the appropriate thickness of the film is obtained. The used solvent is usually volatile and evaporates at the same time. Parameters such as the viscosity and concentration of the solution, as well as the solvent, influence the film thickness.

Deposition, spin-up, spin-off, and evaporation are the four basic phases in this process. Using a pipette, the solution is cast onto the substrate in the first stage. The centrifugal motion will disseminate the solution across the substrate, whether the substrate is spinning already (dynamic spin coating) or spinning after deposition (static spin coating).

The substrate is then rotated at the specified speed, either immediately or after a slower spreading stage. Most of the solution has been ejected from the substrate at this point. The fluid may spin at a different rate than the substrate at first, but as drag balances rotational accelerations, the rotation speeds will eventually match, resulting in the fluid being level. As the fluid is now dominated by viscous forces, it begins to thin. Due to interference effects, the film frequently changes color when the fluid is hurled off (see video below). This indicates that the film is mostly dry when the color stops changing. Because the fluid must form droplets near the edge to be thrown off, edge effects might occur.

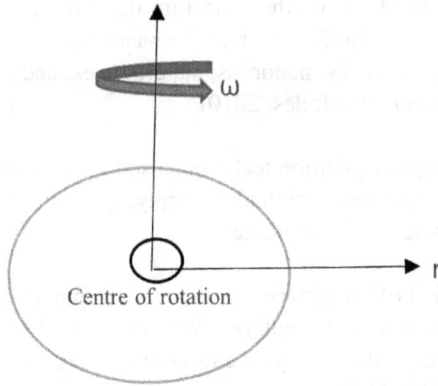

FIGURE 2.6 Schematic of the Emslie, Bonner, and Peck model showing the rotating disk approximation.

Finally, fluid outflow ceases, and thinning is dominated by solvent evaporation. The rate of solvent evaporation is determined by the volatility of the solvent, the vapor pressure, and the surrounding environment. Nonuniformities in the evaporation rate, such as at the substrate's edge, will result in nonuniformities in the film.

Spin Coating Thickness of a Spin Coater: In general, the thickness of a spin-coated film is proportional to the inverse of the spin speed squared, as shown in equation 2.3, where h_f is the final film thickness and $\sqrt{E}$ is the angular velocity/spin speed. The equation can also be used to calculate the spin curve.

$$h_f = \frac{1}{\sqrt{\omega}}$$

(2.3)

Attempts have been made to predict spin coating final film thickness without using experiments by different researchers. Equation 2.4 shows the Emslie, Bonner, and Peck model used on an endless rotating disk, a nonvolatile, viscous fluid. The model made various assumptions, including neglecting the effects of evaporation (which will vary depending on how volatile the solvent is) and ignoring the potential of non-Newtonian behavior.

$$\frac{\partial h}{\partial t} + \frac{\rho \omega^2 r}{\eta} h^2 \frac{\partial h}{\partial r} = -\frac{2\rho \omega^2 h^3}{3\eta}$$

(2.4)

where t is the process start time, r is the distance from the center of rotation, ρ is the density, η is the viscosity, and h is the fluid layer thickness (rather than the dry thin film). Angular velocity is ω, rate of change in thickness is represented by $\partial h/\partial t$, and the rate of spreading is represented by $\partial h/\partial r$.

Equation 2.5 is obtained if the film is uniform, and the start time t becomes zero and there is no evaporation and hence is not used to compute the final dry film thickness.

$$h = \frac{h_0}{\left(1 + \frac{4\rho\omega^2}{3\eta} h_0^2 t\right)^{\frac{1}{2}}} \tag{2.5}$$

The modification of the Emslie, Bonner, and Peck model to include evaporation rate of the solvent was done by Meyerhofer in 1978 and shown in equation 2.6.

$$\frac{dh}{dt} = -\frac{2\rho\omega^2 h^3}{3\eta} - E \tag{2.6}$$

where E stands for the uniform solvent evaporation rate, which is expressed in units of solvent volume evaporated per unit area per unit time.

Spin Speed of Spin Coater: The range of spin speeds available is critical since it determines the thickness range that can be obtained with a particular solution. From roughly 1000 rpm and up, spin coating can readily make uniform films, but with care and attention, good film quality may be achieved down to around 500 or 600 rpm in most cases (and even lower in some cases). Most spin coaters have a maximum speed of 6000–8000 revolutions per minute (although specialist coaters may go to 12,000 rpm or higher). As a result, a typical range of working spin coating rpms (say, 600–6000 rpm) might cover a factor of ten, resulting in a maximum variance in film thickness of roughly factor $\sqrt{(10)} = 3.2$.

This reliance on the square root of spin speed has both benefits and drawbacks. The downside is that it limits the range of thicknesses that can be attained with a particular solution (around a factor of 3–4). On the other hand, it has the advantage of allowing exact film thickness control within this range. The maximum thickness that a given material/solvent combination may create is also determined by the highest concentration of the material that can be dissolved in the solvent. Thicknesses of >1 m can be produced with high-solubility materials (100 mg/ml or greater). Meanwhile, the greatest thickness for some low solubility conjugated polymers (a few mg/ml) may be around 20 nm or so. The thickness of a film is approximately linearly influenced by the concentration of the substance in the ink at reduced concentrations, but as concentrations increase, this will influence the viscosity of the ink, resulting in a non-linear connection.

Duration of Spin Coating: Spin coating's goal is to maintain the substrate spinning until the film is completely dry. As a result, it is dependent not only on the solvent's boiling point and vapor pressure but also on the ambient conditions (temperature and humidity) in which the spin coating is performed. A spin coating time of 30 seconds is usually more than adequate for most solvents (such as water, ethanol, IPA, acetone,

```
┌─────────────────┐
│   Spin Coating  │
└─────────────────┘
          │
          ▼
┌─────────────────────┐                    ┌─────────────────┐
│  Dynamic Dispense   │                    │ Static Dispense │
└─────────────────────┘                    └─────────────────┘
```

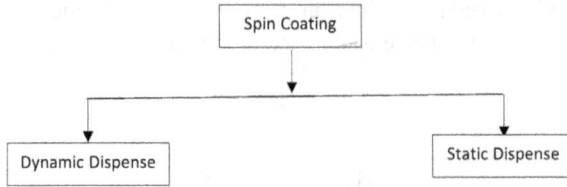

FIGURE 2.7 Schematic of types of spin coating based on dispense.

methanol, xylene, butanol, chloroform, toluene, and chlorobenzene) and is therefore recommended as a starting point for most procedures.

There are two types of spin coating processes based on the dispense process, namely dynamic and static as seen in Figure 2.7.

Dynamic dispensing is chosen because it is a more regulated process that results in less substrate fluctuation. Because there is less time for the solvent to evaporate before spinning begins, the ramp speed and dispensing time are less significant (so long as the substrate has been allowed time to reach the desired rpm). In general, a dynamic dispense consumes less ink, albeit this is dependent on the surface's wetting qualities. When employing low spin rates below 1000 rpm or particularly viscous solutions, the downside of a dynamic dispense is that it becomes progressively difficult to acquire complete substrate coverage. Because there isn't enough centripetal force to drag the liquid across the surface, and because the rotation speed is slower, there's a higher risk the ink may be dispensed before the substrate has completed a full rotation (at 600 rpm the substrate is rotating once every 0.1 seconds, commensurate with a fast pipette drop). As a result, it is recommended that a static dispense at 500 rpm or lower, with either technique working in the 500–1000 rpm range, should be used.

Application: Spin coating is frequently utilized in the microfabrication of functional oxide layers on glass or single crystal substrates using sol–gel precursors, where it may produce homogeneous thin films with nanoscale thicknesses (Hanaor et al., 2011). It's widely used in photolithography to deposit 1 micrometer-thick layers of photoresist. Photoresist is typically spun for 30–60 seconds at 20–80 revolutions per second. It's also commonly employed in the creation of polymer-based planar photonic structures.

Merit and Demerit: The consistency of the film thickness is one advantage of spin-coating thin films. Thicknesses do not change by more than 1% due to self-leveling. Spin coating thicker polymer and photoresist films, on the other hand, can result in quite large edge beads with physical constraints to planarization (Arscott, 2020). The simplicity and efficiency with which a process may be set up, together with the thin and homogeneous coating that can be achieved at varied thicknesses, makes spin coating excellent for both research and rapid prototyping. Second, the ability to have high spin speeds leads to quick drying durations (because of the high airflow), which leads to great consistency at both the macroscopic and nano length scales, and typically eliminates the need for postdeposition heat treatment. In addition, compared to other approaches, spin coating is a very low-cost way to batch process individual substrates, as many of them require both more expensive equipment and high-energy processes.

The main drawback of spin coating is that it is a batch (single substrate) process with a poor throughput when compared to roll-to-roll techniques such as Slot-Die coating. In addition, the actual material used in a spin coating process is often relatively low (about 10% or less), with the remainder hurled over to the side and wasted. This isn't normally a problem in research settings, but it's a waste in large-scale manufacturing. Fast drying times can cause some nanotechnologies (small molecule OFETs, for example) to operate poorly since they require time to self-assemble and/or crystallize. Despite these disadvantages, spin coating is frequently used as a starting point and standard for most academic and industrial operations that require a thin and homogenous coating.

2.3.2 Physical Vapor Deposition

PLD, sputtering (DC and RF), molecular beam epitaxy, thermal evaporation. They are here discussed in detail below.

2.3.2.1 Sputtering Techniques

Vapor species may be created by kinetic ejection from the surface of a material (called target or cathode) by bombardment with energetic and nonreactive ions. The ejection process, known as sputtering, takes place because of momentum transfer between the impinging ions and the atoms of the target surface. The sputtered atoms are condensed on a substrate to form a film. Since the number of sputtered atoms is proportional to the number of ions, the sputtering process provides a very simple and precise control of the rate of film deposition. A large number of sputtering variants have been developed over the years. These variants and processes are described below:

2.3.2.1.1 Glow Discharge Sputtering

The simplest arrangements to produce ions are provided by a normal glow discharge created at a residual pressure of about 10^{-2} torr of the required gas (generally Ar) by applying 1–3 kV DC between a cathode (target) and an anode (on which the substrate is placed) separated by about 5 cm. The thickness d of the cathode dark space (across which most of the applied voltage drops) is inversely proportional to the gas pressure p. Because of collisions with gas atoms, the sputtered atoms reach the substrate with randomized directions and energies.

2.3.2.1.2 Magnetron Sputtering

Arrangements in which the applied electric and magnetic fields are perpendicular to each other are called magnetron in sputtering systems. In a planar cathode system, the magnetic field is applied parallel to the cathode to confine the primary electron motion to the vicinity of the cathode and thus increase the ionization efficiency and prevent the electron bombardment of the film. Permanent magnets are placed behind the cathode in various geometries in such a way that the cathode surface has at least one region where the locus of the magnetic field lines is parallel to the cathode surface in a closed path. The discharge plasma is constrained near the cathode surface by endless toroidal trapping regions bounded by a tunnel-shaped magnetic field. The tunnel shape and thus the electron paths depend on the magnet geometry and arrangement. The usefulness of this technique in depositing several solar cell materials such as TiO_2, ITO, CdS, and Cu_2S has been demonstrated recently.

2.3.2.1.3   Radio Frequency (RF) Sputtering

Sputtering at low pressures (10^{-3} torr) is also possible by enhancing gas ionization with the help of an inductively coupled external rf field. If the cathode is an insulator material, dc sputtering is not possible owing to the building up of positive (Ar^+) surface charges. However, a high-frequency alternating potential may be used to neutralize the insulator surface periodically with plasma electrons, which have a much higher mobility than the positive ions. Whether or not the cathode surface develops a positive bias, which is responsible for sputtering, depends on the amplitude and frequency of rf and the geometry of the cathode. The rf technique can be used with any sputtering geometry in glow discharge or magnetron modes. It is an indispensable technique for deposition of thin films of semiconductors and insulators.

2.3.2.1.4   Ion Beam Sputtering

Sputter deposition under controlled high-vacuum conditions can be achieved by using an ion beam source. In the primary ion beam deposition process, the ions of the required material are produced and condensed on a surface to form a thin film. In the secondary ion beam deposition process, the Ar^+ ions from a beam source are used to sputter a target in vacuum and condense the sputtered species on a substrate. Both techniques have undergone major technology developments and have now become standard but expensive tools for utilizing the benefits of a sputtering process under vacuum deposition conditions. Ion deposition has been used to deposit films of ITO on Si for SIS solar cells.

2.3.2.1.5   Ion Plating

A combination of thermal evaporation onto a substrate (cathode) which is simultaneously bombarded with positive ions (e.g., Ar^+) from a glow discharge or an ion source is called ion plating. A better and meaningful ion plating technique involves ionization of the vapor by bombardment with accelerated electrons from a thermal source and depositing the ions onto a substrate with or without postionization acceleration.

2.3.2.1.6   Reactive Sputtering

This technique can be accomplished by introducing the reactant in gas form into the inert gas plasma. Whether the chemical reaction takes place on the cathode, in the plasma, or at the anode depends on the pressure and chemical activity of the reacting species under given surface and temperature conditions. Among the major applications of this technique is the preparation of controlled composition oxide films for MIS and SIS solar cells and for antireflective (AR) coatings. It is possible with this technique to create patterned thin films for solar cell grid structures.

2.4 DC/RF MULTITARGET MAGNETRON SPUTTERING EQUIPMENT

The equipment available at the center consists of vacuum deposition chamber, control system, mechanical pump, water chiller, Ar gas cylinder, and air compressor as it can be shown in Figure 2.1.

The equipment is a multitarget whereby more than one target can be loaded for deposition. This makes it easy to deposit multilayer thin film without venting the chamber and it also serves as a source of power to sputter as direct current and radio frequency. The schematic diagram shown in Figure 2.8 presents detailed information about the equipment.

2.5 MAINTENANCE DC/RF MAGNETRON SPUTTERING EQUIPMENT

Always check the oil gauge of the mechanical pump. Always check the water level of the chiller. Ensure the mechanical pump, air compressor, and chiller are properly connected before starting the system. Ensure proper electrical connection to avoid surge.

2.6 PRECAUTIONS

Wear nitride hand glove. Never touch the electrical components of the system when the system is on. In case of any emergency press the emergency button.

2.7 PROCEDURE FOR THIN FILM DEPOSITION USING DC/RF MAGNETRON SPUTTERING SYSTEM

Acquire the desired target of required size and specification. Mount the target on the target holder. Mount the substrate on the substrate holder. Adjust the target holder to desired distance between the target and the substrate. Close the chamber and pump down to high vacuum. Introduced the Argon sputtering gas at a control rate. Introduce the desired reactive gas at a control rate for reactive sputtering. Start sputtering by turning on the power (DC or RF). Adjust the voltage or current to increase or decrease the sputtering power for DC sputtering. Adjust the radio frequency to increase or decrease the sputtering power for RF sputtering. Allow the sputtering for few minutes before opening the substrate shutter. This is necessary to remove contaminants on the surface of the target before the deposition. Monitor the time of deposition. After deposition, follow the same sequence to vent the chamber. Shut down the system properly.

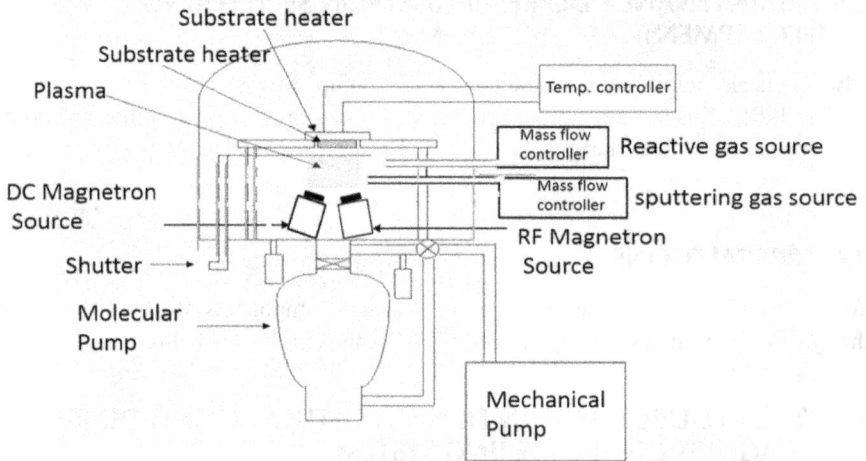

FIGURE 2.8   (a) Picture and (b) schematic of DC/RF magnetron sputtering system.

Pulsed Laser Deposition: PLD makes it simple to make multicomponent films with the desired stoichiometric ratio. It has a rapid deposition rate, a short test period, and requires a low substrate temperature. PLD produces consistent films. The procedure is straightforward and adaptable, with plenty of room for growth and compatibility.

The history of laser-assisted film growth began in 1960, shortly after Maiman's technical achievement of the first laser. Three years after Breech and Cross explored the laser-vaporization and excitation of atoms from solid surfaces, Smith and Turner used a ruby laser to deposit the first thin films in 1965. The deposited films, however, were still inferior to those produced by other methods such as CVD and molecular beam epitaxy. In the early 1980s, a few research groups (mostly from the former Soviet

Union) produced impressive accomplishments using laser technology to manufacture thin film structures. Dijkkamp, Xindi Wu, and T. Venkatesan achieved a breakthrough in 1987 when they were able to laser implant a thin film of YBa2Cu3O7, a
high-temperature superconductive material, that was superior to films produced using
other methods. Pulsed laser deposition has since been used to manufacture high-
quality crystalline films, such as doped garnet thin films, for use in planar waveguide
lasers (Grant-Jacob et al., 2016, Beecher et al., 2017). Ceramic oxides (Koinuma
et al., 1991), nitride films (Vispute et al., 1997), ferromagnetic films (Yoshitake et al.,
2003), metallic multilayers (Lu et al., 2021), and different superlattices have all been
deposited (Ye et al., 2021). PLD became a very competitive method for the formation of thin, well-defined films with complex stoichiometry in the 1990s when new
laser technologies, such as lasers with high repetition rate and short pulse durations,
became available.

Although the basic setup is uncomplicated in comparison to many other deposition processes, the physical phenomena of laser-target interaction and film formation
are highly complicated. When a laser pulse is absorbed by a target, the energy is
transformed first to electronic excitation, then to thermal, chemical, and mechanical
energy, causing evaporation, ablation, plasma production, and even exfoliation (Ma
and Chen, 2016). Before depositing on the typically heated substrate, the ejected
species expand into the surrounding vacuum in the form of a plume including many
energetic species such as atoms, molecules, electrons, ions, clusters, particulates, and
molten globules.

Pulsed Laser Deposition Chamber: There are a variety of ways to construct a
PLD deposition chamber. The laser evaporates the target material, which is usually a revolving disc attached to a support as shown in Figure 2.9. It can, however,
be sintered into a cylindrical rod with rotational motion and up-and-down translational movement along its axis. This unique design allows not only for the use of a

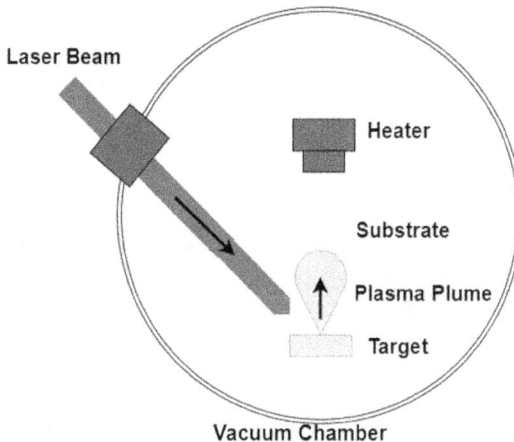

FIGURE 2.9 Schematic of pulsed laser deposition chamber.

synchronized reactive gas pulse but also for the use of a multicomponent target rod to make films of various multilayers.

The following are some of the elements that influence the rate of deposition of pulsed laser deposition: laser pulse energy, target material, laser repetition rate, substrate temperature (Grant-Jacob et al., 2017), distance between target and substrate, and type of gas and chamber pressure (oxygen, argon, etc.) (Scharf and Krebs, 2002)

Pulsed Laser Deposition Process: The ablation of the target material by laser irradiation; the production of a plasma plume with high energy ions, electrons, and neutrals; and the crystalline growth of the film itself on the heated substrate are all very complex mechanisms. The PLD process can be broken down into four stages: laser absorption on the target surface and laser ablation of the target material and creation of a plasma, dynamic of the plasma, deposition of the ablation material on the substrate, and nucleation and growth of the film on the substrate surface. The crystallinity, homogeneity, and stoichiometry of the resultant film are all dependent on each of these stages.

2.8   CHEMICAL VAPOR DEPOSITION (CVD)

CVD is a vacuum deposition technology for producing high-quality, high-performance solid materials and thin films. Typical CVD involves exposing the wafer (substrate) to one or more volatile precursors, which react and/or degrade on the substrate surface to form the desired deposit (Binti Hamzan et al., 2021). Volatile by-products are frequently formed, which are eliminated by the reaction chamber's gas flow. CVD is frequently used in micromachining to produce materials in a variety of morphologies, including monocrystalline, polycrystalline, amorphous, and epitaxial. Silicon (dioxide, carbide, nitride, oxynitride), carbon (fiber, nanofibers, nanotubes, diamond, and graphene), fluorocarbons, filaments, tungsten, titanium nitride, and a variety of high-k dielectrics are among these materials. The principle of CVD is based on the production of a gaseous species containing the coating element within a coating retort or chamber (Kempster, 1992). Alternatively, the gaseous species could be produced outside of the coating retort and delivered via a delivery system. Figure 2.10 shows the schematic of the CVD process.

CVD can be classified based on the operating conditions, physical characteristics of the vapor, substrate heating, plasma method, and nature of the vapor as shown in Figure 2.11.

The operating condition includes atmospheric pressure CVD (APCVD), low-pressure, ultrahigh vacuum, and subatmospheric CVD. APCVD involves pressure in the atmosphere, and CVD at subatmospheric pressures is known as low-pressure CVD (LPCVD). Reduced pressures tend to eliminate undesirable gas-phase reactions and improve wafer film uniformity. Ultrahigh vacuum CVD (UHVCVD) is a type of CVD that takes place at extremely low pressures, often below 10^6 Pa (108 torr). It's worth noting that a lower divide between high and ultra-high vacuum is typical in other industries, with 10^7 Pa being the most common. CVD at subatmospheric

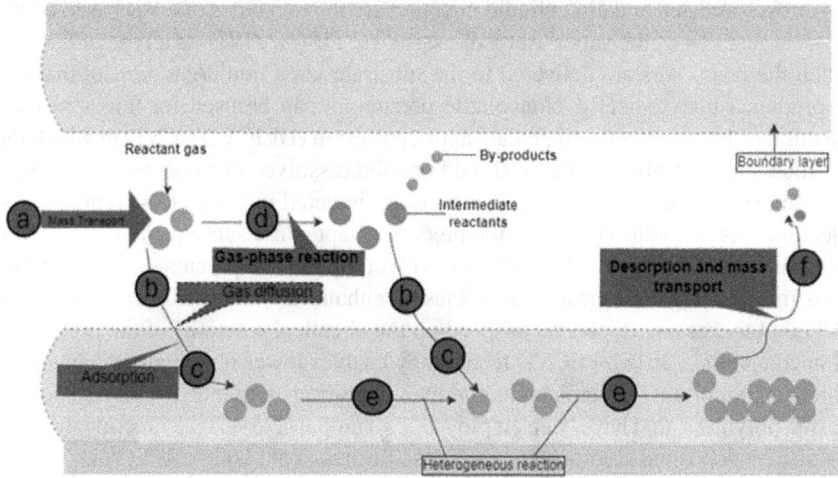

FIGURE 2.10 Schematic of chemical vapor deposition.

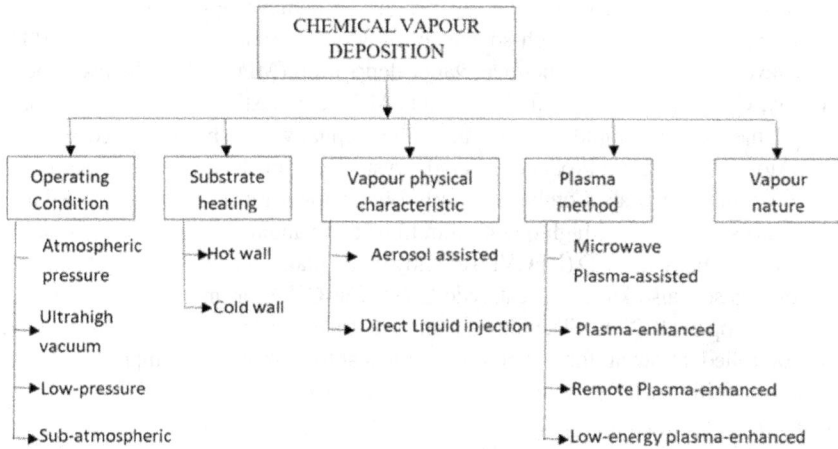

FIGURE 2.11 Chemical vapor deposition classification.

pressures is known as subatmospheric CVD (SACVD). Fills high aspect ratio Si structures with silicon dioxide using tetraethyl orthosilicate (TEOS) and Ozone (SiO$_2$) (Shareef et al., 1995).

Also, CVD is classified based on the substrate heating as hot wall and cold wall. The hot wall CVD method uses an external power source to heat the chamber, and the substrate is heated by radiation from the heated room walls. Cold wall CVD, on the other hand, is a kind of CVD in which only the substrate is heated directly, either by induction or by sending current through the substrate or a heater in contact with it. The walls of the chamber are at room temperature.

Aerosol-assisted and direct liquid injection are the two types of CVD based on the physical features of the vapor. Aerosol-assisted CVD (AACVD) is a type of CVD in which the precursors are delivered to the substrate via a liquid/gas aerosol that can be produced ultrasonically. Nonvolatile precursors can be used for this approach. Injection of liquid directly chemical vapor deposition (DLICVD)–CVD in which the precursors are dissolved in liquid (liquid or solid dissolved in a convenient solvent). In a vaporization chamber, liquid solutions are injected into injectors (typically car injectors). As in traditional CVD, the precursor vapors are subsequently transferred to the substrate. This method can be used with liquid or solid precursors. This method helps in achieving high growth rates. Plasma-enhanced CVD (PECVD) is a type of CVD that makes use of plasma to speed up the chemical reactions of the precursors (Hamedani et al., 2016). PECVD technology enables lower-temperature deposition, which is important in semiconductor manufacturing. Organic coatings, such as plasma polymers, that have been employed for nanoparticle surface functionalization can also be deposited at lower temperatures (Tavares et al., 2008). RPECVD (remote plasma-enhanced CVD) is like PECVD, except the wafer substrate is not directly in the plasma discharge zone. Processing temperatures can be reduced to room temperature by removing the wafer from the plasma area. Low-energy plasma-enhanced CVD uses a high-density, low-energy plasma to achieve epitaxial deposition of semiconductor materials at high speeds and low temperatures (Isella et al., 2004). Microwave plasma-assisted chemical vapor deposition (MPCVD) is the last type.

Lastly, CVD has been classified to include ALD, combustion, hot filament, hybrid-physical, metalorganic, rapid thermal, vapor-phase epitaxy, and photo-initiated and laser CVD. Atomic-layer CVD Produces layered, crystalline films by depositing consecutive layers of various materials. Combustion CVD, often known as flame pyrolysis, is a flame-based process for forming high-quality thin films and nanomaterials in an open atmosphere. Heated filament CVD (HFCVD) employs a hot filament to chemically decompose the source gases, also known as catalytic CVD (Cat-CVD) or, more generally, started CVD (Schropp et al., 2000). The filament and substrate temperatures are thus independently controlled, allowing for cooler temperatures at the substrate for improved absorption rates and higher temperatures at the filament for breakdown of precursors to free radicals (Lau et al., 2000). HPCVD stands for hybrid physical-chemical vapor deposition, which entails both the chemical breakdown of a precursor gas and the vaporization of a solid source. MOCVD (metalorganic CVD) is a CVD method that uses metalorganic precursors. Rapid thermal CVD (RTCVD) is a procedure that rapidly heats the wafer substrate using heating lamps or other ways. Instead of heating the gas or the chamber walls, just the substrate is heated, which helps to reduce undesirable gas-phase reactions that can contribute to particle formation. Epitaxy in the vapor phase (VPE). UV light is used to induce chemical reactions in photo-initiated CVD (PICVD). Given that plasmas release a lot of UV radiation, it's like plasma processing. PICVD may be conducted at or near atmospheric pressure under appropriate conditions (Dion and Tavares, 2013). Laser chemical vapor deposition (LCVD)—in semiconductor applications, this CVD technique uses lasers to heat spots or lines on a substrate. Lasers are utilized to rapidly break down the precursor gas process temperatures that can surpass 2000°C to build up a solid structure in MEMS and fiber fabrication, like how laser sintering-based 3D printers construct solids from powders.

Chemical Vapor Deposition Usage: CVD is frequently used to produce conformal coatings and modify substrate surfaces in ways that other surface modification techniques cannot. CVD is particularly useful for producing incredibly thin layers of material in the ALD technique. Such videos can be used in a variety of ways. Some integrated circuits (ICs) and solar devices contain gallium arsenide. In photovoltaic devices, amorphous polysilicon is employed. Wear resistance is conferred by some carbides and nitrides (Savale, 2018). CVD polymerization, possibly the most flexible of all applications, enables super-thin coatings with a variety of desirable properties, including lubricity, hydrophobicity, and weather resistance, to mention a few (Asatekin et al., 2010). Metalorganic frameworks, a type of crystalline nanoporous material, have recently been shown to be achievable by this method (Stassen et al., 2016). Gas sensing and low-k dielectrics are expected to be used for these films, which have recently been scaled up as an integrated cleanroom process depositing large-area substrates (Cruz et al., 2019). Membrane coatings, such as those used in desalination or water treatment, benefit from CVD processes because they can be uniform (conformal) and thin enough not to clog membrane pores.

2.9 CURRENT TRENDS IN THIN FILMS

The world is evolving, with most skills and technologies presumed to expire after five years. Thin films have evolved over the years as seen in the usage, applications, and properties. Thin films current trend is seen in applications such as clean energy (hydrogen, small modular reactor, wind blades, solar), sensors, corrosion inhibition, storage, wear, and abrasion, among others (Kilinc et al., 2022, Herth and Rauch, 2022, Fayomi et al., 2019). An alternate deposition source that could achieve a high deposition rate for industrial usage was electromagnetic levitation.

2.10 CONCLUSION

Modern research, technology, and industrial uses all depend heavily on the thin film production method. There have been various attempts to deliver superior thin films by treating the materials' surfaces. The three greatest, widely utilized deposition processes, physical vapor deposition, solution-based deposition, and chemical vapor deposition, are used in both scientific research and industry applications. It has been well acknowledged that the columnar microstructure that is frequently present in films formed by PVD and CVD has an impact on a variety of film properties. Numerous deposition mechanisms with high material throughput and superior film characteristics have emerged in recent years.

REFERENCES

Abel, S., Tesfaye, J. L., Gudata, L., Lamessa, F., Shanmugam, R., Dwarampudi, L. P., Nagaprasad, N., & Krishnaraj, R. (2022). Investigating the influence of bath temperature on the chemical bath deposition of nanosynthesized lead selenide thin films for photovoltaic application. *Journal of Nanomaterials, 2022,* 1–6.

Al-Assadi, Z. I., & Al-Assadi, F. I. (2021). Enhancing the aesthetic aspect of the solar systems used as facades for building by designing multi-layer optical coatings. *Technium, 3*(11), 1–10.

Albu, D. F., Lungu, J., Popescu-Pelin, G., Mihăilescu, C. N., Socol, G., Georgescu, A., Socol, M., Bănică, A., Ciupina, V., & Mihailescu, I. N. (2022). Thin film fabrication by pulsed laser deposition from TiO_2 targets in O_2, N_2, He, or Ar for dye-sensitized solar cells. *Coatings, 12*, 293.

Aoba, T. (2004). Solubility properties of human tooth mineral and pathogenesis of dental caries. *Oral Diseases, 10*(5), 249–257.

Arscott, S. (2020). The limits of edge bead planarization and surface levelling in spin-coated liquid films. *Journal of Micromechanics and Microengineering, 30*, 025003.

Asatekin, A., Barr, M. C., Baxamusa, S. H., Lau, K. K., Tenhaeff, W., Xu, J., & Gleason, K. K. (2010). Designing polymer surfaces via vapor deposition. *Materials Today, 13*, 26–33.

Bailey, N., Rockett, T., Flores, S., Reyes, D., David, J., & Richards, R. (2022). Effect of MBE growth conditions on GaAsBi photoluminescence lineshape and localised state filling. *Scientific Reports, 12*, 1–8.

Beecher, S. J., Grant-Jacob, J. A., Hua, P., Prentice, J. J., Eason, R. W., Shepherd, D. P., & Mackenzie, J. I. (2017). Ytterbium-doped-garnet crystal waveguide lasers grown by pulsed laser deposition. *Optical Materials Express, 7*, 1628–1633.

Binti Hamzan, N., Ng, C. Y. B., Sadri, R., Lee, M. K., Chang, L.-J., Tripathi, M., Dalton, A., & Goh, B. T. (2021). Controlled physical properties and growth mechanism of manganese silicide nanorods. *Journal of Alloys and Compounds, 851*, 156693.

Bonnet, D. (2001). Cadmium telluride solar cells. In M. D. Archer, & R. Hill (Eds.), *Clean electricity from photovoltaics* (pp. 245–269). Imperial College Press.

Chaudhary, K. T. (2021). Thin film deposition: Solution based approach. In A. E. Ares (Ed.), *Thin films*. IntechOpen.

Cheng, H.-C., Chen, C.-F., & Lee, C.-C. (2006). Thin-film transistors with active layers of zinc oxide (ZnO) fabricated by low-temperature chemical bath method. *Thin Solid Films, 498*, 142–145.

Cruz, A. J., Stassen, I., Krishtab, M., Marcoen, K., Stassin, T., Rodríguez-Hermida, S., Teyssandier, J., Pletincx, S., Verbeke, R., & Rubio-Giménez, V. (2019). Integrated cleanroom process for the vapor-phase deposition of large-area zeolitic imidazolate framework thin films. *Chemistry of Materials, 31*, 9462–9471.

Dan, A., Bijalwan, P. K., Pathak, A. S., & Bhagat, A. N. (2022). A review on physical vapor deposition-based metallic coatings on steel as an alternative to conventional galvanized coatings. *Journal of Coatings Technology and Research, 19*(2), 403–438, 1–36.

Danglad-Flores, J., Eickelmann, S., & Riegler, H. (2018). Deposition of polymer films by spin casting: A quantitative analysis. *Chemical Engineering Science, 179*, 257–264.

Dion, C. D., & Tavares, J. R. (2013). Photo-initiated chemical vapor deposition as a scalable particle functionalization technology (a practical review). *Powder Technology, 239*, 484–491.

El-Shamy, A. M., Elsayed, E., & Rashad, M. (2022). Preparation and characterization of ZnO thin film on anodic Al_2O_3 as a substrate for several applications. *Egyptian Journal of Chemistry, 65*(10), 119–129.

Elsheikh, A. H., Sharshir, S. W., Ali, M. K. A., Shaibo, J., Edreis, E. M., Abdelhamid, T., Du, C., & Haiou, Z. (2019). Thin film technology for solar steam generation: A new dawn. *Solar Energy, 177*, 561–575.

Emslie, A. G., Bonner, F. T., & Peck, L. G. (1958). Flow of a viscous liquid on a rotating disk. *Journal of Applied Physics, 29*, 858–862.

Ezekoye, B., Offor, P., Ezekoye, V., & Ezema, F. (2013). Chemical bath deposition technique of thin films: A review. *International Journal of Scientific Research, 2*, 452–456.

Fan, H.-P., Yang, X.-X., & Lu, F.-H. (2022). Air-based deposition of titanium-aluminum oxynitride thin films by reactive magnetron sputtering. *Surface and Coatings Technology, 436*, 128287.

Fayomi, O., Akande, I., Abioye, O., & Fakehinde, O. (2019). New trend in thin film composite coating deposition: A mini review. *Procedia Manufacturing, 35*, 1007–1012.

Fazal, T., Iqbal, S., Shah, M., Bahadur, A., Ismail, B., Abd-Rabboh, H. S., Hameed, R., Mahmood, Q., Ibrar, A., & Nasar, M. S. (2022). Deposition of bismuth sulfide and aluminum doped bismuth sulfide thin films for photovoltaic applications. *Journal of Materials Science: Materials in Electronics, 33*, 42–53.

Glaser, J. A. (2011). *Kirk–Othmer chemical technology and the environment.* Springer.

Göde, F., Gümüş, C., & Zor, M. (2007). Investigations on the physical properties of the poly-crystalline ZnS thin films deposited by the chemical bath deposition method. *Journal of Crystal Growth, 299*, 136–141.

Grant-Jacob, J. A., Beecher, S. J., Parsonage, T. L., Hua, P., Mackenzie, J. I., Shepherd, D. P., & Eason, R. W. (2016). An 11.5 W Yb: YAG planar waveguide laser fabricated via pulsed laser deposition. *Optical Materials Express, 6*, 91–96.

Grant-Jacob, J. A., Beecher, S. J., Riris, H., Anthony, W. Y., Shepherd, D. P., Eason, R. W., & Mackenzie, J. I. (2017). Dynamic control of refractive index during pulsed-laser-deposited waveguide growth. *Optical Materials Express, 7*, 4073–4081.

Guire, M. R. D., Bauermann, L. P., Parikh, H., & Bill, J. (2013). Chemical bath deposition. In T. Schneller, R. Waser, M. Kosec, & D. Payne (Eds.), *Chemical solution deposition of functional oxide thin films.* Springer.

Hamedani, Y., Macha, P., Bunning, T. J., Naik, R. R., & Vasudev, M. C. (2016). Plasma-enhanced chemical vapor deposition: Where we are and the outlook for the future. *Chemical Vapor Deposition-Recent Advances and Applications in Optical, Solar Cells and Solid State Devices, 4*, 243–280.

Hanaor, D., Triani, G., & Sorrell, C. (2011). Morphology and photocatalytic activity of highly oriented mixed phase titanium dioxide thin films. *Surface and Coatings Technology, 205*, 3658–3664.

Herth, E., & Rauch, J.-Y. (2022). ICPECVD-dielectric thin-films CMOS-compatible: Trends in eco-friendly deposition. *International Journal of Precision Engineering and Manufacturing-Green Technology, 9*, 933–940.

Hodes, G. (2007). Semiconductor and ceramic nanoparticle films deposited by chemical bath deposition. *Physical Chemistry Chemical Physics, 9*, 2181–2196.

Hone, F. G., & Abza, T. (2019). Short review of factors affecting chemical bath deposition method for metal chalcogenide thin films. *International Journal of Thin Film Science and Technology, 8*, 43.

Isella, G., Chrastina, D., Rössner, B., Hackbarth, T., Herzog, H.-J., König, U., & Von Känel, H. (2004). Low-energy plasma-enhanced chemical vapor deposition for strained Si and Ge heterostructures and devices. *Solid-State Electronics, 48*, 1317–1323.

Jilani, A., Abdel-Wahab, M. S., & Hammad, A. H. (2017). Advance deposition techniques for thin film and coating. *Modern Technologies for Creating the Thin-Film Systems and Coatings, 2*, 137–149.

Kaiser, N. (2002). Review of the fundamentals of thin-film growth. *Applied Optics, 41*, 3053–3060.

Kempster, A. (1992). The principles and applications of chemical vapour deposition. *Transactions of the IMF, 70*, 68–75.

Kilinc, N., Sanduvac, S., & Erkovan, M. (2022). Platinum-Nickel alloy thin films for low concentration hydrogen sensor application. *Journal of Alloys and Compounds, 892*, 162237.

Koinuma, H., Nagata, H., Tsukahara, T., Gonda, S., & Yoshimoto, M. (1991). Ceramic layer epitaxy by pulsed laser deposition in an ultrahigh vacuum system. *Applied Physics Letters, 58*, 2027–2029.

Konh, M., Wang, Y., Chen, H., Bhatt, S., Xiao, J. Q., & Teplyakov, A. V. (2022). Selectivity in atomically precise etching: Thermal atomic layer etching of a CoFeB alloy and its protection by MgO. *Applied Surface Science, 575*, 151751.

Lau, K. K., Caulfield, J. A., & Gleason, K. K. (2000). Structure and morphology of fluorocarbon films grown by hot filament chemical vapor deposition. *Chemistry of Materials, 12*, 3032–3037.

Liebig, J. (1856). Ueber versilberung und vergoldung von glas. *Justus Liebigs Annalen der Chemie, 98*, 132–139.

Lokhande, C., Gondkar, P., Mane, R. S., Shinde, V., & Han, S.-H. (2009). CBD grown ZnO-based gas sensors and dye-sensitized solar cells. *Journal of Alloys and Compounds, 475*, 304–311.

Lorenz, P., Zieger, G., Dellith, J., & Schmidt, H. (2022). Electron beam co-deposition of thermoelectric BiSb thin films from two separate targets. *Thin Solid Films, 745*, 139082.

Lu, Y., Huang, G., Wang, S., Xi, L., Qin, G., Tan, J., & Tan, C. (2021). Pulsed laser deposition of multilayer diamond-like carbon film grown on stainless steel. *Diamond and Related Materials, 120*, 108615.

Lukong, V., Ukoba, K., & Jen, T. (2022). Heat-assisted sol–gel synthesis of tio2 nanoparticles structural, morphological and optical analysis for self-cleaning application. *Journal of King Saud University-Science, 34*, 101746.

Ma, C., & Chen, C. (2016). Pulsed laser deposition for complex oxide thin film and nanostructure. In Y. Lin, & X. Chen (Eds.), *Advanced nano deposition methods* (pp. 1–31). Wiley-VCH. https://doi.org/10.1002/9783527696406.ch1

Mahana, D., Mauraya, A. K., Pal, P., Singh, P., & Muthusamy, S. K. (2022). Comparative study on surface states and CO gas sensing characteristics of CuO thin films synthesised by vacuum evaporation and sputtering processes. *Materials Research Bulletin, 145*, 111567.

Martínez-Benítez, A., Oseguera-Galindo, D. O., Escobar-Alarcón, L., Meléndez-Lira, M., Quiñones-Galván, J. G., Pérez-Centeno, A., & Santana-Aranda, M. A. (2022). On the chemical bath deposition of the ternary compound ZnxHg1-xS. *Optical Materials, 124*, 111983.

Mcpeak, K. M., Le, T. P., Britton, N. G., Nickolov, Z. S., Elabd, Y. A., & Baxter, J. B. (2011). Chemical bath deposition of ZnO nanowires at near-neutral pH conditions without hexamethylenetetramine (HMTA): Understanding the role of HMTA in ZnO nanowire growth. *Langmuir, 27*, 3672–3677.

Moreno-Regino, V., Castañeda-De-La-Hoya, F., Torres-Castanedo, C., Márquez-Marín, J., Castanedo-Pérez, R., Torres-Delgado, G., & Zelaya-Ángel, O. (2019). Structural, optical, electrical and morphological properties of CdS films deposited by CBD varying the complexing agent concentration. *Results in Physics, 13*, 102238.

Myny, K. (2018). The development of flexible integrated circuits based on thin-film transistors. *Nature Electronics, 1*, 30–39.

Nair, P., Nair, M., Garcıa, V., Arenas, O., Pena, Y., Castillo, A., Ayala, I., Gomezdaza, O., Sanchez, A., & Campos, J. (1998). Semiconductor thin films by chemical bath deposition for solar energy related applications. *Solar Energy Materials and Solar Cells, 52*, 313–344.

Osherov, A., Ezersky, V., & Golan, Y. (2007). The role of solution composition in chemical bath deposition of epitaxial thin films of PbS on GaAs (1 0 0). *Journal of Crystal Growth, 308*, 334–339.

Oviroh, P. O., Akbarzadeh, R., Pan, D., Coetzee, R. A. M., & Jen, T.-C. (2019). New development of atomic layer deposition: Processes, methods and applications. *Science and Technology of Advanced Materials, 20*, 465–496.

Patra, C., & Das, D. (2022). Room temperature synthesized highly conducting B-doped nanocrystalline silicon thin films on flexible polymer substrates by ICP-CVD. *Applied Surface Science, 583*, 152499.

Ren, M., Yu, H.-L., Zhu, L.-N., Li, H.-Q., Wang, H.-D., Xing, Z.-G., & Xu, B.-S. (2022). Microstructure, mechanical properties and tribological behaviors of TiAlN-Ag composite coatings by pulsed magnetron sputtering method. *Surface and Coatings Technology, 436*, 128286.

Savale, P. (2018). Comparative study of various chemical deposition methods for synthesis of thin films: A review. *Asian Journal of Research in Chemistry, 11*(1), 195–205.

Scharf, T., & Krebs, H. (2002). Influence of inert gas pressure on deposition rate during pulsed laser deposition. *Applied Physics A, 75*, 551–554.

Schropp, R., Stannowski, B., Brockhoff, A., Van Veenendaal, P., & Rath, J. (2000). Hot wire CVD of heterogeneous and polycrystalline silicon semiconducting thin films for application in thin film transistors and solar cells. *Materials Physics and Mechanics, 1*, 73–82.

Seshan, K. (2012). *Handbook of thin film deposition*. William Andrew.

Shafi, M. A., Bouich, A., Fradi, K., Guaita, J. M., Khan, L., & Mari, B. (2022). Effect of deposition cycles on the properties of ZnO thin films deposited by spin coating method for CZTS-based solar cells. *Optik, 258*, 168854.

Shareef, I., Rubloff, G., Anderle, M., Gill, W., Cotte, J., & Kim, D. (1995). Subatmospheric chemical vapor deposition ozone/TEOS process for SiO_2 trench filling. *Journal of Vacuum Science & Technology B: Microelectronics and Nanometer Structures Processing, Measurement, and Phenomena, 13*, 1888–1892.

Singh, M. P., Chandi, P. S., Singh, R. C. (2007). Synthesis of nano-crystalline tin oxide powder through fine crystallization in liquid phase. *Journal of Optoelectronics and Advanced Materials, 9*(10), 3275.

Stassen, I., Styles, M., Grenci, G., Gorp, H. V., Vanderlinden, W., Feyter, S. D., Falcaro, P., Vos, D. D., Vereecken, P., & Ameloot, R. (2016). Chemical vapour deposition of zeolitic imidazolate framework thin films. *Nature Materials, 15*, 304–310.

Switzer, J. A., & Hodes, G. (2010). Electrodeposition and chemical bath deposition of functional nanomaterials. *MRS Bulletin, 35*, 743–750.

Tavares, J., Swanson, E. J., & Coulombe, S. (2008). Plasma synthesis of coated metal nanoparticles with surface properties tailored for dispersion. *Plasma Processes and Polymers, 5*, 759–769.

Ukoba, K., Eloka-Eboka, A., & Inambao, F. (2018a). Review of nanostructured NiO thin film deposition using the spray pyrolysis technique. *Renewable and Sustainable Energy Reviews, 82*, 2900–2915.

Ukoba, K. O., Eloka-Eboka, A. C., & Inambao, F. L. (2018b). Experimental optimization of nanostructured nickel oxide deposited by spray pyrolysis for solar cells application. *International Journal of Applied Engineering Research, 13*(6), 3165–3175.

Vispute, R., Talyansky, V., Trajanovic, Z., Choopun, S., Downes, M., Sharma, R., Venkatesan, T., Woods, M., Lareau, R., & Jones, K. (1997). High quality crystalline ZnO buffer layers on sapphire (001) by pulsed laser deposition for III–V nitrides. *Applied Physics Letters, 70*, 2735–2737.

Vyas, S. (2020). A short review on properties and applications of ZnO based thin film and devices. *Johnson Matthey Technology Review, 64*(2), 202–218.

Wilson, S., Hunt, R., & Duffy, B. (2000). The rate of spreading in spin coating. *Journal of Fluid Mechanics, 413*, 65–88.

Ye, B., Miao, T., Zhu, Y., Huang, H., Yang, Y., Shuai, M., Zhu, Z., Guo, H., Wang, W., & Zhu, Y. (2021). Pulsed laser deposition of large-sized superlattice films with high uniformity. *Review of Scientific Instruments, 92*, 113906.

Yoshitake, T., Nakagauchi, D., & Nagayama, K. (2003). Ferromagnetic iron silicide thin films prepared by pulsed-laser deposition. *Japanese Journal of Applied Physics, 42*, L849.

Zaremba, O. T., Goldt, A. E., Khabushev, E. M., Anisimov, A. S., & Nasibulin, A. G. (2022). Highly efficient doping of carbon nanotube films with chloroauric acid by dip-coating. Materials Science and Engineering: *B, 278*, 115648.

Zou, X., Fan, H., Tian, Y., Zhang, M., & Yan, X. (2015). Chemical bath deposition of Cu_2O quantum dots onto ZnO nanorod arrays for application in photovoltaic devices. *RSC Advances, 5*, 23401–23409.

3 Thin Films Properties, Characterization Techniques, and Applications

3.1 THIN FILMS PROPERTIES

A material's overall performance and usage are dependent on its properties. The properties will determine the material application and functions. A thorough understanding of thin films properties is vital for correct usage and utilization. Thin films are exceedingly thermally stable and tough, but they are delicate. Organic materials, on the other hand, have good thermal stability and are durable, but they are soft.

The properties of material such as chemical properties, mechanical properties, electrical properties, and other properties shown in Figure 3.1 influence the growth condition of the material. Growth condition is needed for the performance of a material especially in some environment.

3.2 THIN FILMS CHARACTERIZATION AND TECHNIQUES

Material characterization is the measurement and determination of a material's physical, chemical, mechanical, and microstructural properties. It is a comprehensive and general procedure for probing and measuring the structure and properties of a material. It is a crucial step in the field of materials science, without which no scientific understanding of engineering materials would be possible (Sam et al., 2009). Some definitions confine the term's application to procedures that examine the microscopic structure and properties of materials (Leng, 2009). However, others apply to any materials analysis process, including macroscopic techniques like mechanical testing, thermal analysis, and density computation (Zhang et al., 2008).

The scale of structures seen in materials characterization spans from angstroms, as in imaging individual atoms and chemical bonds, to centimeters, as in imaging coarse grain structures in metals. While many characterization techniques, such as fundamental optical microscopy, have been used for decades, new techniques and methodologies are continually being developed. The invention of the electron microscope and secondary ion mass spectrometry in the 20th century revolutionized the field, allowing for the imaging and analysis of structures and compositions on much smaller scales than previously possible, resulting in a massive increase in our understanding of why different materials exhibit different properties and behaviors

DOI: 10.1201/9781003364481-3

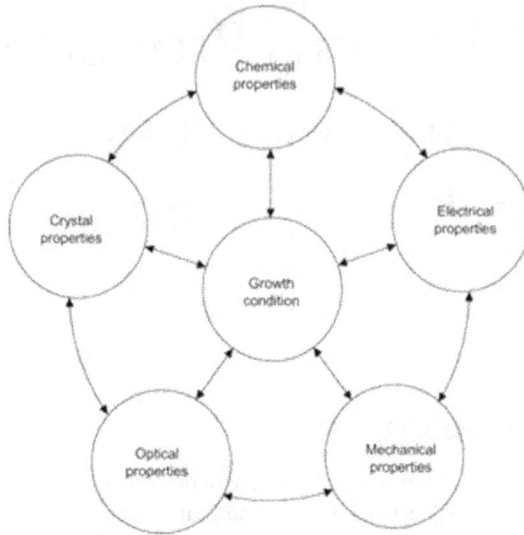

FIGURE 3.1 A schematic of the relationship between growth condition and thin films properties.

(Ismail & Nurdin, 1998). In the last 30 years, atomic force microscopy has enhanced the greatest feasible resolution for investigation of specific substances.

One of the first characteristics of a film's nature that is mentioned is its thickness (Ohring, 2001). The reason for this is that the qualities of thin films are usually determined by their thickness. The use of films in optical applications has historically prompted the development of systems for precisely measuring film thickness. Other key film features, such as structure and chemical composition, on the other hand, were only rudimentarily defined until recently. In some applications, the actual film thickness isn't so important, at least not within certain parameters. This is true for decorative, metallurgical, and protective films and coatings. Microelectronic applications, on the other hand, frequently demand exact and repeatable film metrology, including thickness and lateral dimensions.

The various types of films and their applications have resulted in a plethora of methods for measuring film thickness. They are classified into optical and mechanical approaches and are mostly nondestructive but can occasionally be destructive. For most procedures, it is necessary to remove films from the deposition chamber for measurement. However, for real-time monitoring of film thickness during growth, several technologies are either commercially available or can be experimentally customized.

The tools and process of identifying, isolating, or quantifying substances or materials, as well as characterizing their physical properties, are known as characterization and analytical procedures. Different reviews have been made about the standard or minimal analytical techniques to employ in biooil (Michailof et al., 2016), rice husk (Park et al., 2003), particulate matter (Galvão et al., 2018), drug-related complexes in aqueous solution (Mura, 2014) and nonaqueous solution (Mura, 2015), and humic materials (Helal et al., 2011). However, there have been attempts to perform detailed review of thin films electrical characterization (Wager and Keir, 1997) and surface characterization (Chason and Mayer, 1997); thermoelectric (Bahk et al., 2013), structural (Alvarez-Fernandez et al., 2020), and optical characterization (Dietz, 2001); and morphology (Song et al., 2018). Thin films characterization is performed to study the physical, chemical, and related properties of the material being studied. The characterization to be performed depends on the purpose and final usage of the device. These characterizations are grouped under morphology, structural, elemental, optical, absorption, and electrical.

3.2.1 SPECTROSCOPY

Spectroscopy is a class of characterization techniques that employs a variety of concepts to expose a material's chemical composition, composition variation, crystal structure, and photoelectric characteristics. The following are some examples of spectroscopic techniques.

Table 3.1 shows the spectroscopy used for EDS and other types of spectroscopies and their underlying principle.

3.2.2 OPTICAL RADIATION

Ultraviolet-visible spectroscopy (UV-vis), thermoluminescence (TL), Fourier transform infrared spectroscopy (FTIR), photoluminescence (PL) operate using an X-ray. An X-ray is an electromagnetic radiation with a wavelength of 1 Å. These devices are operated by allowing high-energy electrons accelerated by electrons to strike a metallic target. Spectroscope using X-rays include X-ray diffraction (XRD) and small-angle X-ray scattering (SAXS). Others include energy-dispersive X-ray

TABLE 3.1
Types of Spectroscopies

Method	Type
X-ray and Photoelectron (Egerton, 2005a)	X-ray photoelectron spectroscopy, energy-dispersive X-ray spectroscopy, atomic, photoelectron, emission, and EXAFS
Radiowave	ESR/EPR, terahertz, NMR, ferromagnetic resonance
Vibrational (Krivanek et al., 2014)	FT-IR, Raman, resonance Raman, rotational, rotational–vibrational, vibrational, vibrational circular dichroism, nuclear resonance vibrational spectroscopy
Nucleon	Gamma and Mössbauer
UV–Vis–NIR	Ultraviolet–visible, fluorescence, vibronic, near-infrared, resonance-enhanced multiphoton ionization (REMPI), Raman optical activity spectroscopy, Raman spectroscopy, and laser-induced

spectroscopy (EDX or EDS), wavelength dispersive X-ray spectroscopy (WDX or WDS), electron energy loss spectroscopy (EELS), X-ray photoelectron spectroscopy (XPS), Auger electron spectroscopy (AES), and X-ray photon correlation spectroscopy (XPCS)

3.2.2.1 Mass Spectrometry

This is grouped according to the mode and the secondary ion mass. Modes of mass spectrometry include electron ionization (EI), thermal ionization mass spectrometry (TIMS), and a laser energy-absorbing matrix is used in the matrix-assisted laser desorption/ionization (MALDI). It is a method to ionize big molecules with little fragmentation. It has been used to analyze numerous organic compounds, including polymers, dendrimers, and other macromolecules, as well as biomolecules (biopolymers like DNA, proteins, peptides, and carbohydrates), which have a tendency to be brittle and fragment when ionized using more traditional ionization techniques. Although MALDI normally produces significantly fewer multicharged ions, it is similar to electrospray ionization (ESI) in that both procedures are relatively gentle (low fragmentation) ways of getting ions of big molecules in the gas phase.

3.2.2.2 Nuclear Spectroscopy

This includes nuclear magnetic resonance spectroscopy (NMR), Mössbauer spectroscopy (MBS), and Perturbed angular correlation (PAC).

Other types include photon correlation spectroscopy/dynamic light scattering (DLS), terahertz spectroscopy (THz), electron paramagnetic/spin resonance (EPR, ESR), small-angle neutron scattering (SANS), and Rutherford backscattering spectrometry (RBS).

3.2.2.3 Macroscopic Properties Testing

The following are used to analyze the macroscopic properties of samples. They are differential thermal analysis (DTA), mechanical testing, dielectric thermal analysis (DEA or DETA), thermogravimetric analysis (TGA), ultrasound method, differential scanning calorimetry (DSC), and impulse excitation technique (IET). However, mechanical testing is further divided into creep, compressive, tensile, torsional, fatigue, toughness, and hardness testing. Ultrasound techniques include resonant ultrasound spectroscopy and time-domain ultrasound (Truell et al., 2013).

3.3 MORPHOLOGY

Morphology refers to the features of the exterior appearance (shape, structure, color, pattern, and size) as well as the form and structure of inside parts of a material or animal, referred to as internal morphology (or eidonomy or anatomy). Surface morphology and topography are essential polymeric materials (from nano- to macroscale) derived from their chemical nature/structure and manufacturing processes (Popelka et al., 2020).

3.3.1   History and Origin of Morphology

The word "morphology" comes from the Ancient Greek words "morph," which means "shape," and "lógos," which means "word, study, research."(Bailly, 1951). While the concept of form in biology, as opposed to function, dates to Aristotle, in 1790, Johann Wolfgang von Goethe did research on morphology, anatomy, and optics. He also developed a new manner of inquiry called morphology, which studied the form and transformation of organisms. He was one of the founders of this science, along with Karl Friedrich Burdach and Lorenz Oken. Independently of Johann Wolfgang von Goethe (1749–1832), Burdach coined the term "morphology" in 1800. The word was first used by Goethe in his diary in 1796 in Jena. Burdach also coined the term "biology," in 1800 (Bopp, 1993). Lorenz Oken, Georges Cuvier, Étienne Geoffroy Saint-Hilaire, Richard Owen, Karl Gegenbaur, and Ernst Haeckel are among the most influential morphology theorists (Di Gregorio, 2005; Myers and Richards, 2009). Cuvier and E.G. Saint-Hilaire had a famous argument in 1830 that is considered to have exemplified the two primary variations in biological theory—whether the animal structure was due to function or evolution.

3.3.2   Equipment Used for Morphology Characterization

There are various characterization equipment used for testing and analyzing the morphology of a sample. These include scanning electron microscope (SEM) and TEM, among others. Electron microscopes, SEMs, and scanning transmission electron microscopes all use electron beam excitation. X-ray fluorescence (XRF) spectrometers use X-ray beam excitation. The X-ray energy is converted into voltage signals via a detector. This is then transferred to a pulse processor that measures the signals. This sends them to an analyzer for display and analysis. The most popular detector used to be a Si(Li) detector chilled with liquid nitrogen to cryogenic temperatures. The SDDs (silicon drift detectors) with Peltier cooling are now common in modern systems. Figure 3.2 shows the different microscopes as adapted based on classification by Aryal (2022).

3.3.3   Scanning Electron Microscope

It is a type of electron microscope that uses a focused beam of electrons to scan the surface of a sample to obtain images. When electrons interact with atoms in a sample, they produce a variety of signals that carry information about the sample's surface topography and composition. A raster scan pattern is used to scan the electron beam, and the beam's position is combined with the intensity of the detected signal to create an image. Secondary electrons generated by atoms stimulated by the electron beam are detected by a secondary electron detector (Everhart–Thornley detector) in the most common SEM mode. The number of secondary electrons that may be detected, and hence the signal intensity, is influenced by specimen topography, among other factors.

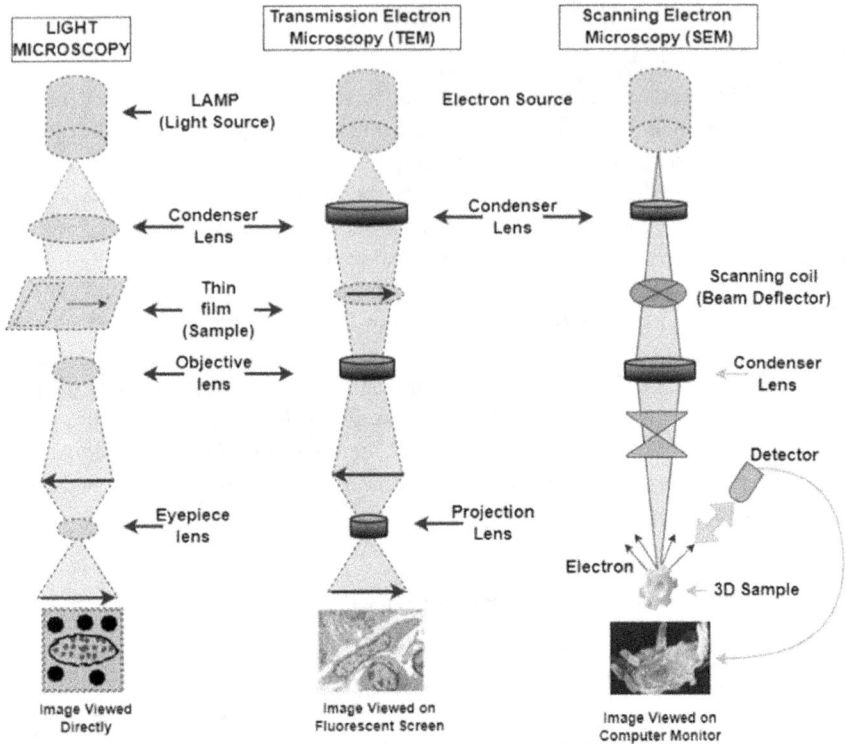

FIGURE 3.2 Schematic of different electron microscope.

3.3.4 History and Origin of SEM

Dennis McMullan was one of the pioneers of the scanning electron microscope (SEM), a type of electron microscope that produces images of a sample by scanning the surface with a focused beam of electrons. He started his PhD project on the construction of an SEM in 1948 at Cambridge University under the supervision of Charles Oatley (McMullan, 1995, 2008). Although Max Knoll utilized an electron beam scanner to create a photo with a 50-mm object field width demonstrating channeling contrast, Manfred von Ardenne (von Ardenne, 1937) invented a high-resolution microscope in 1937 scanning a very small raster with a demagnified and finely focused electron beam. von Ardenne used scanning of the electron beam to try to outperform the transmission electron microscope (TEM) in terms of resolution and to overcome significant chromatic aberration issues that come with real-time imaging in the TEM. He went on to talk about the several detection modes, capabilities, and theory of SEM, as well as the development of the first high-resolution SEM. Zworykin's group reported further work (Zworykin, 1945), accompanied by Cambridge groups developed by Charles Oatley in the 1950s and early 1960s (Wells, 2004), all of which culminated in Cambridge Scientific Instrument Company promoting the first commercial device as the "Stereoscan" in 1965, which was produced by DuPont. Figure 3.3 shows the first SEM and high-resolution SEM of the ZEISS model.

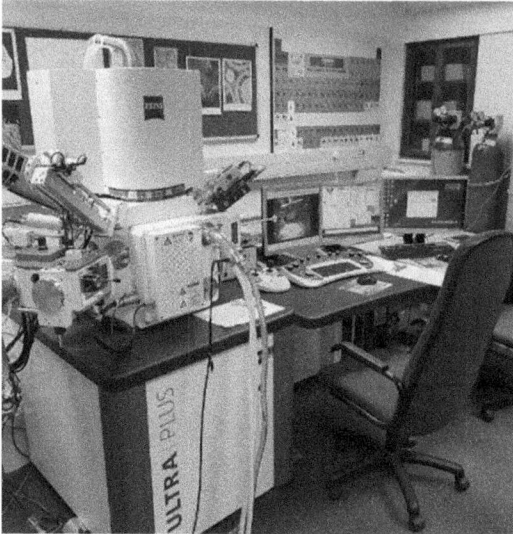

FIGURE 3.3   Schematic of ZEISS Ultra Plus scanning electron microscope.

3.3.5   COMPONENTS OF AN SEM

Stage, electron gun, scan coils, viewing chamber, condenser lens, condenser aperture, secondary electron, backscattered electron, and objective aperture are the key components of an SEM. The cathode and anode make the electron gun. Secondly, the number of electrons that go down the column is controlled by the condenser lens. The objective lens concentrates the beam on a specific area of the sample. The electron beam is deflected with the use of a deflection coil. The secondary electrons are drawn to SED. Backscattered electrons and X-rays are detected by additional sensors. Figure 3.4 gives the schematic of the key components of an SEM.

Operating Principle of Scanning Electron Microscope: In an SEM, accelerated electrons carry a lot of kinetic energy. This is dissipated as various signals caused by electron-sample interactions as the incident electrons decelerate in the solid sample as shown in Figure 3.5.

The signals include backscattered electrons, secondary electrons, cathodoluminescence (visible light), photons, transmitted electron, heat, and absorbed current called specimen current. Backscattered and secondary electrons create SEM images. Diffracted backscattered electrons create crystal structures and orientations of samples. Secondary electrons mean free path is restricted owing to the low energies of 50 eV. Secondary electrons are highly localized and can analyze samples to 1 nm. Backscattered electrons (BSE) are best for highlighting compositional contrasts in multiphase samples (i.e., rapid phase discrimination). Backscattered electronscans reveal the distribution of distinct elements in the sample but not their identities. Colloidal gold immuno-labels with a diameter of 5 or 10 nm can be imaged with BSE imaging, which would be difficult or impossible to detect in secondary

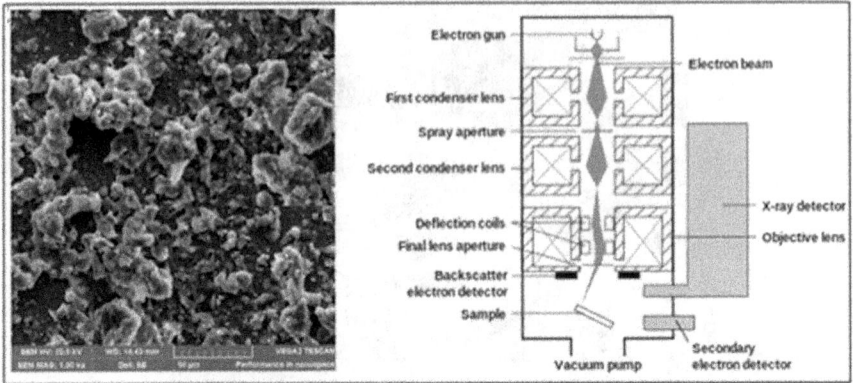

FIGURE 3.4   A schematic of major components of SEM.

Source: Lukong, Ukoba, & Jen, 2021.

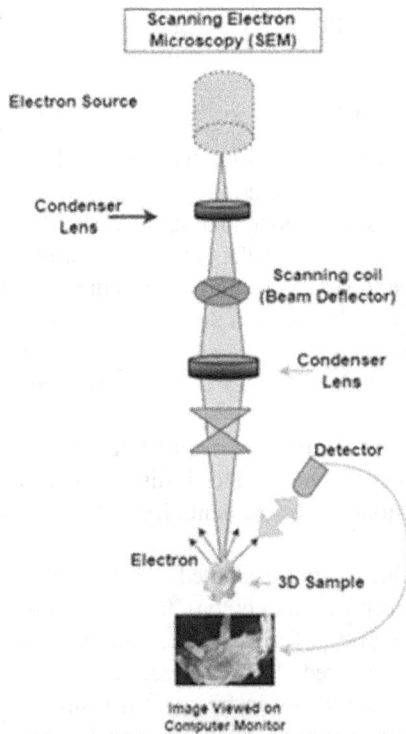

FIGURE 3.5   Schematic of working of an SEM with SEM image.

electron pictures (Suzuki, 2002). Inelastic collisions of incoming electrons with electrons in discrete orbitals (shells) of atoms in the sample produce X-rays. When excited electrons return to lower energy states, they produce X-rays with a specific wavelength (related to the difference in energy levels of electrons in different shells for a given element). As a result, each element in a sample "stimulated" by the electron beam produces distinct X-rays. SEM is a "nondestructive" test because X-rays created by electron interactions do not cause the sample to lose volume, allowing the same materials to be analyzed multiple times. SEM micrographs have a significant depth of field due to the relatively narrow electron beam, giving them a distinctive three-dimensional look important for analyzing a sample's surface structure (Goldstein et al., 1981). The micrograph of pollen displayed above exemplifies this. Magnifications range from about 10 times (equivalent to a powerful hand lens) to more than 500,000 times, or about 250 times the magnification limit of the greatest light microscopes.

In a nutshell, the essential premise is that a suitable source, such as a tungsten filament or a field emission cannon, produces an electron beam. To produce a thin beam of electrons, the electron beam is accelerated with a high voltage and passed through a system of apertures and electromagnetic lenses. The beam then scans the specimen's surface, causing electrons to be emitted from the specimen and captured by a detector that is properly positioned.

3.3.6 MERIT OF SEM

It provides detailed three-dimensional and topographical imagery as well as a wide range of data. This works quite quickly. Data can be generated in digital form using modern SEMs. Most SEM samples require only a few steps of preparation.

3.3.7 DEMERIT OF SEM

SEMs are huge and expensive. Operating an SEM necessitates specialized training. SEMs can only work with solid samples. The electrons that scatter from beneath the sample surface pose a limited danger of radiation exposure in SEMs. SEM requires steady power supply and fluctuation causes issues for the machine.

3.4 APPLICATION OF SEM

The SEM is widely used in high images of shapes of objects (SEI) and to show fluctuation in chemical characteristics (Swapp, 2014): acquiring elemental maps or spot chemical analyses with EDS; secondly, phase discrimination is based on mean atomic number (commonly related to relative density) with BSE; and lastly, compositional maps based on differences in trace element "activitors" (typically transition metal and rare earth elements). SEMs are also commonly used to determine phases using qualitative chemical analyses and/or crystalline structure. The SEM may also be used to precisely measure very small features and objects down to 50 nm in size. Backscattered electron (BSE) images can be utilized to distinguish phases in multiphase mixtures quickly.

3.5 SAMPLE PREPARATION AND COLLECTION

Depending on the nature of the samples and the data sought, sample preparation can be simple or complex for SEM analysis. Acquisition of a sample that will fit into the SEM chamber and some accommodation to prevent charge build-up on electrically insulating samples are the only requirements for minimal preparation. A thin layer of conducting substance, such as carbon, gold, or another metal or alloy, is applied to most electrically insulating samples. The best material for conductive coatings depends on the type of data to be collected: carbon is best for elemental analysis, whereas metal coatings are best for high-resolution electron imaging. In an instrument capable of "low vacuum" operation, an electrically insulating material can be studied without a conductive covering.

3.5.1 UNDERSTANDING OR READING AN SEM MICROGRAPH

An SEM micrograph reveals the surface of sample. It shows the shape of the sample, how they are arranged, and if there is a crack or pinhole. Reporting it simply means describing the arrangement, size, and shapes of the scanned surface.

Apart from obtaining the SEM micrograph, the SEM equipment can be used to obtain the thickness of a sample by using stylus profilometer. It can also be used to identify the elements present in a sample by using a component called energy-dispersive X-ray (EDX) spectroscopy and particle size by using a software such as ImageJ.

3.6 ELEMENTAL COMPOSITION (EDX)

Most SEM equipment is equipped with technology to perform elemental composition analysis of a sample. This tool is called energy-dispersive X-ray spectroscopy (EDS, EDX, EDXS, or XEDS). It is a tool that gives the element present in a sample by composition. It is also used for elemental color mapping, in which the element present is colored to show the distribution in the scanned area of the sample. It is sometimes called energy dispersive X-ray analysis (EDXA or EDAX) or energy-dispersive X-ray microanalysis (EDXMA). The composition of a sample, such as thin films, is determined using X-ray energy dispersive spectroscopy (EDX). Not only can it measure relative amounts of each atom, but we can also visualize the arrangement of atoms in samples. It is based on an interaction between an X-ray source and a sample. Its characterization skills are largely attributable to the fundamental premise that each element has a unique atomic structure, resulting in a distinct collection of peaks on its electromagnetic emission spectrum (which is the main principle of spectroscopy) (Goldstein, 2003). Moseley's law predicts peak positions with a level of precision far above the experimental resolution of a standard EDX device.

A beam of electrons is focused into the sample being investigated to promote the emission of distinctive X-rays from it. At rest, an atom in the sample comprises ground-state (or unexcited) electrons bound to the nucleus in discrete energy levels or electron shells. The incident beam may excite an electron in an inner shell, causing it to be ejected from the shell and leaving an electron hole in its place. The hole is subsequently filled with an electron from an outer, higher-energy shell, and the energy difference between the higher-energy shell and the lower-energy shell may be emitted as an X-ray. An energy-dispersive spectrometer can determine the amount and energy of X-rays emitted by a specimen. EDS permits the elemental makeup of the sample to be assessed since the energies of the X-rays are typical of the difference in energy between the two shells and of the atomic structure of the emitting element. EDS is composed of X-ray beam, analyzer, pulse detector, and X-ray detector. Figure 3.6 gives the operation of the EDX and an image of TiO_2 EDX.

EDS detectors are utilized with AEM in two different ways. The standard detector (Be-window type) detects an X-ray via a Be window of about 7 m thickness, whereas the UTW (ultrathin window)–type detector detects an X-ray through a thin plastic film covered with about 0.1-m-thick aluminum (Harada and Ikuhara, 2013). The elements heavier than Na (Z = 11) are observable in the Be-window type because Be film absorbs X-rays, whereas the elements heavier than C (Z = 6) are detectable in the UTW type. In other words, the most important aspect of the UTW type is the capacity to analyze light elements such as C, N, O, and so on. The Be-window type has a strong detectability since it can increase the detector's effective area. EDS can do both qualitative and quantitative analyses, and it also can perform elemental analysis with ease. The greatest disadvantage, however, is that the energy resolution is only approximately 150 eV. As a result, spectrum overlap occurs, necessitating careful interpretation. In EDS analysis, there are primarily two ways. One is the point analysis approach, in which an X-ray spectrum is collected after the electron probe is halted at one place on the object using TEM/STEM. The elemental mapping method, which uses STEM to detect the electron probe on the sample in two dimensions, modulates the illumination pertaining to the intensity of a specific characteristic X-ray, synchronizes with a scanning signal, and presents a two-dimensional image of the characteristic X-ray intensity on a liquid crystal monitor. The most modern AEMs with FEG and a Cs-corrector may focus the electron probe diameter on the specimen to as small as 0.1 nm. As a result, atomic-level point analysis and two-dimensional elemental mapping are conceivable, and they are frequently employed in material science study.

3.6.1 Understanding or Reading an EDX Micrograph

Figure 3.7a shows the EDX micrograph with composition and Figure 3.7b shows the EDX mapping. The symbols of each element present in the sample are represented. Thereafter, a table showing the weight percentage (wt%) of each of the elements is given. This shows the weight of the element that is contained in that segment scanned by the machine. The elements' weight percentage adds up to 100%. For the EDX mapping, the element present in the scanned piece of the sample is represented by different colors. These colors are assigned before scanning. A red color was assigned to manganese (Mn) in Figure 3.7b. The color variation can be seen in the final piece scanned.

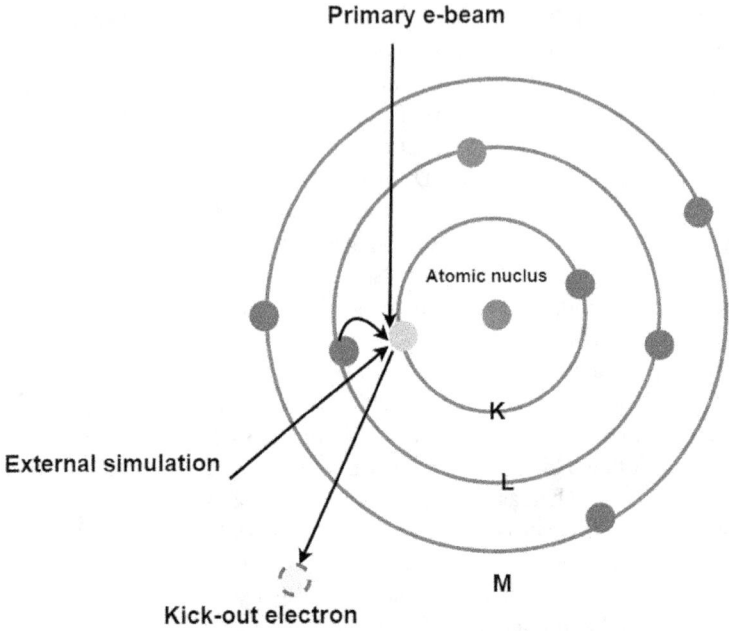

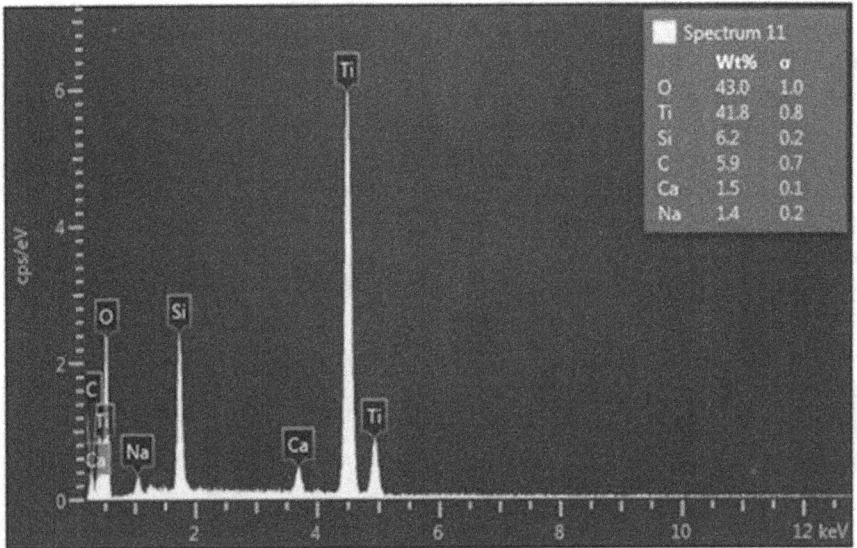

FIGURE 3.6 Schematic of (a) EDX operation and (b) EDX image of TiO_2.

Source: Lukong, Ukoba, Yoro, & Jen, 2022.

3.6.2   HISTORY OF EDS

The first usage of EDS was in the 1960s. This was made possible because microana-lyzer had solid state detectors imbedded into them. Wavelength-dispersive spectrom-eter (WDS) was used in place of EDS before the discovery.

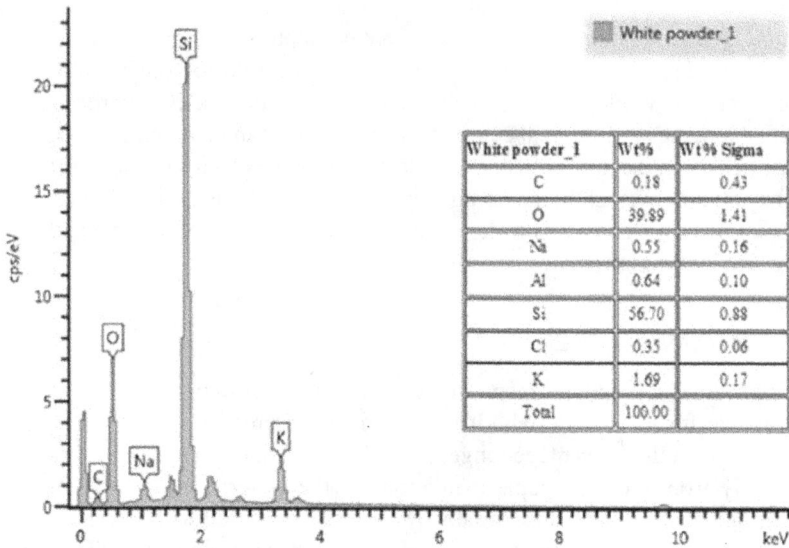

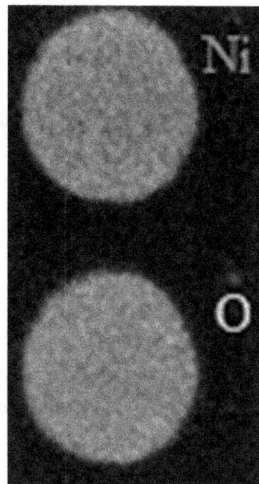

White powder_1	Wt%	Wt% Sigma
C	0.18	0.43
O	39.89	1.41
Na	0.55	0.16
Al	0.64	0.10
Si	56.70	0.88
Cl	0.35	0.06
K	1.69	0.17
Total	100.00	

FIGURE 3.7   EDX micrograph: (a) elemental composition of nanosilica produced from palm kernel shell ash; (b) EDX mapping photographs for Ni and O.

Sources: Imoisili, Ukoba, & Jen, 2020; Nwanya, Botha, Ezema, & Maaza, 2021.

3.6.3 ADVANTAGE OF EDS

The advantages of EDS include elemental exposure for all but the lightest elements (carbon and above are detectable, boron is troublesome); a fairly rapid elemental analysis approach (in most circumstances); elemental coverage for all but the lightest elements (carbon and above are detectable, boron is difficult); quantitative elemental data; the capacity to scan areas and isolated places (raster scanning); a wide spatial range ranging from 1 mm^2 to submicron-squared scale. Image data generated by an electron microscope is connected to elemental spectra. The data can be used to create elemental maps, or "dot maps." Variable excitation voltages and modeling tools such as Monte Carlo simulations can be used to obtain depth information. Only the top handful of microns of the substance under inquiry are used to collect information (surface sensitive). While many people think of this as a destructive approach, especially when it comes to electrical components, it isn't in many circumstances (e.g., in most cases electronic components are not damaged by the electron beam). WDS has several advantages over EDS, including increased sensitivity by around 1–3 orders of magnitude for most elements, a wider elemental range down to beryllium, and quantitative analysis that is more accurate than EDS.

3.6.4 DISADVANTAGE OF EDS

Nitrogen provides a very faint response in most detector systems, making its detection unreliable for most materials. It is a relatively insensitive method with lower detection limits in the percentage range; only elemental data is generated; generated data are only from the top couple of microns of the material under investigation, complicating bulk analyses; it is a relatively insensitive method with lower detection limits in the percentage range; only elemental data is generated; quantitative analysis of heterogeneous materials frequently yields erroneous results; nonconductive samples may need to be coated with a conductive film, usually resulting in destructive analysis; samples must be submitted to vacuum conditions; chamber dimensions often limit the size of samples that can be analyzed (large chamber systems do get around this limitation, but they are the exception, not the rule); nonconductive samples may need to be coated with a conductive film, usually resulting in destructive analysis.

3.6.5 LIMITATION OF EDS

Gases cannot be studied since samples must be exposed to vacuum conditions, and liquids are limited to those with very low volatility and will not contaminate the system (some labs will not perform any liquids analysis out of contamination concerns) (Wolfgong, 2016a). It is critical to emphasize that only rudimentary information is provided, and many times incorrect assumptions about the identity of unknown materials are made as a result. When carbon is found, for example, it is conventional to presume that organic stuff is present. This is a hypothesis based on elemental data, and it's often a fair one, although there are many examples of inorganic carbon, including carbonate corrosion products. Lead carbonate, for example, is frequently the primary cause of tin/lead solder corrosion. It would be inaccurate to attribute carbon observed on such solder to organic contamination, such as flux residue, in this scenario. Furthermore, some elements (boron, beryllium, and nitrogen) are difficult to detect, while others (hydrogen, helium, and lithium) are not detectable at all, resulting in material misidentifications (Wolfgong, 2016b). This is mentioned because when using these approaches, extreme caution should be used when assigning exact materials identification.

3.6.6 TRENDS IN EDS

An improved EDS detector, known as the silicon drift detector, is gaining popularity. A high-resistivity silicon chip is used to push electrons to a small collecting anode in the SDD. The benefit comes from the anode's exceptionally low capacitance, which allows for faster response time and higher throughput. The SDD has several advantages: counting and processing at a high rate and reducing the amount of time spent dead (time spent on processing X-ray event). At high count rates, this detector has better resolution than standard Si(Li) detectors. More exact X-ray maps or particle data are obtained in seconds, as well as faster analytical capabilities. It can operate and store at increased temperature in absence of liquid nitrogen. Because the SDD chip's capacitance is independent of the detector's active area, significantly bigger SDD chips can be used (40 mm^2 or more). This enables a higher collecting rate of counts. Large-area chips can have other advantages. Minimizing SEM beam current allows for better imaging under analytical settings, as well as lower specimen damage (Kosasih et al., 2021) and enhanced spatial resolution for high-speed maps.

Traditional silicon-based methods suffer from poor quantum efficiency due to a drop in detector stopping power when the X-ray energy of interest is more than 30 keV. High-density semiconductor detectors, such as cadmium telluride (CdTe) and cadmium zinc telluride (CdZnTe), are more efficient at higher X-ray energy and can operate at room temperature. At 100 keV, single-element devices and more recently pixelated imaging detectors like the high energy X-ray imaging technology (HEXITEC) system can achieve energy resolutions of the order of 1%. A new form of EDS detector, based on a superconducting microcalorimeter, has become commercially available in recent years. The simultaneous detection capabilities of EDS are combined with the great spectral resolution of WDS in this new technique. There are two parts to the EDS microcalorimeter: an absorber and a superconducting transition-edge sensor (TES) thermometer. The former absorbs X-rays released by the sample and converts the energy into heat, while the latter measures the temperature change because of the heat inflow. Historically, the EDS microcalorimeter has had a variety of flaws, including poor count rates and limited detector regions. The count rate is limited by the calorimeter's electrical circuit's dependency on the time constant. To keep the heat capacity low and enhance thermal sensitivity, the detector area must be tiny (resolution). The deployment of arrays of hundreds of superconducting EDS microcalorimeters has enhanced the count rate and detector area, and the importance of this technology is expanding.

EDS can identify major and minor elements at amounts high exceeding 10 wt% (major) and less than 10 wt% (minor) (concentrations between 1 and 10 wt%). Because the detection limit for bulk materials is 0.1 wt%, EDS is unable to identify trace elements (concentrations below 0.01 wt%) (Makhlouf and Aliofkhazraei, 2015). The technique of energy dispersive spectroscopy (EDS) is primarily employed for qualitative material analysis, but it can also provide semiquantitative results. SEM instrumentation is typically equipped with an EDS system to allow chemical analysis of features seen in the SEM monitor. In failure analysis scenarios where spot analysis is critical to reaching a reliable conclusion, simultaneous SEM and EDS analysis is useful. Secondary and backscattered electrons utilized in image formation for morphological analysis, as well as X-rays used for identification and quantification of compounds present at measurable amounts, are among the signals produced by an SEM/EDS system.

3.6.7 Particle Size Analysis

Particle size analysis is a technique for determining the particle size distribution in a material. Solid samples, suspensions, emulsions, and aerosols can all benefit from particle size analysis. Particle size measurement can be done in a variety of ways. It evaluates the particle size variation as well as the average or mean particle size in a liquid or powder sample. Measurement of particle size is commonly accomplished with particle size analyzers (PSAs) (Khanam et al., 2016). These are based on a variety of technologies, including high-resolution image processing, Brownian motion analysis, particle gravitational settling, and light scattering (Rayleigh and Mie scattering). The following equipment can be used to measure the particle size of sample especially nanoparticles: dynamic light scattering (E), nanoparticle tracking analysis (SP), disc centrifugation (E), tunable resistive pulse sensing (SP), electron microscopy (SP), and atomic force microscopy (SP) (Hristov et al., 2017). Another approach is to use an image analysis software like ImageJ with image picture of SEM or TEM (Yokoyama et al., 2008).

3.6.8 Determining the Correct Particle Size Analysis

There are a variety of approaches for performing particle size. Although, they differ in the final output. The range of particle size often influences the technique used. Therefore, it is critical to select the technique that is most relevant to the application. The particle size is inferred from a measurement, rather than a direct measurement of particle diameter, for instance, particle motion, electrical resistance, and light scattering. This allows an instrument to quickly measure a particle size distribution, but it does necessitate some type of assumptions or calibration about the particles nature. The assumption of spherical particles is frequently used, resulting in a result with an equivalent spherical diameter. When comparing the findings of different equipment, the particle size distribution measured to differ is common. The approach that is most appropriate to apply is usually the one that is aligned with the data's end usage. For example, while deciding whether to test a chemical compound using laser diffraction or dynamic light scattering (Lotya et al., 2013), one should consider the sample type (solid or liquid), expected size range, chemical stability, amount of sample available, and the sector applications (Foerter-Barth and Teipel, 2000). A sedimentation approach for size is highly pertinent when developing a sedimentation tank. Nonetheless, this technique is not always viable, necessitating the employment of a different technique (Primavera et al., 2014). Expert systems are built online to aid in choosing (and eradicating) particle sizes analysis equipment (Ganguli and Bandopadhyay, 2002).

3.6.9 Light Scattering Particle Size Analysis

Particle size analysis based on light scattering has a wide range of applications since it provides for very simple optical characterization of samples, allowing for improved product quality control in a variety of industries such as pharmaceutical, food, cosmetics, and polymer production (Syvitski, 1991). Many advances in scattering light methods for characterizing particle have been made in recent years. DLS (dynamic light scattering) is now a standard technology in the industry for particles to lower micrometer ranges in the lower nanometer (Stetefeld et al., 2016). In the academic sector, it is the most extensively utilized light scattering means for characterizing particle (Franks et al., 2019; Xu, 2015). When suspended particles are irradiated with laser, this approach analyzes the changes of scattered light to estimate the Brownian motion velocity, which may be utilized to calculate the particles size hydrodynamic using the Stokes–Einstein relationships. The DLS is a quick, noninvasive procedure that is also repeatable and precise (Mu et al., 2020). Furthermore, because the approach is based on measuring light scattering as a function of time, it is known to be absolute; therefore DLS equipment need not be calibrated (Xu, 2015). One of its drawbacks is that it cannot resolve extremely polydisperse samples properly, and the big particles present can compromise accuracy of size. Other scattering approaches have surfaced, such as NTA (nanoparticle tracking analysis), which uses picture recording to detect individual particle movement via scattering (Kim et al., 2019). NTA, like DLS, uses the diffusion coefficient to determine the hydrodynamic size of particles, but it is capable of circumventing some of DLS' restrictions (Kim et al., 2019).

PSAs (particle size analyzers) rely on laser diffraction (LD) or static light scattering (Blott et al., 2004). They are the most prominent and commonly utilized instruments for evaluating hundreds of nanometer particles to many millimeters, and the above techniques mentioned are best fit for evaluating particles usually in the submicron regions. Nonelectromagnetic systems based on wave propagation, for example, ultrasonic analyzers, use a similar scattering principle. A laser beam is utilized in irradiating dilute particle suspended in LD PSAs. The forward-directed light scattered by the particle is lens focused onto a vast array of concentric photodetectors ring. The scattering angle of the laser beam increases as the size of the particle decreases. Using Mie or Fraunhofer scattering model, one can estimate the particle size distribution by the angle-dependent scattered intensity measurement (Vargas-Ubera et al., 2007). The refractive index prior knowledge of the measured particles, and the dispersant, is essential in the latter scenario.

3.6.10 Transmission Electron Microscope

Transmission electron microscopes (TEMs) are microscopes that visualize samples and produce a magnified image using a particle stream of electrons. TEMs can magnify objects by up to 2 million times. Consider how little a cell is to get a better picture of just how small that is. Transmission electron microscopy is a microscopy technique in which an image is formed by passing an electron beam through a specimen. A suspension on a grid or an ultrathin segment less than 100 nm thick is the most common specimen. With its diverse imaging modes and analytical capabilities, transmission electron microscopy (TEM) has become an invaluable instrument for the chemical and structural characterization of all forms of samples at the nanoscale (Sciau, 2016).

The ability of transmission electron microscopy to create images of atomic arrangements at limited locations within materials is its distinctive feature (Pennycook, 2017). It shows the microstructure, or the differences in structure from one region to the next. It also shows the boundaries connecting them. When macroscopic properties are regulated or impacted by defects or interfaces, such as in the development of advanced structural materials with complex second-phase microstructures or electronic materials that rely on precise control of interfaces and multilayers, TEM plays a key role. The information obtained from X-ray or neutron diffraction is highly complementary.

Because electrons are charged particles, unlike X-rays or neutrons, they can be accelerated and accurately focused by electromagnetic fields, thus making the TEM plays a unique role. In the manner of an optical microscope, the scattered beams can be gathered by a lens and refocused to generate a true real-space image, with each point in the image corresponding to a precise point in the object. Electrons have a far stronger interaction with matter than protons, and electron diffraction can be conducted on nanometer-sized materials. The electron wavelength varies from 0.004 to 0.001 nm for accelerating voltages of 100–1000 kV, orders of magnitude less than atomic spacings in materials.

These approaches can pinpoint the average structure of complicated samples, but not the structure of a single nanostructure or local region. TEM, or its scanning equivalent, STEM, is playing an increasingly important role in basic condensed matter physics research as condensed matter physics evolves toward the study of ever more complicated materials and interest in nanoscale physics and devices grows (Pennycook, 2017).

3.6.11 History of TEM

In 1873 Ernst Abbe proposed that the capability to resolve details in an object is limited by the imaging light wavelength, or a few hundred nanometers for visible light microscope. In 1931, Ernst Ruska and Max Knoll at the University of Berlin combined these qualities to create the first transmission electron microscope (TEM), for which Ruska received the Nobel Prize in Physics in 1986. Köhler and Rohr pioneered improvements in ultraviolet (UV) microscopes, which enhanced determining power by two factors (Davidson, 2013; KASTEN, 1989). However, because of the absorption of UV by glass, this necessitated the use of expensive quartz lenses. Due to the wavelength constraint, it was thought that obtaining an image with submicrometer information was impossible (Ruska, 1987). Plücker discovered that magnetic fields deflect "cathode rays" (electrons) in 1858 (Hawkes, 2002). Ferdinand Braun utilized this effect to create basic CRO (cathode-ray oscilloscope) measuring equipment in 1897 (Süsskind, 1980). Riecke discovered that cathodes ray might be used as a source of light in 1891 (Braun, 2000). In 1891, Riecke discovered that magnetic fields could focus cathode rays that allow simple electromagnetic lens design. Hans Busch published in 1926 that extended this hypothesis and demonstrated that the lens maker equations could be applied to electron given the right assumptions (Jain, 1977; Kalita et al., 2017; Robinson, 1986). Adolf Matthias, Professor of High Voltage Technology and Electrical Installations at the Technical University of Berlin, appointed Max Knoll to lead a team of researchers to enhance the CRO design in 1928 (Schulze, 2007). Different doctoral students, notably Bodo von Borries and Ernst Ruska, were part of the team. The team focused their research on CRO column location and lens design, as well as optimizing building better CROs parameters and fabricating components of the electron optical to generate images with low magnifications (almost 1:1). The group succeeded in magnifying mesh grids images put over the anodes aperture in 1931. The apparatus, which arguably created the first electron microscope, utilized two magnetic lenses to obtain high magnification. Reinhold Rudenberg, Siemens' scientific director, patented an electrostatic lens electron microscope in the same year (Rudenberg and Rudenberg, 2010).

From the invention of the first transmission electron microscopes in the 1930s to the imaging of crystal lattices at 0.2–0.3 nm resolution in the 1980s, it took about 50 years (Iijima, 1980). Over the next two decades, resolution gradually improved to 0.1–0.2 nm. Electron microscopy is currently undergoing a revolution. A succession of nonround magnetic lenses can be employed to rectify the aberrations of the (round) objective lens thanks to the development of solid-state electronics, notably computer and charge-coupled device (CCD) detectors. The increase in resolution witnessed in recent years is comparable to that seen in recent decades, a remarkable achievement that has pushed TEM into the subangstrom regime for the first time in history.

3.7 SAMPLE PREPARATION OF TEM

TEM sample preparation can be a time-consuming process. For a standard TEM, samples should be fewer than 100 nm thick (Cheville and Stasko, 2014). Unlike X-ray radiation or neutron, electron in the beam readily interacts with the material, an effect which grows in proportion to the atomic number squared (Z2). The thickness of high-quality samples will be equivalent to the electron mean free paths passing all the way through them, which may be few tens of nanometers. The type of information to be collected from a TEM specimen is dependent on the material under investigation and the type of information to be obtained from the specimen.

The material type also determines the method employed in preparing the sample. Tissue sectioning, mechanical processing (milling), sample staining, etching (ion and chemical), nanowire-assisted transfer, replication, and ion milling are some of the TEM preparation methods. Deposition of dilute samples with the sample onto films on support grid can be used to swiftly prepare materials with dimensions sufficiently tiny for electrons transparent, for instance, powder compounds, nanotubes, small organisms, or viruses. To endure the higher vacuum in the specimen chamber and allow cutting tissue into electrons transparent thin section, biological specimens can be embedded in resin. The biological samples can be stained with negative staining substance for bacteria and viruses, such as uranyl acetate, or with heavy metals, such as osmium tetroxide, in the case of embedding sections. After embedding in vitreous ice, samples can be kept at liquid nitrogen temperatures (Amzallag et al., 2006). Although specimens in metallurgy and material science can normally resist high vacuum, they must still be etched or prepared as thin foils so that some piece of the sample is thin enough for the beam to pass through. The scattering cross-section of the atoms that make up the material may impose limitations on the thickness of the substance.

For sectioning tissue, biological tissues are frequently encased in resin blocks and ultramicrotome thin to less than 100 nm. As the resin block goes over a diamond or glass knife edge, it fractures (Porter and Blum, 1953). This approach is utilized to create thin, plainly distorted sample that allows tissue ultrastructure to be examined. Inorganic sample, for example, aluminum, can be embedded also in resins and ultrathin sectioned utilizing sapphire, coated glass, or greater angle diamond blades in this fashion (Phillips, 1961). Tissue samples must be covered with small layers of conducting materials, for instance, carbon, in preventing charge build-up at the surface of the sample when seen in the TEM.

Sample staining is used to improve contrast in biological tissue TEM samples by using high atomic number stains. The stain absorbs or scatters a portion of the beam electrons that would otherwise be projected onto the imaging device. Prior to TEM observation, heavy metal compounds, for instance, lead, osmium, gold (in immunogold labeling), or uranium, can be utilized to deposit electron-dense atoms in or on the sample in desired cellular or protein regions. Understanding how heavy metals bind to various biological tissues and cellular structures is required for this process (Alberts et al., 2003).

Dimpling is a specimen preparation procedure that results in a specimen with a thinned central area and a thick enough outer rim to allow for easy handling. Traditionally, ion milling has been the last step in the specimen preparation process. The application of high voltage accelerates charged argon ions to the specimen surface in this process. As a result of momentum transfer, ion impact on the specimen surface removes material.

For imaging TEM samples, mechanical polishing is also used to prepare the samples. To guarantee consistent thickness of the sample across the region of interest, high-quality polishing is required. In the last phases of polishing, a diamond or cubic boron nitride polishing compound may be used to eliminate any scratches that may create contrast fluctuations due to varied sample thickness. Final stage thinning may require additional precise procedures like ion etching, even after careful mechanical milling.

Chemical etching can be used to prepare some samples, particularly metallic specimens. In preparing samples for TEM observation, chemical etchant, for example, acids, is used to thin it. Devices for controlling the thinning process may include system to detect when the specimen is thinned to an appropriate optical transparency level and can allow the operator to regulate the current or voltage running through the sample.

Etching of ion is a type of sputtering in removing small amounts of material. This is utilized in polishing samples that have been polished using other methods. Ion etchings generate plasma streams that are directed to the specimen surface by passing inert gases via electric fields. For gases like argon, the acceleration energies are usually a few kilovolts. The samples can be rotated to ensure that the sample surface is polished evenly. Such technologies have a sputtering tens rate of micrometers per hour, restricting the process to only tremendously fine polishing. Recent research has shown that ion etching with argon gas may file down MTJ stack structure to specific layers, which can subsequently be resolved atomically. The MgO layer within MTJs comprises a considerable number of grain boundaries, which may be reducing the device's characteristics, according to TEM pictures obtained in plain views instead of cross-sections (Bean et al., 2017).

3.7.1 OPERATING PRINCIPLE OF TEM

The TEM works on the same principles as a light microscope, except instead of light, it employs electrons. Because electrons have a significantly shorter wavelength than light, the resolution of TEM images is many orders of magnitude better than that of light microscope images. As a result, TEMs can disclose the tiniest details of interior structure, down to individual atoms in some situations.

3.7.2 ELECTRON SOURCE

The TEM is made up of an emission source or cathode, which can be a tungsten filament or needle, or a single crystal of lanthanum hexaboride (LaB6) (Egerton, 2005b). The gun is linked to a high-voltage source (usually 100–300 kV), and given enough current, it will start emitting electrons into the vacuum via thermionic or field electron emission. The electron source of a thermionic source is mounted typically in Wehnelt cylinders to enable initial focus of the emitted electron into the beams and to stabilize the current via passive feedback circuits. Instead, field emissions source controls the electric field shapes and strength at the sharp tips with an electrostatic electrode known as extractor, a gun lens, and a suppressor, each with a distinct voltage. The "electron cannon" is the name given to the combination of the cathodes and this initial electrostatic lens element. The beam is frequently accelerated by a series of electrostatic plates after leaving the gun until it enters the condenser lens system and reaches its final voltage, which is the next section of the microscope. The electron beam is then further focused by the TEM's higher lenses to the required size and placement on the material (Rose, 2008).

The manipulation of the electron beam is accomplished using two physical effects. Electrons will move according to the left-hand rule when they encounter a magnetic field, allowing electromagnets to manipulate the electron beam. Magnetic fields can be used to create a magnetic lens with changeable focusing power, with the lens shape resulting from the distribution of magnetic flux. Electrostatic forces can also cause the electrons to be deflected at a constant angle. The development of a shift in the beam path by coupling two deflections in opposite directions with a small intermediate gap allows for beam TEM shift, which is critical for STEM. The phenomena, together with the utilized electron imaging systems, allow for considerable beam path control for TEM performance. Unlike an optical microscope, the optical arrangement of a TEM may be rapidly modified since lenses in the beam axis can be activated, have their strength changed, or be completely disabled just by rapid electrical toggling that is restricted by phenomena such as the lenses magnetic hysteresis.

3.7.3   SPECIMEN STAGE

Airlocks are built into TEM specimen stage designs to allow the specimen holder to be inserted into the vacuum with minimum vacuum loss in other regions of the microscope. A standard-size sample grid or self-supporting specimen is held in the specimen holders. The standard TEM grid size is 3.05 mm in diameter. It has 100 m thickness with a mesh size of 100 m. The sample is deposited onto a meshed region with a diameter of around 2.5 mm. Copper, molybdenum, gold, and platinum are common grid materials. The sample holder, linked with specimen stage, holds this grid. Depending on the type of experiment being conducted, a wide range of stage and holder configurations are available. Aside from 3.05 mm grids, the type of experiment determines the stage holder and configuration. However, 2.3 mm grids and 3.05 mm grids are mainly used. The thickness of electron transparent specimens is typically less than 100 nm. Although, this number varies depending on the accelerating voltage.

3.7.4   ELECTRON GUN

The electron gun is composed of Wehnelt cap, extract anode, biased circuit, and filament. Electron can be "pumped" from the electrons gun to the TEM columns and the anode plates by connecting filament to negative component power supply, completing the circuits. The guns are made to fire electron beam out of the assembly at specific angles, called the gun divergence semiangles. When the Wehnelt cylinder is built with a higher negative charge than the filament, electrons that depart the filament in a diverging pattern are forced into a converging pattern with a minimum size of the gun crossover diameter under appropriate operation. A cross-section of an electron gun assembly with electron extraction is shown in Figure 3.8.

Richardson's law can be used to connect the thermionic emission, J, current density to the emitting materials work function as seen in equation 3.1.

$$J = AT^2 exp\left(\frac{-\Phi}{kT}\right) \tag{3.1}$$

A is Richardson's constant, T is the temperature of the material and Φ is the work function (Williams and Carter, 1996).

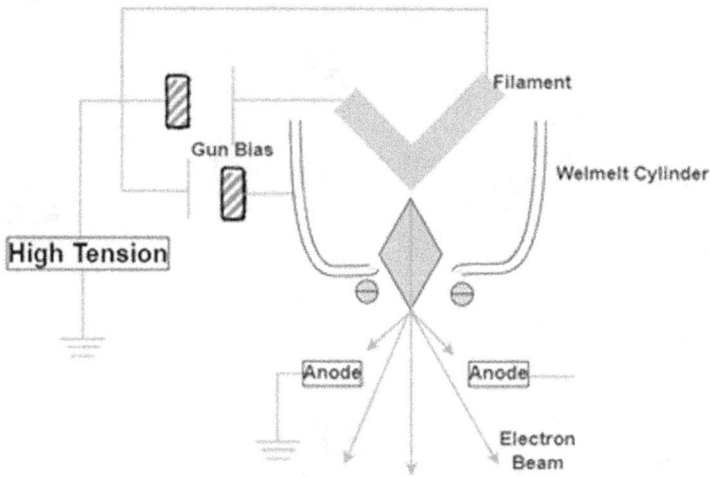

FIGURE 3.8 Cross-sectional diagram of an electron gun assembly, illustrating electron extraction.

Equation 3.1 illustrates that to generate an appropriate current density, the emitter must be heated, taking care not to harm the emitter by applying too much heat. For this reason, the gun filament must be made of elements with a high melting point, such as tungsten, or materials with a low work function, such as LaB6 (Buckingham, 1965). Furthermore, to create thermionic emission, both tungsten thermionic and lanthanum hexaboride sources must be heated, which can be done with the help of a tiny resistive strip. To avoid thermal shock and temperature gradients from destroying the filament, a delay in the application of current to the tip is frequently enforced. The wait is normally a few seconds for LaB6, and substantially less for tungsten.

3.7.5 IMAGING

The condenser lens concentrates the electron beam from the electron gun into a compact, thin, coherent beam. The condenser aperture limits this beam by excluding high-angle electrons. The beam then strikes the specimen and depending on the thickness and electron transparency of the material, parts of it are transferred. The objective lens focuses this transmitted portion into an image on a phosphor screen or a charge-coupled device (CCD) camera. By blocking off high-angle diffracted electrons, optional objective apertures can improve contrast. The image is then expanded all the way down the column through the intermediate and projection lenses. When a picture hits a phosphor screen, light is produced, allowing the viewer to see it as shown in Figure 3.9. The darker areas of the picture reflect areas of the sample through which fewer electrons are communicated, while the lighter areas represent areas of the sample through which more electrons were transmitted.

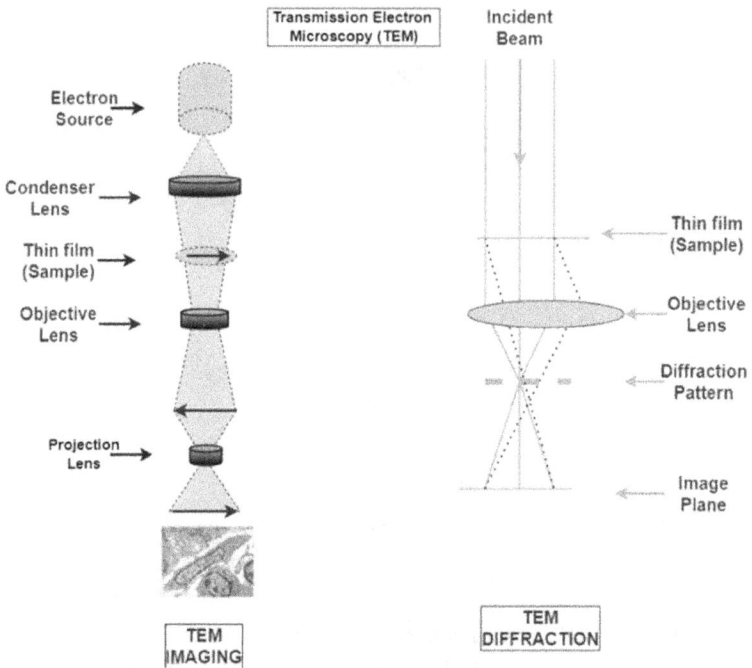

FIGURE 3.9   Schematic of imaging and diffraction of TEM.

3.7.6 DIFFRACTION

A basic diagram depicting the passage of an electron beam in a TEM from immediately above the specimen to the phosphor screen is Figure 3.10. The electrons are scattered as they pass through the sample due to the electrostatic potential created by the constituent elements (Williams and Carter, 1996). They pass through the specimen and then through the electromagnetic objective lens, which concentrates all the scattered electrons from one location on the specimen into a single point in the picture plane. A dotted line is also depicted where the electrons distributed in the same direction by the sample are gathered into a single point. This is the objective lens's back focal plane, where the diffraction pattern is generated.

3.7.7 LIMITATION OF TEM

The TEM technique has several disadvantages. The TEM analysis technique is time-consuming with limited throughput sample because many samples necessitate substantial specimen preparations to create samples thin enough to be electron see-through. During the preparation procedure, the specimen's structure may also alter. Also, because the field of view is narrow, it's possible that the region evaluated isn't representative of the entire sample. The electron beam has the potential to destroy the sample, especially with biological material. The significant inherent aberrations of a round magnetic lens were discovered early in the history of the microscope

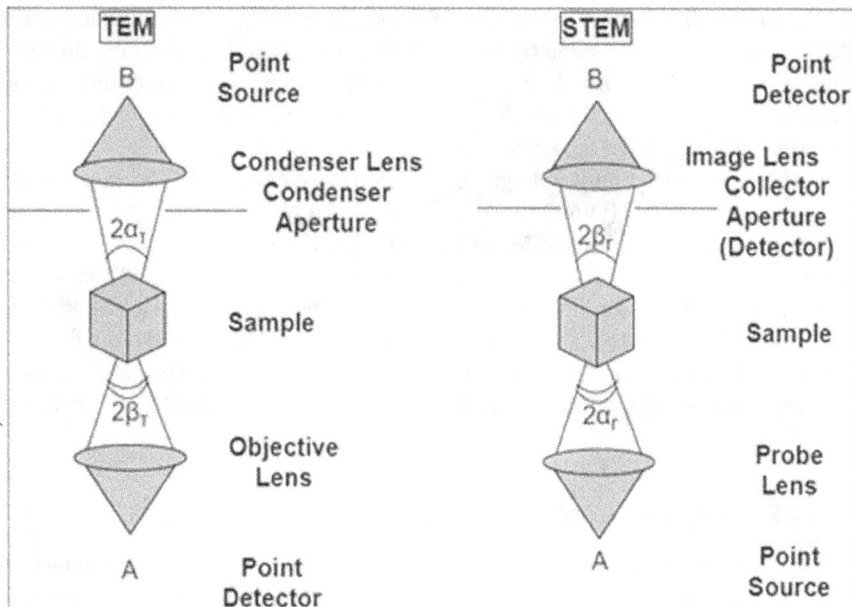

FIGURE 3.10   Schematic ray diagram illustrating the optical reciprocity between TEM (left) and STEM (right). The convergence angle in TEM, α_T, becomes the collection angle in STEM, β_r.

to be a serious constraint (Linck et al., 2017). The primary aberration is spherical aberration, which results in a ray deviation of $Cs\alpha 3$, where α is the objective lens's semiangle (Tanaka, 2008). Cs is the same order as the focal length in electron lenses. This means that only very small apertures could be utilized, and since microscope resolution is defined as $0.61\lambda/\sin\alpha$, the best resolution would be limited to ~50λ by diffraction(Sawada et al., 2005).

3.7.8   MODIFICATION OF TEM

The limitation of TEM can be improved upon in various ways. The methods include scanning TEM, low-voltage electron microscope, in situ TEM (mechanical, environmental and pressure), cryo-TEM, corrected aberration TEM, and dynamic and Ultrafast TEM.

When combined with proper detectors, TEM can be converted into STEM (scanning transmission electron microscope) by adding device that rasters convergent beam across the material to generate the images. Scanning coils are used to deflect the beam, which is then collected using a current detector, such as a Faraday cup, which works as a direct electron counter. The transmitted component of the beam can be assessed by connecting the electron count to scanning beam position (also called the "probe"). Beam tilting or the usage of annular dark field detectors can be used to get the nontransmitted components. A STEM is TEM with the electrons source and observation points swapped in relation to the electron beam's travel direction. The STEM apparatus uses the same optical setup as a TEM, but it reverses the direction of electron passage (or time) during operation. STEM employs a variety of detectors with adjustable collecting angle depending on which electron the user wishes to capture instead of using holes to regulate electrons detected, like in TEM.

The electron accelerating voltages of LVEM (low-voltage electron microscope) is between 5 and 25 kV (Drummy et al., 2004). Some of them can be a single compact apparatus that combines SEM, TEM, and STEM. Low voltage improves visual contrasts that are critical for biological specimen. The need for staining is significantly reduced, if not eliminated, because of the increased contrast (Mukherjee, 2013). In STEM, TEM, and SEM modes, resolutions of a few nanometers are feasible. Because of the low energy of the electron beam, permanent magnets can be utilized as lenses, allowing for the use of a small column not using cooling (Nebesářová and Vancová, 2007).

3.7.9 SELECTED AREA ELECTRON DIFFRACTION

It is a technique that is used in conjunction with S/TEM to assess the lattice parameters, orientation, crystal structure, and crystallinity of a sample by analyzing the electron diffraction pattern created by the electron beam's interaction with the sample atoms. It's often utilized for phase identification, structural intergrowth determination, and growth direction determination, among other things. SAED's lattice parameters are accurate to around 5%, and kinematically forbidden reflections are frequently present due to repeated diffraction. Selected area electron diffraction (SAED) is also known as electron diffraction.

Under parallel electron illumination, a SAED is obtained. A convergent beam electron diffraction (CBED) is obtained in the case of the convergent beam (Morniroli et al., 2006). SAED uses a broad beam that illuminates a large sample region. The specified aperture area in the plane picture is utilized to analyze only a certain sample region. This contrasts with nanodiffraction, which uses condensed beams to a small probe to achieve site-selectivity (Fultz and Howe, 2013). When orienting a sample for high-resolution microscopy or setting up dark-field imaging conditions, SAED is crucial. The Fourier transform can be used to convert high-resolution electron microscope pictures into an artificial diffraction pattern.

3.7.10 HOW TO UNDERSTAND AND ANALYZE SAED

The diffraction pattern obtained in the reverse space of the lattice planes is referred to as SAED. It is used to calculate the d-spacing of crystal planes by calculating the radius of the spots observed in the SAED pattern from the brilliant center. The d-spacing can then be calculated using the camera constant.

Figure 3.11 shows the SAED of copper oxide. The single Bragg's reflection is represented by smattered bright spot. The rings of the shiny spot represent the crystallinity. In this case, of this ring, multiple concentric rings represent polycrystalline.

FIGURE 3.11 A schematic of a SAED.

Source: Nwanya, Botha, Ezema, & Maaza, 2021.

A double exposure is needed to be able to visualize the zero-diffraction spot. Thereafter, distance between central point and each ring is measured. Spot result from diffraction of a single crystal and multiple crystals form rings. The zone axis of the patterns can be determined.

The diffraction patterns of several randomly oriented crystallites superimpose to generate an image of concentric rings if the illuminated area specified by the aperture covers many of them. For nanoparticles, powders, or polycrystalline samples, the ring diffractogram is typical. Each ring's diameter corresponds to the interplanar distance of a plane system in the sample. This diffractogram gives statistical information, such as general crystallinity or texture, rather than information regarding individual grains or sample orientation. Despite having enough crystallinity to generate smooth rings, textured materials have a nonuniform intensity distribution over the ring circumference. Ring diffractograms can also be used to distinguish among amorphous and nanocrystalline phases (De Graef, 2003).

Not all the features shown in the diffraction image are desirable. To protect the camera, the transmitted beam is frequently excessively powerful and must be shaded by a beam-stopper. In most cases, the beam-stopper also obscures some of the useful information. The backdrop intensity steadily increases as you get closer to the center of the rings, diminishing the contrast of the diffraction rings. Modern analytical software enables the reduction of such undesired image elements, which, when combined with other characteristics, enhances image clarity and aids in image interpretation.

3.7.11   PRINCIPLE OF SAED

A thin crystalline sample is illuminated in a transmission electron microscope by a parallel stream of accelerated electrons to hundreds of kiloelectron volts. If the sample is thinned enough, even metallic samples become transparent to electrons at these energies (typically less than 100 nm). The high-energetic electron behaves as wave with few thousandths wavelengths nanometers owing to the durability of wave-particle. The relativistic wavelength can be calculated using equation 3.2.

$$\lambda = \frac{hc}{\sqrt{eV\left(2m_oC^2 + eV\right)}} \qquad (3.2)$$

where m_0 is the rest mass of an electron, h is Planck's constant, c is the speed of light, e is the elementary charge, and V is an electric potential that accelerates electrons (also called acceleration voltage). For example, a wavelength of 2.508 pm is produced by a 200,000 kV acceleration voltage.

The electrons are diffracted on the crystal lattice, which acts as a diffraction grating, because the distance between atoms in crystals is around a hundred times bigger. Part of the electrons are scattered at specific angles (diffracted beams) due to diffraction, while others pass through the substance without changing their direction (transmitted beams). The electron beam ordinarily incident on the atomic lattice can be regarded as a planar wave, which is retransmitted as a spherical wave by each atom to calculate the diffraction angles. The spherical wave from the diffracted beams number under angles θn is given by equation 3.3

$$d\sin . n = n\lambda \qquad (3.3)$$

where d is the distance between atoms (if only one row of atoms is assumed as in the illustration aside) or a distance between atomic planes parallel to the beam, and θn is the order of diffraction (in a real 3D atomic structure).

Normally, an incident electron beam is transmitted ($n = 1$) and diffracted (n equal to 1 and 2) at certain angles.

Each set of initially parallel beams intersects in the rear focus plane after being bent by the microscope's magnetic lens, generating the diffraction pattern. The transmitted beams cross in the optical axis exactly. The diffracted beams cross at a specific distance from the optical axis (equivalent to the interplanar distance of the diffracting planes) and at a specific azimuth (corresponding to the orientation of the planes diffracting the beams). This permits a pattern of bright spots to emerge, which is typical of SAED (Wang et al., 2022). Using the CrysTBox simulation engine, the relationship between spot and ring diffraction was depicted on 1 to 1000 grains of MgO.

The user can choose the sample area from which the diffraction pattern will be acquired, which is why SAED is called "selected."

3.8 ATOMIC FORCE MICROSCOPY

Atomic force microscopy (AFM), also known as scanning force microscopy (SFM), is a form of scanning probe microscopy (SPM) that has achieved resolution on the order of fractions of a nanometer, which is more than 1000 times greater than the optical diffraction limit (Peng et al., 2021). Atomic force microscopy (AFM) is a form of scanning probe microscopy (SPM) that has been shown to have resolution on the order of fractions of a nanometer, which is more than 1000 times greater than the optical diffraction limit. The data is acquired by using a mechanical probe to "feel" or "touch" the surface. The image obtained depend on the tip orientation to the sample (Xu and Arnsdorf, 1994). Piezoelectric elements enable precise scanning by facilitating tiny but accurate and precise movements on (electronic) command. The naming is because it measures force interactions at atomic scale of samples. However, nuclear force is not used in the testing. It is a versatile tool for measuring the morphology (Recek et al., 2015) and studying mechanical properties of a sample (Ren et al., 2015).

3.8.1 OPERATION OF AN AFM

The AFM operates by imaging the surface and probing the sample (Bolshakova et al., 2004). An AFM creates images by scanning a tiny cantilever across a sample's surface. The cantilever's pointed tip makes contact with the surface, bending it and altering the amount of laser light reflected into the photodiode. The response signal is restored by adjusting the height of the cantilever, resulting in the measured cantilever height tracing the surface. A detailed schematic representation of AFM operation is depicted in Figure 3.12.

Force measurement, topographic imaging, and manipulation are the three main capabilities of the AFM. AFMs can be used to measure the forces between the probe and the sample as a function of their mutual separation in force measurement. This

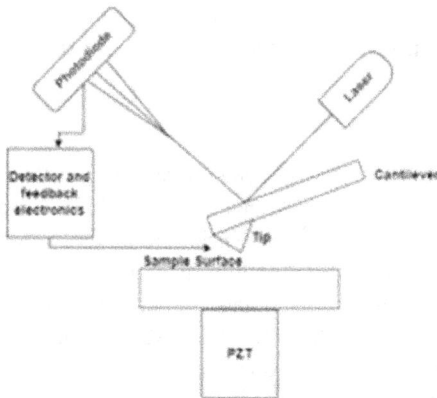

FIGURE 3.12   Operation of an AFM microscope.

can be used to do force spectroscopy and evaluate the sample's mechanical properties, such as its Young's modulus, which is a measure of stiffness (Ren et al., 2015). The response of the probe to the pressures imposed by the sample can be utilized to create a high-resolution image of the three-dimensional shape (topography) of a sample surface. This is accomplished by raster scanning the sample's position in relation to the tip and recording the probe's height, which corresponds to a constant probe–sample interaction (for more information, see Topographic image). A pseudocolor plot of the surface topography is widely used. Although Binnig, Quate, and Gerber's initial publication on atomic force microscopy hypothesized the prospect of achieving atomic resolution in 1986, significant experimental challenge had to be conquered before Ohnesorge and Binnig demonstrated atomic resolution of step edges and defects in liquid (ambient) condition in 1993 (Ohnesorge and Binnig, 1993). Giessibl had to wait a bit longer to reveal true atomic resolution of the silicon 7 × 7 surface—atomic images of this surface obtained by STM had proven the scientific world of scanning tunneling microscopy's amazing spatial resolution (Carmichael, 1995). The forces between the tip and the sample can also be used to manipulate the sample's properties in a controlled manner. Atomic manipulation, scanning probe lithography, and local cell stimulation are examples of this. Other features of the material can be monitored locally and displayed as an image, frequently with very high resolution, while topographical photos are being acquired. Mechanical traits like stiffness and adhesion strength, as well as electrical properties like conductivity and surface potential, are examples of such properties. The bulk of SPM techniques are AFM adaptations that employ this modality (Santos and Castanho, 2004).

AFM differs from rival technologies like optical microscopy and electron microscopy in that it does not need lenses or beam irradiation. As a result, it has no spatial resolution limitations owing to diffraction and aberration, and there is no need to prepare a place for guiding the beam (by producing a vacuum) or stain the sample. Scanning probe microscopy techniques [which encompasses AFM, scanning tunneling microscopy (STM), near-field scanning optical microscope (SNOM/NSOM), stimulated emission depletion (STED) microscopy, and scanning electron microscopy and electrochemical AFM (EC-AFM)] are all examples of scanning microscopy. Although SNOM and STED illuminate the sample with visible, infrared, or even terahertz light, their resolution is not limited by the diffraction limit.

3.8.2 HISTORY OF AFM

IBM scientists pioneered the AFM in 1982 (Biderman and Bui, 2021). The scanning tunneling microscope (STM), a forerunner to the AFM, was invented by Gerd Binnig and Heinrich Rohrer at IBM Research–Zurich in the early 1980s, earning them the Nobel Prize in Physics in 1986 (Binnig et al., 1986). Binnig invented the atomic force microscope, and Binnig, Quate, and Gerber performed the first experimental application in 1986 (Wang and Hao, 2022). In 1989, the first commercially available atomic force microscope was released (Shi et al., 2013). One of the most important technologies for photographing, measuring, and manipulating materials at the nanoscale is the AFM. It has since been used in chemistry and material science (Wang et al., 2021).

3.9 STRUCTURAL (XRD, XRF)

Structure of a material is known as the way the material is arranged internally. It describes the role of the element of the material in the final structure of the material. It is a major determinant in the overall properties, performance, and application of the material. The type, crystal structure, number, shape, and topological organization of phases and defects in a crystalline material, such as point defects, dislocations, stacking faults, or grain boundaries, determine the microstructure. Understanding the properties of nanomaterials necessitates a thorough understanding of their structure, from the atomic/molecular (local) to the crystal structure (long range order) and microstructure (mesoscopic scale and defect structure). In many situations, X-ray, electron, and neutron diffraction are utilized to characterize the crystal structure, which has been found to be size dependent. The precision of crystal structure determination is limited by the line broadening at very small crystallite sizes, i.e., difficulty discriminating between tetragonal and cubic zirconia. In general, all diffraction reflexes are present in CVS nanopowders, and the background is quite low. This suggests that the individual particles have a high degree of crystallinity and a low defect density. Nanocrystalline silicon carbide is an exception, as it exhibits stacking defects and twinning, particularly when manufactured at low temperatures. Not only the phase composition, lattice parameters, and atom locations in the unit cell may be derived via Rietveld refinement, but also crystallite size and microstrain.

3.9.1 X-RAY DIFFRACTOMETER

It is an analytical, highly versatile chemical technique for understanding the phase and element of a material. It is a nondestructive technique used to observe the chemical composition of a material through identification of the crystalline phases inherent in the material. The science of detecting the atomic and molecular structure of a crystal, in which the crystalline structure causes a beam of incident X-rays to diffract in many different directions, is known as X-ray crystallography. A crystallographer can create a three-dimensional picture of the density of electrons within the crystal by measuring the angles and intensities of these diffracted beams. The mean locations of the atoms in the crystal, as well as their chemical bonds, crystallographic disorder, and other information, can be deduced using this electron density. A pictorial representation of XRD equipment is shown in Figure 3.13.

3.9.2 HOW DOES IT WORK?

When a wave runs into several obstacles that are spaced evenly apart, diffraction happens. The unique phase relationship between two or maybe more waves that were dispersed by the obstacles is what causes diffraction. The pH, pressure, and temperature play a crucial role in determining the XRD patterns of a material (Ramohlola et al., 2020).

Electromagnetic waves with short wavelengths, in the region of 0.05 to 0.25 nm, are utilized to create X-rays for diffraction (wavelength on the order of the atomic spacings for solids). A voltage of around 35–50 kV must be applied between a cathode and an anode target metal that are both enclosed in a vacuum in order to create X-rays

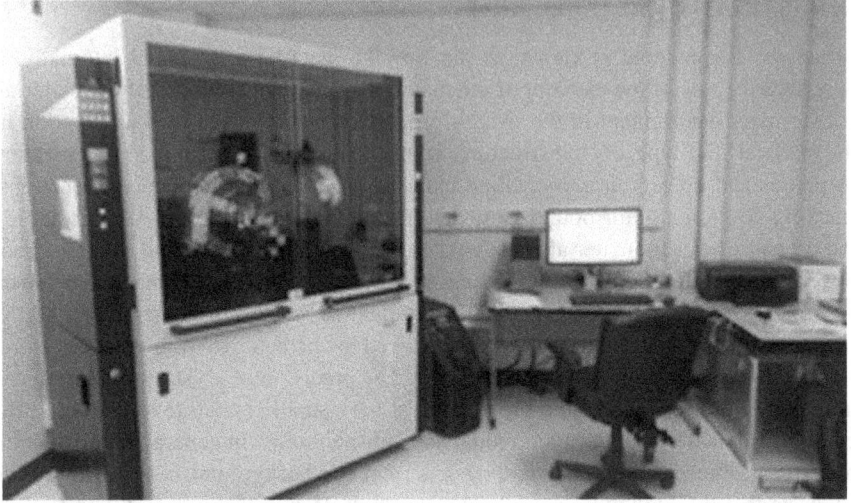

FIGURE 3.13 X-ray diffractometer.

for diffraction purposes. Electrons are emitted via thermionic emission when the tungsten filament of the cathode is heated, and they are then propelled through the vacuum by the significant voltage differential between the anode and cathode. An electrically charged particle (electron or ion) that is released by a substance at a high temperature is known as a thermon.

An electronic X-ray detector in the diffractometer immediately measures the intensity of the diffracted beam. There are several types of detectors (also known as counters), but all of them transform the incoming X-rays into surges or pulses of electric current that are then supplied into different electronic components for processing. When an X-ray beam enters a detector, its intensity is precisely proportional to the number of current pulses that are counted by the electronics in one unit of time.

In X-ray diffractometers, three different detector types are typically utilized, viz. proportional, scintillation, and semiconductor. A proportional detector is likely employed in a powder works equipment. Scintillation is still utilized in some new types of diffractometers, but not frequently. Semiconductor detectors have a lot of benefits, like being extremely effective.

It is possible to examine crystalline materials in bulk, powder, sheet, or thin film forms. It is crucial that the specimen is a good representation of the components. A small coating of crystalline powder is put on a planar substrate, which is frequently a nondiffracting material like a glass microscope slide, and subjected to an X-ray beam for a powder specimen. The amount of powder in the sample is only a few milligrams. The powder's grains should be no larger than 50 m. The specimen should ideally consist of many tiny, unevenly orientated grains that are equiaxed. The peaks in the diffraction pattern expand when the grain size is less than 1 m. Two powder mixtures must be fully combined if they are to be described.

XRD pattern is made up of several peaks of varying strengths. Peak intensity is plotted against measured diffraction angle 2θ in the spectrum. Each peak or reflection in the diffraction pattern represents an individual set of planes in the specimen from which X-rays were diffracted. The intensity is inversely correlated with the number of X-ray photons of a given energy that the detector has counted for each angle 2θ. Major focus is on the relative peak intensities and the differences in their integrated intensities (area under the peak). The peak places must be found using the pattern. The shapes and dimensions of the unit cells in the materials' crystal structures determine where the peaks appear in an X-ray diffraction pattern. The number of peaks depends on the crystal structure's symmetry. The number of peaks will rise as symmetry decreases.

Example:
In its diffraction patterns, hexagonal crystal structure exhibits more peaks than cubic crystal structure does. These non-cubic-structured materials can easily be identified from diffraction patterns from cubic materials at a look. The peaks of simple cubic and BCC structures are evenly spaced. The peaks alternately show up as a pair and a single peak in the FCC structure. The peaks are instead more unevenly spaced in diamond cubic form. The positions of the atoms are obtained in the crystal by examining the intensity of the reflections in single-phase materials. The width of a single peak, which is frequently defined as the whole width at half the maximum height, can be used to estimate the size of the crystallites and whether or not lattice distortions are present.

The "fingerprint" of a material is the diffraction pattern. A single isolated peak's position does not describe a substance or its solid state structure. The quantity of a specific phase present in the specimen is correlated with the intensity of the diffraction peaks. A powder pattern is a representation of a 3D structure in 1D. Powder pattern is excellent for phase identification since it describes the substance by its structure. Because XRD patterns' intensities correlate with the amount of the substance being studied, they can be utilized to estimate the quantitative ratio of phases in mixtures. The majority of contemporary XRD devices use computers to do peak search in addition to computing the 2θ angles and integrated intensities for each peak. The majority of commercial software enables quick matching and identification by allowing users to compare conventional patterns (from databases) with experimentally obtained patterns. Other features of some software programs include the ability to calculate crystallite size, lattice strain, and lattice parameters more precisely.

3.9.3　Interpretation of XRD Result

XRD is used to obtain the crystal size, crystal structure, and crystal strain. The nature of the material is obtained from Bragg's peak. An amorphous material has a very broad humped peak. However, a crystalline material contains a sharp peak. The positioning of atoms in a lattice structure is revealed by the peak intensity. The crystallite size and strain is obtained by the peak width.

3.9.4　Some Equations and Calculations Useful in XRD

There are certain variables that are obtained via computations. Most are computed directly or indirectly from the XRD patterns obtained. They are discussed below:

3.9.5 AVERAGE CRYSTALLITE DIMENSION

The size of crystallite particle is calculated using the Debye–Scherrer formula shown in Equation 3.4. Debye–Scherrer describes the enlargement of a peak in a diffraction pattern to the size of submicrometer crystallites in a solid. It was created in 1918 and uses measurements of the whole width at half maximum of peaks (β) in radians placed at any two points in the pattern to determine the nanocrystallite size (L) using XRD radiation of wavelength λ (nm). K's shape factor ranges from 0.62 to 2.08 but is typically estimated to be around 0.89. Homogenous material has a k value of 0.94 and heterogeneous has a k value of 0.89. The Bragg reflection is used to obtain the Debye–Scherrer ring.

$$D = \frac{k\lambda}{\beta \cos\theta} \qquad (3.4)$$

With λ representing the X-ray wavelength of Cu-k radiation (nm), β is the full width of a maximum of half (FWHM) obtained in the radians, θ represents the Bragg diffraction angle, and k is a forming constant (0.9).

As an example, Figure 3.14 shows an XRD micrograph. The micrograph is an XRD diffraction pattern for TiO_2 thin films after aging sol for 24 h, 48 h, and 72 h.

3.9.6 FOURIER TRANSFORM INFRARED SPECTROSCOPY (FTIR)

It is a subfield of spectroscopy. It is used to investigate the interaction of radiation—including electromagnetic radiation, light, and particle radiation—with matter. Among the numerous varieties of spectroscopy, FTIR is one that produces an infrared spectrum of a substance's absorption, emission, photoconductivity, or Raman scattering, whether it be a gas, liquid, or solid.

3.10 ELECTRICAL TECHNIQUE

This is used to determine the electrical properties of a material or samples (thin films). It is key for electrical applications. Electrical techniques are used to evaluate the carrier concentration, width of depletion, resistivity, interface states, oxide charge, deep-level impurities, carrier lifetime, contact resistance, mobility, and barrier height, among others. Instruments such as Langmuir probe, resistivity and Hall effect, and deep-level transient spectroscopy are used for determining electrical properties.

3.10.1 LANGMUIR PROBE

Irving Langmuir invented the Langmuir probe shown in the figure below. The Langmuir probe is used for measuring the electron temperature, electric potential, and electron density.

The working principles involve introducing electrodes into plasma with a fixed or variable electric potential between the different electrodes or between them and the surrounding vessel. The I-V characteristic of the Debye sheath, or the current density flowing to a surface in a plasma as a function of the voltage drop across the sheath, is

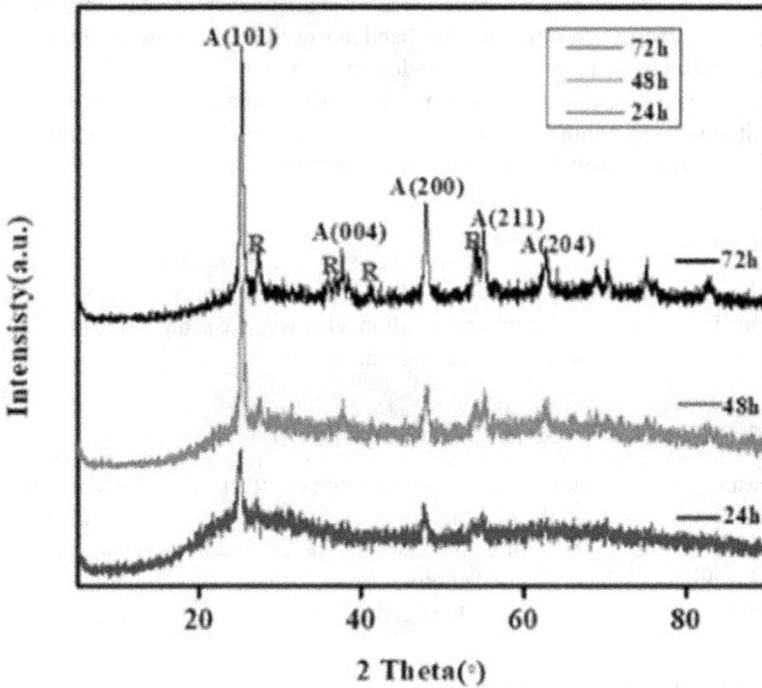

FIGURE 3.14 XRD diffraction pattern of the deposited TiO2 thin films after aging sol for 24 h, 48 h, and 72 h.

Source: Lukong, Ukoba, and Jen, 2021.

the basis of Langmuir probe theory. This research shows how the I-V characteristic can be used to calculate the electron density, electron temperature, and plasma potential. In certain circumstances, a more thorough study can provide data on the electron energy distribution function (EEDF), ion temperature, or ion density.

3.10.2 DEEP LEVEL TRANSIENT SPECTROSCOPY

It is an instrument used for measuring charge carrier traps. It performed the measurement by using the defect parameters to evaluate the materials concentration.

The working principle involves measuring the equilibrium state of the reverse bias. Thereafter, the major charge carrier is captured by the n-type material. The trapped charge carrier is emitted proceeded by measuring the capacitance transient with respect to time (Capan and Brodar, 2022).

3.10.3 PHOTOLUMINESCENCE

Markus E. Beck, Janice C. Lee, and Erel Milshtein invented the nondestructive test of photoluminescence for photovoltaic films. It is an approach frequently used to describe the optoelectronic characteristics of semiconductors and other materials. A laser with energy greater than the bandgap excites electrons in the material's valence band to move them into the conductance band.

A laser beam measures the light given by a photovoltaic thin film as it falls from the excited state to ground state using photoexcitation. The material impurities and imperfection is read from the luminescence spectrum.

3.10.4 Operating Principle

When light hits a sample, it is absorbed, and the extra energy is transferred to the substance by a process called photoexcitation. One way the sample releases this extra energy is by luminescence, or light emission.

3.10.5 Reading the Photoluminescence Result

By measuring the radiation intensity as a function of either the excitation wavelength or the emission wavelength, photoluminescence spectra are created. By observing emission at a fixed wavelength while changing the excitation wavelength, an excitation spectrum is obtained. Emitted photons are read off and analyzed if a photoexcited thin film's energy is more than the bandgap.

3.11 FOUR-POINT PROBE

Frank Wenner invented the four-point probe about a hundred years ago. It is used to evaluate the resistivity value of materials with a layer of an electronic property. The linear array of four electrodes in the four-point probe is evenly spaced apart. The two outer electrodes in the typical design feed a current I through the sample. A high-impedance voltmeter measures the voltage drop V that results from this between the two inner electrodes.

3.12 FUTURE TRENDS OF THIN FILMS

Thin films are moving toward being at the forefront of the Fourth Industrial Revolution. The majority of the 4IR will be thin film based. This includes the ability to screen-print solar panels on wearable devices including clothes and other devices. It will also play a role in bendable and chargeable devices.

3.13 ROLE OF THIN FILMS IN SUSTAINABLE DEVELOPMENT AND CLIMATE CHANGE

Climate change is changing the world and the way things are done. Countries and companies are striving to become ESG compliance. ESG means environmental, social, and corporate governance. It is developed as a framework for organizations' strategy plans to breed value with consideration of the impact of activities on the environment, society, and economy.

TABLE 3.2
Summary of Thin Film Characterization Techniques

Technique	Examples
Scanning Probe Microscopy	Scanning tunneling microscopy (STM), atomic force microscopy
Optical Techniques	X-ray diffraction (XRD)
Diffraction	Optical microscopy, ellipsometry
Reflection/Transmission	Photoluminescence, photoconductivity, Fourier transform infrared
Absorption/Emission	spectroscopy (FTIR), Raman spectroscopy, ultraviolet photoelectron spectroscopy (UPS), X-ray photoelectron spectroscopy (XPS)
Ion Beam Techniques	Rutherford backscattering spectrometry (RBS)
	Elastic recoil detection (ERD)
	Ion Induced X-ray spectroscopy (IIX)
	Nuclear reaction analysis (NRA), Channeling, secondary ion mass spectrometry (SIMS)
Analytical Electron Microscopy	Auger electron spectroscopy (AES), energy dispersive X-ray analysis (EDX), electron energy loss spectroscopy (EELS)
Electrical Techniques	Resistivity and Hall effect, capacitance-voltage (C-V), deep-level transient spectroscopy (DLTS), Langmuir probe
Electron Microscopy	Low energy electron diffraction (LEED), reflection high energy electron diffraction (RHEED), scanning electron microscopy (SEM), transmission electron microscopy (TEM)

Sustainability looks at ways humans and the planet can interact and coexist for longer periods without the other going into oblivion in a profitable way. The ESG is the pillar of sustainability. Waste reduction and green energy are some ways of ensuring the sustainability of the planet and humanity. Thin film has a role to play in the sustainability of man and the environment.

3.14 CONCLUSION

A thorough understanding of the property of a material helps in the right utilization and performance of the material. This chapter discussed the properties of thin films and different ways of testing and analyzing thin films with key examples and explanation. This includes the working principle, applications, sample preparation, and pros and cons of the characterization techniques. The chapter covers the morphology; structural, chemical, electrical, and optical properties; and characterization techniques in detail. The trends of thin films and discussion on sustainability were also covered vis-à-vis thin films and climate change. It is believed that the chapter will provide the needed knowledge for beginners to advance thin films readers and researchers.

REFERENCES

Alberts, B., Johnson, A., Lewis, J., Raff, M., Roberts, K., & Walter, P. (2003). Molecular biology of the cell. *Scandinavian Journal of Rheumatology, 32*(2), 125–125.

Alvarez-Fernandez, A., Reid, B., Fornerod, M. J., Taylor, A., Divitini, G., & Guldin, S. (2020). Structural characterization of mesoporous thin film architectures: a tutorial overview. *ACS Applied Materials & Interfaces, 12*(5), 5195–5208.

Amzallag, A., Vaillant, C., Jacob, M., Unser, M., Bednar, J., Kahn, J. D., Dubochet, J., Stasiak, A., & Maddocks, J. H. (2006). 3D reconstruction and comparison of shapes of DNA minicircles observed by cryo-electron microscopy. *Nucleic Acids Research, 34*(18), e125–e125.

Aryal, S. (2022). *Differences between light microscope and electron microscope.* https://micro biologyinfo.com/differences-between-light-microscope-and-electron-microscope/

Bahk, J.-H., Favaloro, T., & Shakouri, A. (2013). Thin film thermoelectric characterization techniques. In G. Chen et al. (eds.), *Annual review of heat transfer* (vol. 16, pp. 1–52). Begell House Inc.

Bailly, A. (1951). *Abrégé du dictionnaire grec-français.* Hachette.

Bean, J. J., Saito, M., Fukami, S., Sato, H., Ikeda, S., Ohno, H., … McKenna, K. P. (2017). Atomic structure and electronic properties of MgO grain boundaries in tunnelling mag-netoresistive devices. *Scientific Reports, 7*(1), 1–9.

Biderman, N., & Bui, H. S. (2021). Atomic force microscopy (AFM) as a surface characteriza-tion tool for hair, skin, and cosmetic deposition. In K. L. Mittal, H. S. Bui (eds.), *Surface Science and Adhesion in Cosmetics* (pp. 245–278). Wiley.

Binnig, G., Quate, C. F., & Gerber, C. (1986). Atomic force microscope. *Physical Review Letters, 56*(9), 930.

Blott, S. J., Croft, D. J., Pye, K., Saye, S. E., & Wilson, H. E. (2004). Particle size analysis by laser diffraction. *Geological Society, London, Special Publications, 232*(1), 63–73.

Bolshakova, A. V., Kiselyova, O. I., & Yaminsky, I. V. (2004). Microbial surfaces investigated using atomic force microscopy. *Biotechnology Progress, 20*(6), 1615–1622.

Bopp, M. (1993). Karl Mägdefrau, Geschichte der Botanik, Gustav Fischer Verlag, Stuttgart, Jena, New York (1992), 2. bearb. Auflage 1992 VIII, 359 S., 160 Abb.,(ISBN 3-437-20489-0). Price: 78,-DM.

Braun, F. (2000). Nobel Prize for Physics. *The Biographical Dictionary of Scientists: Lebedev to Zworykin, 2,* 1020.

Buckingham, J. (1965). Thermionic emission properties of a lanthanum hexaboride/rhenium cathode. *British Journal of Applied Physics, 16*(12), 1821.

Capan, I., & Brodar, T. (2022). Majority and minority charge carrier traps in n-type 4H-SiC studied by junction spectroscopy techniques. *Electronic Materials, 3*(1), 115–123.

Carmichael, S. W. (1995). Atomic resolution with the atomic force microscope. *Microscopy Today, 3*(4), 6–7.

Chason, E., & Mayer, T. (1997). Thin film and surface characterization by specular X-ray reflectivity. *Critical Reviews in Solid State and Material Sciences, 22*(1), 1–67.

Cheville, N., & Stasko, J. (2014). Techniques in electron microscopy of animal tissue. *Veterinary Pathology, 51*(1), 28–41.

Davidson, M. W. (2013). Pioneers in optics: James Bradley and August Köhler. *Microscopy Today, 21*(5), 50–52.

De Graef, M. (2003). *Introduction to conventional transmission electron microscopy.* Cambridge University Press.

Di Gregorio, M. A. (2005). *From here to eternity: Ernst Haeckel and scientific faith* (Vol. 3). Vandenhoeck & Ruprecht.

Dietz, N. (2001). Real-time optical characterization of thin film growth. *Materials Science and Engineering: B, 87*(1), 1–22.

Drummy, L. F., Yang, J., & Martin, D. C. (2004). Low-voltage electron microscopy of polymer and organic molecular thin films. *Ultramicroscopy, 99*(4), 247–256.

Egerton, R. F. (2005a). An introduction to microscopy *Physical Principles of Electron Microscopy* (pp. 1–25). Springer.

Egerton, R. F. (2005b). *Physical principles of electron microscopy* (Vol. 56). Springer.

Foerter-Barth, U., & Teipel, U. (2000). Characterization of particles by means of laser light diffraction and dynamic light scattering *Developments in Mineral Processing* (Vol. 13, pp. C1–1). Elsevier.

Franks, K., Kestens, V., Braun, A., Roebben, G., & Linsinger, T. P. (2019). Non-equivalence of different evaluation algorithms to derive mean particle size from dynamic light scattering data. *Journal of Nanoparticle Research, 21*(9), 1–10.

Fultz, B., & Howe, J. (2013). High-resolution STEM and related imaging techniques *Transmission Electron Microscopy and Diffractometry of Materials* (pp. 587–615). Springer.

Galvão, E. S., Santos, J. M., Lima, A. T., Reis Jr, N. C., Orlando, M. T. D. A., & Stuetz, R. M. (2018). Trends in analytical techniques applied to particulate matter characterization: a critical review of fundaments and applications. *Chemosphere, 199*, 546–568.

Ganguli, R., & Bandopadhyay, S. (2002). Expert system for equipment selection. *International Journal of Surface Mining, Reclamation and Environment, 16*(3), 163–170.

Goldstein, J. (2003). *Scanning electron microscopy and X-ray microanalysis*. Springer.

Goldstein, J. I., Newbury, D. E., Echlin, P., Joy, D. C., Fiori, C., & Lifshin, E. (1981). Quantitative X-ray microanalysis. In: *Scanning electron microscopy and X-ray microanalysis* (pp. 305–392). Springer.

Harada, Y., & Ikuhara, Y. (2013). The Latest Analytical Electron Microscope and its Application to Ceramics. In S. Somiya (ed.), *Handbook of Advanced Ceramics* (p. 3–21). Elsevier Inc. Chapters.

Hawkes, P. (2002). Signposts in electron optics. In: *Advances in imaging and electron physics* (Vol. 123, pp. 1–28). Elsevier.

Helal, A. A., Murad, G., & Helal, A. (2011). Characterization of different humic materials by various analytical techniques. *Arabian Journal of Chemistry, 4*(1), 51–54.

Hristov, D. R., Ye, D., de Araújo, J. M., Ashcroft, C., DiPaolo, B., Hart, R., … Dawson, K. A. (2017). Using single nanoparticle tracking obtained by nanophotonic force microscopy to simultaneously characterize nanoparticle size distribution and nanoparticle–surface interactions. *Nanoscale, 9*(13), 4524–4535.

Iijima, S. (1980). High resolution electron microscopy of some carbonaceous materials. *Journal of Microscopy, 119*(1), 99–111.

Imoisili, P. E., Ukoba, K. O., & Jen, T.-C. (2020). Green technology extraction and characterisation of silica nanoparticles from palm kernel shell ash via sol–gel. *Journal of Materials Research and Technology, 9*(1), 307–313.

Ismail, H., & Nurdin, H. I. (1998). Tensile properties and scanning electron microscopy examination of the fracture surface of oil palm wood flour/natural rubber composites. *Iranian Polymer Journal, 7*, 53–58.

Jain, J. (1977). An introduction to electron microscope. *IETE Journal of Education, 18*(2), 65–82.

Kalita, O. C. P., Doley, P., & Kalita, A. (2017). Uses of transmission electron microscope in microscopy and its advantages and disadvantages. *Life Sciences Leaflets, 85*, 8–13.

Kasten, F. H. (1989). The origins of modern fluorescence microscopy and fluorescent probes. In E. Kohen (ed.), *Cell structure and function by microspectrofluorometry* (pp. 3–50). Elsevier.

Khanam, T., Syuhada Wan Ata, W. N., & Rashedi, A. (2016). *Particle size measurement in waste water influent and effluent using particle size analyzer and quantitative image analysis technique.* Paper presented at the Advanced Materials Research.

Kim, A., Bernt, W., & Cho, N.-J. (2019). Improved size determination by nanoparticle tracking analysis: influence of recognition radius. *Analytical Chemistry, 91*(15), 9508–9515.

Kim, A., Ng, W. B., Bernt, W., & Cho, N.-J. (2019). Validation of size estimation of nanoparticle tracking analysis on polydisperse macromolecule assembly. *Scientific Reports, 9*(1), 1–14.

Kosasih, F. U., Cacovich, S., Divitini, G., & Ducati, C. (2021). Nanometric chemical analysis of beam-sensitive materials: A case study of STEM-EDX on Perovskite Solar Cells. *Small Methods, 5*(2), 2000835.

Krivanek, O. L., Lovejoy, T. C., Dellby, N., Aoki, T., Carpenter, R., Rez, P., … Lagos, M. J. (2014). Vibrational spectroscopy in the electron microscope. *Nature, 514*(7521), 209–212.

Leng, Y. (2009). *Materials characterization: introduction to microscopic and spectroscopic methods.* John Wiley & Sons.

Linck, M., Ercius, P. A., Pierce, J. S., & McMorran, B. J. (2017). Aberration corrected STEM by means of diffraction gratings. *Ultramicroscopy, 182*, 36–43.

Lotya, M., Rakovich, A., Donegan, J. F., & Coleman, J. N. (2013). Measuring the lateral size of liquid-exfoliated nanosheets with dynamic light scattering. *Nanotechnology, 24*(26), 265703.

Lukong, V., Ukoba, K., & Jen, T. (2021). Analysis of sol aging effects on self-cleaning properties of TiO$_2$ thin film. *Materials Research Express, 8*(10), 105502.

Lukong, V., Ukoba, K., Yoro, K., & Jen, T. (2022). Annealing temperature variation and its influence on the self-cleaning properties of TiO$_2$ thin films. *Heliyon, 8*(5), 1–9, e09460.

Makhlouf, A. S. H., & Aliofkhazraei, M. (2015). *Handbook of materials failure analysis with case studies from the oil and gas industry.* Butterworth-Heinemann.

McMullan, D. (1995). Scanning electron microscopy 1928–1965. *Scanning, 17*(3), 175–185.

McMullan, D. (2008). The early development of the scanning electron microscope *Biological Low-Voltage Scanning Electron Microscopy* (pp. 1–25). Springer.

Michailof, C. M., Kalogiannis, K. G., Sfetsas, T., Patiaka, D. T., & Lappas, A. A. (2016). Advanced analytical techniques for bio-oil characterization. *Wiley Interdisciplinary Reviews: Energy and Environment, 5*(6), 614–639.

Morniroli, J.-P., Marceau, R., Ringer, S., & Boulanger, L. (2006). LACBED characterization of dislocation loops. *Philosophical Magazine, 86*(29–31), 4883–4900.

Mu, W., Jiang, L., Zhang, J., Shi, Y., Gray, J. E., Tunali, I., … Zhao, X. (2020). Non-invasive decision support for NSCLC treatment using PET/CT radiomics. *Nature Communications, 11*(1), 1–11.

Mukherjee, S. (2013). Analytical techniques for clay studies. In: *The science of clays* (pp. 69–110). Springer.

Mura, P. (2014). Analytical techniques for characterization of cyclodextrin complexes in aqueous solution: A review. *Journal of Pharmaceutical and Biomedical Analysis, 101*, 238–250.

Mura, P. (2015). Analytical techniques for characterization of cyclodextrin complexes in the solid state: A review. *Journal of Pharmaceutical and Biomedical Analysis, 113*, 226–238.

Myers, P., & Richards, R. J. (2009). The tragic sense of life: Ernst Haeckel and the struggle over evolutionary thought. *The Review of Politics, 71*(3), 505.

Nebesářová, J., & Vancová, M. (2007). How to observe small biological objects in low voltage electron microscope. *Microscopy and Microanalysis, 13*(S03), 248–249.

Nwanya, A. C., Botha, S., Ezema, F. I., & Maaza, M. (2021). Functional metal oxides synthesized using natural extracts from waste maize materials. *Current Research in Green and Sustainable Chemistry, 4*, 100054.

Ohnesorge, F., & Binnig, G. (1993). True atomic resolution by atomic force microscopy through repulsive and attractive forces. *Science, 260*(5113), 1451–1456.

Ohring, M. (2001). *Materials science of thin films.* Elsevier.

Park, B.-D., Wi, S. G., Lee, K. H., Singh, A. P., Yoon, T.-H., & Kim, Y. S. (2003). Characterization of anatomical features and silica distribution in rice husk using microscopic and micro-analytical techniques. *Biomass and Bioenergy, 25*(3), 319–327.

Peng, J., Guo, J., Ma, R., & Jiang, Y. (2021). Water-solid interfaces probed by high-resolution atomic force microscopy. *Surface Science Reports, 77*(1), 100549.

Pennycook, S. J. (2017). The impact of STEM aberration correction on materials science. *Ultramicroscopy, 180*, 22–33.

Phillips, R. (1961). Diamond knife ultra microtomy of metals and the structure of microtomed sections. *British Journal of Applied Physics, 12*(10), 554.

Popelka, A., Zavahir, S., & Habib, S. (2020). Morphology analysis. In M. A. A. AlMaadeed, D. Ponnamma, & M. A. Carignano (eds.), *Polymer science and innovative applications* (pp. 21–68). Elsevier.

Porter, K. R., & Blum, J. (1953). A study in microtomy for electron microscopy. *The Anatomical Record, 117*(4), 685–709.

Primavera, R., Barbacane, R. C., Congia, M., Locatelli, M., & Celia, C. (2014). Laser diffraction and light scattering techniques for the analysis of food matrices. *Advances in Food Safety and Health, 6*, 40–60.

Ramohlola, K. E., Iwuoha, E. I., Hato, M. J., & Modibane, K. D. (2020). Instrumental techniques for characterization of molybdenum disulphide nanostructures. *Journal of Analytical Methods in Chemistry, 2020*, 8896698.

Recek, N., Cheng, X., Keidar, M., Cvelbar, U., Vesel, A., Mozetic, M., & Sherman, J. (2015). Effect of cold plasma on glial cell morphology studied by atomic force microscopy. *PLoS One, 10*(3), e0119111.

Ren, J., Huang, H., Liu, Y., Zheng, X., & Zou, Q. (2015). An atomic force microscope study revealed two mechanisms in the effect of anticancer drugs on rate-dependent Young's modulus of human prostate cancer cells. *PLoS One, 10*(5), e0126107.

Robinson, A. L. (1986). Electron microscope inventors share Nobel Physics Prize: Ernst Ruska built the first electron microscope in 1931; Gerd Binnig and Heinrich Rohrer developed the scanning tunneling microscope 50 years later. *Science, 234*(4778), 821–822.

Rose, H. H. (2008). Optics of high-performance electron microscopes. *Science and Technology of Advanced Materials, 9*(1), 014107.

Rudenberg, H. G., & Rudenberg, P. G. (2010). Origin and background of the invention of the electron microscope: Commentary and expanded notes on memoir of Reinhold Rüdenberg. In P. W. Hawkes (ed.), *Advances in Imaging and Electron Physics* (Vol. 160, pp. 207–286). Elsevier.

Ruska, E. (1987). The development of the electron microscope and of electron microscopy. *Reviews of Modern Physics, 59*(3), 627.

Sam, Z., Lin, L., & Ashok, K. (2009). Materials characterization techniques. *Taylor & Francis Group Cap, 7*, 177–205.

Santos, N. C., & Castanho, M. A. (2004). An overview of the biophysical applications of atomic force microscopy. *Biophysical Chemistry, 107*(2), 133–149.

Sawada, H., Tomita, T., Naruse, M., Honda, T., Hambridge, P., Hartel, P., ... Kirkland, A. (2005). Experimental evaluation of a spherical aberration-corrected TEM and STEM. *Microscopy, 54*(2), 119–121.

Schulze, D. (2007). Pioniere des Elektronenmikroskops. *Practical Metallography, 44*(12), 557–565.

Sciau, P. (2016). Transmission electron microscopy: emerging investigations for cultural heritage materials. In P. W. Hawkes (ed.), *Advances in Imaging and Electron Physics* (Vol. 198, pp. 43–67). Elsevier.

Shi, Y., Gao, S., Lu, M., Li, W., & Xu, X. (2013). *Segmental calibration for commercial AFM in vertical direction.* Paper presented at the Eighth International Symposium on Precision Engineering Measurement and Instrumentation. J. Lin (ed.), 8–11 August, 2012. Chengdu, China.

Song, J., Zhang, M., Yuan, M., Qian, Y., Sun, Y., & Liu, F. (2018). Morphology characterization of bulk heterojunction solar cells. *Small Methods, 2*(3), 1700229.

Stetefeld, J., McKenna, S. A., & Patel, T. R. (2016). Dynamic light scattering: a practical guide and applications in biomedical sciences. *Biophysical reviews, 8*(4), 409–427.

Süsskind, C. (1980). Ferdinand Braun: Forgotten forefather. In L. Marton & C. Marton (eds.), *Advances in Electronics and Electron Physics* (Vol. 50, pp. 241–260). Elsevier.

Suzuki, E. (2002). High-resolution scanning electron microscopy of immunogold-labelled cells by the use of thin plasma coating of osmium. *Journal of Microscopy, 208*(3), 153–157.

Swapp, S. (2014). University of Wyoming.(2013). *Scanning Electron Microscopy (SEM).* https://serc.carleton.edu/research_education/geochemsheets/techniques/SEM.html

Syvitski, J. P. (1991). *Principles, methods, and application of particle size analysis* (Vol. 388). Cambridge University Press Cambridge.

Tanaka, N. (2008). Present status and future prospects of spherical aberration corrected TEM/STEM for study of nanomaterials. *Science and Technology of Advanced Materials, 9*, 1–11.

Truell, R., Elbaum, C., & Chick, B. B. (2013). *Ultrasonic methods in solid state physics.* Academic Press.

Vargas-Ubera, J., Aguilar, J. F., & Gale, D. M. (2007). Reconstruction of particle-size distributions from light-scattering patterns using three inversion methods. *Applied Optics, 46*(1), 124–132.

von Ardenne, M. (1937). *Improvements in electron microscopes.* British Patent (511204).

Wager, J., & Keir, P. (1997). Electrical characterization of thin-film electroluminescent devices. *Annual Review of Materials Science, 27*(1), 223–248.

Wang, J., Jiang, P., Yuan, F. & Wu, X. (2022). Chemical medium-range order in a medium-entropy alloy. *Nature communications, 13*(1), 1021.

Wang, K., Taylor, K. G., & Ma, L. (2021). Advancing the application of atomic force microscopy (AFM) to the characterization and quantification of geological material properties. *International Journal of Coal Geology, 247*, 103852.

Wang, S., & Hao, C. (2022). *Principle and Application of Peak Force Tapping Mode Atomic Force Microscope.* Paper presented at the International Conference on Cognitive based Information Processing and Applications (CIPA 2021), 2, 671–679.

Wells, O. (2004). Building a scanning electron microscope. *Advances in Imaging and Electron Physics, 133*, 127–136.

Williams, D., & Carter, C. (1996). *Transmission electron microscopy, volume 1 Basics.* Plenum Press.

Wolfgong, W. J. (2016a). Chemical analysis techniques for failure analysis: Part 1, common instrumental methods. In A. S. H. Makhlouf & M. Aliofkhazraei (eds.), *Handbook of materials failure analysis with case studies from the aerospace and automotive industries* (pp. 279–307). Elsevier.

Wolfgong, W. J. (2016b). Chemical analysis techniques for failure analysis: Part 2, examples from the lab. In A. S. H. Makhlouf & M. Aliofkhazraei (eds.), *Handbook of materials failure analysis with case studies from the aerospace and automotive industries* (pp. 309–338). Elsevier.

Xu, R. (2015). Light scattering: A review of particle characterization applications. *Particuology, 18*, 11–21.

Xu, S., & Arnsdorf, M. (1994). Calibration of the scanning (atomic) force microscope with gold particles. *Journal of Microscopy, 173*(3), 199–210.

Yokoyama, T., Masuda, H., Suzuki, M., Ehara, K., Nogi, K., Fuji, M., … Hayashi, K. (2008). Basic properties and measuring methods of nanoparticles. In M. Naito, T. Yokoyama, K. Hosokawa, & K. Nogi (eds.), *Nanoparticle technology handbook* (pp. 3–48). Elsevier.

Zhang, S., Li, L., & Kumar, A. (2008). *Materials characterization techniques*. CRC Press.

Zworykin, V. K. (1945). *Electron optics and their electron microscope*. John Wiley & Sons.

4 Demystifying Concept of Atomic Layer Deposition and Its Applications in Fourth Industrial Revolution

4.1 BACKGROUND OF ATOMIC LAYER DEPOSITION

Atomic layer deposition (ALD) is gaining popularity as a thin film deposition process that is uniquely capable of producing homogeneous and conformal films on complex three-dimensional topographies. About 130 countries are actively engaged in ALD research, with the United States leading the pack according to the Web of Science analytics shown in Figure 4.1.

ALD is an improvement in chemical vapor deposition (CVD) technique and was first used by Suntola in Finland in 1974 (Ahvenniemi et al., 2017, Moshe and Mastai, 2013). The team in Finland used it to optimize ZnS films for electroluminescent displays (Puurunen, 2014). Hundreds of ALD chemistries have been discovered during the last few decades for depositing a wide range of materials, predominantly inorganic materials but recently also organic and inorganic–organic hybrid molecules (Miikkulainen et al., 2013). The crystallinity of the generated film (is the material amorphous or crystalline, and if so, which phase(s) are present) is one characteristic that often governs the properties of ALD films in real applications.

It is a process for creating thin films that can be used in a variety of applications. ALD is a type of CVD in which gaseous reactants (precursors) are injected into the reaction chamber and chemical surface reactions are used to produce the desired material. The precursors are pulsed alternately, one at a time, and separated by inert gas purging to avoid gas phase reactions, which is a distinguishing feature of ALD. Figure 4.2 is an adaptation of Saynatjoki's (2012) ALD growth cycle. The reactants' surface reactions are self-terminated, allowing for controlled growth of the desired substance. Even on difficult 3D objects, the unique self-limiting growth mechanism results in excellent conformality and thickness uniformity of the layer. Continued precursor chemistry improvement and reactor design are critical aspects of ALD. ALD has attracted several interests resulting in extensive review of the process (George, 2010, Miikkulainen et al., 2013, Oviroh et al., 2019, Leskelä and Ritala, 2002, Johnson et al., 2014, Shahmohammadi et al., 2022). However, there are limited reviews focusing on the process in its entirety, including new innovations, in recent

 DOI: 10.1201/9781003364481-4

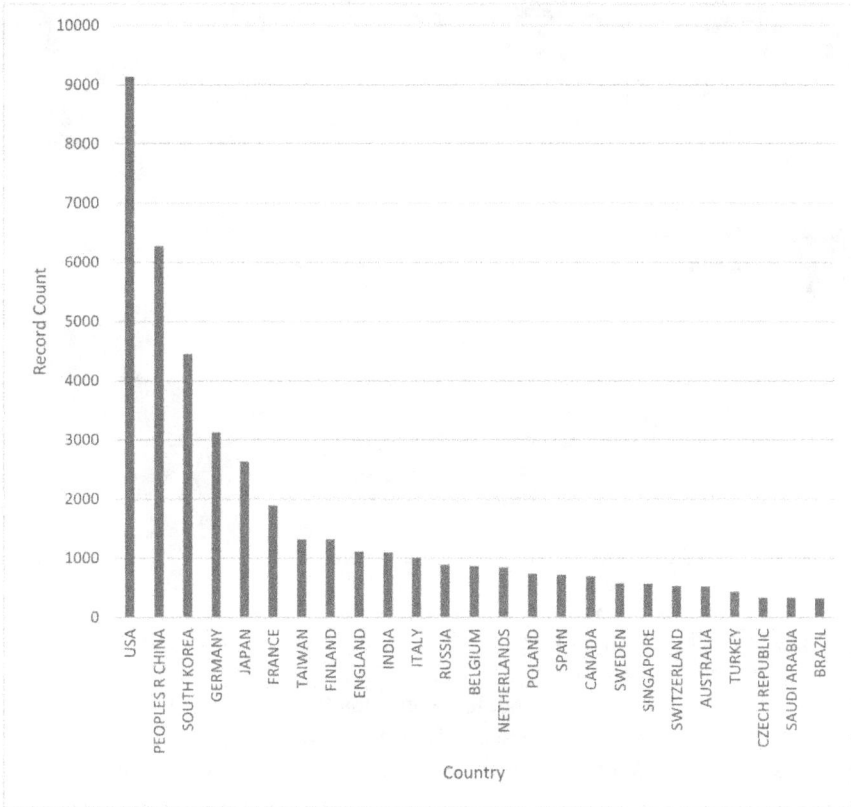

FIGURE 4.1 Atomic layer deposition record count per country.

years as majority of the scarce review are tailored to a specific application such as energy (Gupta et al., 2022).

Chemical vapor deposition (CVD) and atomic layer deposition (ALD) are two important processes in the semiconductor industry (ALD). High-purity materials are deposited as monocrystalline, polycrystalline, amorphous, and epitaxial films with desirable electronic characteristics using these processes. A substrate is heated and exposed to a volatile precursor, which reacts with the substrate surface to form the desired deposit in a conventional CVD experiment. Many of these reactions result in the production of a film, the thickness of which is largely determined by the length of the experiment. Any reaction by-products are volatile, so they can be eliminated by the reaction chamber's gas flow.

There are a few outstanding differences between CVD and ALD. The precursors react simultaneously on the surface or in the gas phase in CVD, and they may also disintegrate, but in ALD, highly reactive precursors react independently via alternating saturating surface reactions, with no self-decomposition. The ALD method is thus governed solely by surface chemistry, whereas the CVD method is influenced by a few fundamental processes. Figure 4.3 shows the coverage metrics of a film on

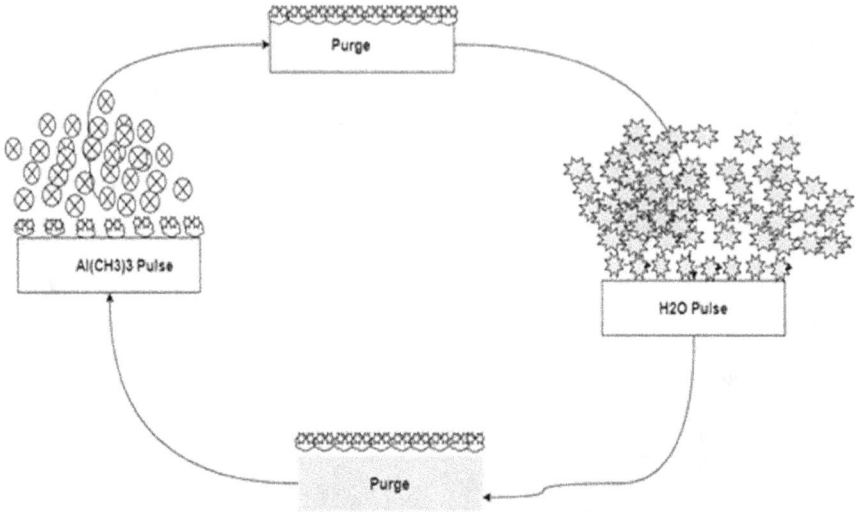

FIGURE 4.2 An ALD growth cycle schematically representation.

a 3D-featured substrate and the schematic comparison of ALD with CVD, PVD, and PECVD adapted from Knoops et al. (2015).

The uniformity metric is used to assess flat surface coverage, whereas the conformality metric is used to assess 3D feature coverage. Another essential parameter is the regulation of film thickness during growth. It's also crucial to note that these measurements can be achieved at low temperatures.

In terms of chemistry, ALD is comparable to CVD, except that the production of a film on a substrate in ALD is self-limiting and involves the reaction of two reagents: a metal precursor and a nonmetal precursor (such as a reducing agent). In the reaction chamber, the metal precursor is injected and allowed to bind to the substrate's surface. Any remaining metal precursor is subsequently purged with an inert gas flow. A second precursor is then added, which reacts with the first on the substrate's surface to generate the desired deposit. This cycle can be repeated several times, yielding films as thin as 0.1A per cycle. ALD is well-suited to the deposition of films on nanoscale surfaces with high aspect ratio features, and it is a hot topic in the microelectronics sector. There are numerous instrumental changes for CVD and ALD that can be utilized to make tailored thin films, but the chemical composition of the precursor component is crucial (Crutchley, 2013).

4.1.1 CHEMISTRY UNDERLYING ALD

The model for ALD procedures is the development of high-k dielectric Al_2O_3 with trimethylaluminum (TMA) and water (H_2O) and is illustrated in equation 4.1 (Klejna and Elliott, 2011).

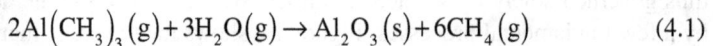

$$2Al\left(CH_3\right)_3 (g) + 3H_2O(g) \rightarrow Al_2O_3 (s) + 6CH_4 (g) \qquad (4.1)$$

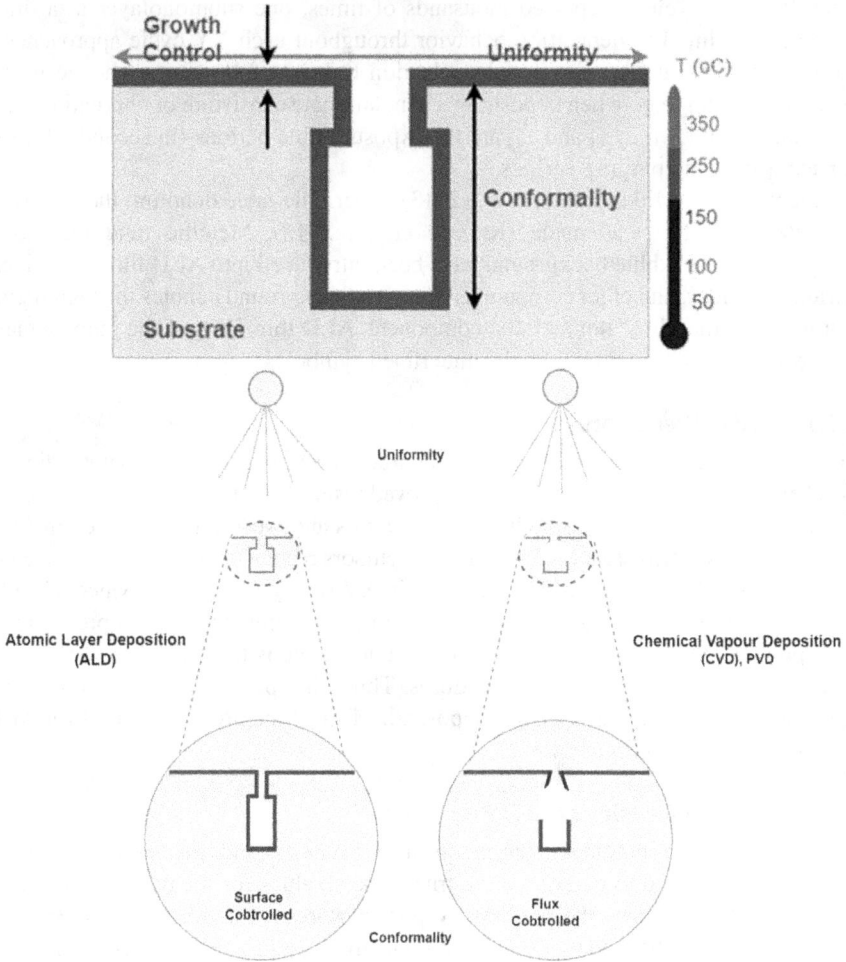

FIGURE 4.3 (a) The coverage metrics of a film on a 3D-featured substrate. (b) Comparison of CVD against ALD.

The TMA (precursor X) and H_2O (precursor Y) react to produce Al_2O_3 amorphous film. The second product, CH_4, does not react with any other species throughout the deposition process and is inert. Each precursor is injected one at a time, separated by a purge period to prevent gas-phase reactions, and the process is illustrated in equation 4.2:

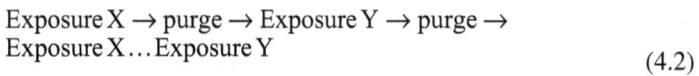

$$\text{Exposure X} \rightarrow \text{purge} \rightarrow \text{Exposure Y} \rightarrow \text{purge} \rightarrow$$
$$\text{Exposure X} \ldots \text{Exposure Y} \tag{4.2}$$

$$T_X \sec \tau_{XP} \sec \tau_Y \sec \tau_{YP} \sec \tau_X \sec$$

where the XY cycle is repeated thousands of times, one submonolayer at a time, forming the film. The deposition behavior throughout each XY cycle approaches a limit-cycle solution after the initial nucleation transient following a change in the precursor system (e.g., when depositing a nanolaminate consisting of alternating thin-film materials). τ_X, τ_Y, τ_{XP}, and τ_{YP} are the exposure time periods (in seconds) for the XY exposures and purging periods.

Adapted from Miikkulainen et al. (2013), a periodic table denoting the materials deposited by ALD was made (Knoops et al., 2015). Metallic (and metalloid) components with a blue background have been introduced into ALD films of oxides, nitrides, carbides, and other compounds. The gray background denotes the ingredients that make up the films' nonmetallic component. ALD thin films of the pure element have been reported, as shown by the underlined symbols.

4.1.1.1 ALD Precursors

The ability to manipulate and optimize any process is predicated on understanding of the chemistry underlying the process. Improved results obtained from ALD are based on the thorough grasp of the chemistry. Precursors are two compounds used in ALD processes (also called "reactants"). These precursors react with a material's surface in a sequential, self-limiting fashion, one at a time. ALD cycle is the sequence of dose-purge-dose-purge of a binary process. Dose time is the time taken for a precursor to be exposed to the surface (Larrabee, 2012). Purge time is time taken to remove the precursor from the chamber between doses. Through repeated exposure to different precursors, a thin coating is slowly deposited. Most of the precursors used for ALD process are shown in Figure 4.4.

4.1.1.2 Characteristics of ALD Precursors

Understanding the precursor thermophysical property and gas delivery system dynamics is required to determine the true dosage values for the precursor delivery system. Table 4.1 shows the predicted vapor pressures for TMA (between 337 and 400 K) and water (between 293 and 343 K) using Antoine's equation coefficients as provided by the National Institute of Standards and Technology (NIST) (Goujon et al.,

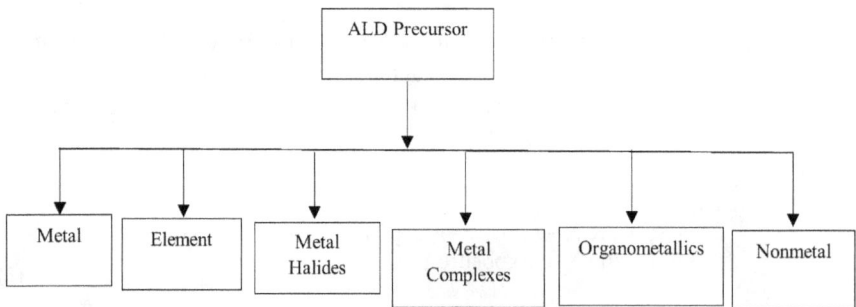

FIGURE 4.4 Different classes of ALD precursors.

2014, Linstorm, 1998). The success of an ALD process is influenced by its properties such as reactivity, thermal stability, and volatility (Koponen et al., 2016). Precursors should possess high reactivity, volatility, and thermal properties with good purging ability. They can be liquid, solid, or gas. To prevent contamination of the thin film, the precursors must be extremely pure (Kanjolia et al., 2008). To avoid undesired side products, they must be reactive on the surface at appropriate temperatures without disintegrating (Malik and O'Brien, 2008).

Precursors must, however, not disintegrate during vaporization and must therefore be thermally stable but volatile. They must be able to dissolve quickly in inert solvents and remain liquid at room temperature. They must also have a preference for and self-limitation of reactivity toward the film surface and substrate. Precursors must chemisorb on contact or react quickly with surface groups while reacting aggressively with one another. In this approach, the saturation state can be reached in a short amount of time (less than 1 second) and a reasonable deposition rate can be ensured. Conventional CVD precursors do not require such an intense reaction. The required ALD reactions should have a big negative ΔG value, but thermodynamic data for just a few precursors is currently available. It appears that not only the qualities of a single precursor molecule matter in ALD but also the mixture of precursors. The volatility of precursor can be improved by using a metal center surrounded by ligands in a process called heteroleptic precursor. This introduces asymmetry, as opposed to a homoleptic molecule where all the ligands are the same.

Most precursors contribute only one element to the deposited film, with the remainder decomposing during the process. Some metal organic precursors, on the other hand, can lead to the unintentional integration of oxygen and carbon into thin films, and this factor should be considered as well. Furthermore, the possibility of precursors prereacting in the vapor phase should be considered. Standard CVD precursors used to be hydrides and metal halides, but now a wide range of metal organic compounds such as metal carbonyls, metal alkoxides, metal alkyls, metal amidinates, metal diketonates, and other metal organic compounds. The names in bold relate to precursor class. The names behind each image are the names of the specific sample compound, followed by the common name in italics, if appropriate.

4.1.1.3 Element Precursors

Zn and Cd, which have been utilized to deposit II–VI semiconductors, are the only metallic elements that have been substantially investigated in ALD. Apart from Hg, the usage of other metallic elements is limited due to their poor volatility. Some II–VI compounds on single crystal substrates show that Zn and Cd react well with elemental chalcogens, and the lack of sterically demanding ligands may allow for a full one monolayer per cycle development.

4.1.1.4 Metal Alkyls Complexes

Metal in transition alkyl complexes are coordination complexes in which a transition metal and an alkyl ligand form a bond. They're everywhere. Metal alkyls, which have metal-carbon linkages, are a typical ALD precursor. The most common metal alkyl compounds utilized in CVD are aluminum and zinc alkyls. Sulfide and oxide films can be made with these alkyl compounds. In the case of aluminum, higher molecular weight alkyls are commonly utilized because they embed less undesired carbon into the films. Trimethylaluminum $(CH_3)_3Al$, is likely the most utilized precursor for laying aluminum oxide due to its near-ideal properties. Diethylzinc $(C_2H_5)_2Zn$ (Fouache and Lincot, 2000) and dimethylzinc $(CH_3)_2Zn$ (Łukasiewicz et al., 2008) are often used to deposit zinc oxide with structure by Basca (Bacsa et al., 2011). The Zn-C bonds have a length of 194.8(5) pm, while the C-Zn-C angle has a length of 176.2 pm. Because they spontaneously fire when exposed to air (pyrophoric activity) and can explode if they encounter water, these metal alkyls must be handled with caution. Less pyrophoric aluminum alkylalkoxides, such as dimethyl aluminum isopropoxide (Ponraj et al., 2013, Cho et al., 2003), can also be used for aluminum oxide ALD. To form the correct film composition, metal alkyls typically react with oxygen, water, sulfur, and other basic reagents. Metal phosphates or silicates can be formed by reacting with alkylphosphates or alkoxysilanols, respectively. A typical alkoxide precursor is titanium isopropoxide (Xie et al., 2007). Alkyl nucleophiles and alkyl electrophiles are the two most common strategies to make metal alkyl complexes. The demerit of alkyl amides is a reactivity that is halfway between that of halides and alkoxides. It also has issue with thermal stability, especially when placed in a hot bubbler for lengthy periods of time.

Because of their volatility and reactivity, metal amides are also regarded as attractive precursors (Niinistö et al., 2004). ZrO_2 films have previously been made by ALD utilizing $Zr(NMe_2)_4$, $Zr(NetMe)_4$, or $Zr(Net_2)_4$ as metal precursors and water or oxygen as the oxygen supply (Kim et al., 2002, Matero et al., 2002). In the amide/water ALD procedures, very low deposition temperatures, even 50°C, can be used to generate conformal ZrO_2 films (Hausmann et al., 2002). In plasma-enhanced ALD, alkyl amides and alkoxides have also been employed as Zr-precursors. In zirconia PEALD investigations, oxygen plasma was used to activate Zr-precursors, resulting in a wider processing window and lower impurity content (Kim et al., 2002, Koo et al., 2002).

Metal alkyl amides have been intensively investigated as nitride thin film CVD precursors. They have also recently been introduced to ALD (Kim, 2003, Park et al., 2001). When high ammonia dosages and lengthy exposure times were used, tetrakis(dimethylamido)titanium (or tetrakis(ethylmethylamido)titanium)-ammonia reactions were shown to be self-limiting at 170–210°C. The deposition rate was twice as fast as monolayer development in TiN in the (Leskelä & Ritala, 2002) direction. Ti–Si–N films can be developed when silane is added to the process, and the composition can be adjusted up to a Si content of 23 at%. percent by varying the number of silane pulses (Min et al., 2000). Silane inhibits growth, and as a result, the rate of growth slows as the silicon content of the films rises. The conformality and barrier characteristics of Ti–Si–N films against Cu diffusion were outstanding.

Metal organic precursors offer a variety of advantages over halides. For example, they can assist in the development of bespoke systems for low-temperature deposition procedures, removing the challenges that come with higher temperatures. Furthermore, metal organic precursors eliminate halogens from the deposition process, which are corrosive.

4.1.1.5 Metal Halides Precursor

Metal halides are formed when metals and halogens combine. Some are ionic, like sodium chloride, whereas others are covalently bonded. Only a few metal halides are discrete molecules, such as uranium hexafluoride, while the majority, such as palladium chloride, take on polymeric forms. According to equation 4.3, all halogens can react with metals to generate metal halides.

$$2M + nX2 \rightarrow 2MXn \tag{4.3}$$

where M stands for metal, X for halogen, and MXn for metal halide.

Metal halides are the first type of precursor used in CVD techniques. They are reasonably priced and widely available. Halide compounds such as fluorides, chlorides, and iodides have been used as ALD precursors (Gordon, 2014, He et al., 2019). Halides are well-known for their high thermal stability (Efimova et al., 2015). In the semiconductor industry, titanium tetrachloride ($TiCl_4$), is used to make titanium nitride (TiN) electrodes for DRAM and hafnium tetrachloride ($HfCl_4$) gate dielectrics for transistors. Halides have a wide range of vapor pressures. The main disadvantage of metal halides is the corrosive nature of the precursors and their reaction byproducts. Etched or corroded substrates, deposited films, ALD apparatus, and vacuum pumps are all possible. The Nb_2O_5 film is partially etched away by a $NbCl_5$ precursor as it forms, resulting in nonuniform thicknesses (Leal et al., 2020). Impurities like halide might persist in the films, leading them to lose their properties. The structure and solution of titanium tetrachloride ($TiCl_4$) was depicted by Shon and associates (Shon et al., 2007).

Metal halides are frequently used as starting materials for various inorganic compounds. Heat, vacuum, or thionyl chloride treatment can all be used to make halide compounds anhydrous. Silver (I) can abstract halide ligands, most commonly as tetrafluoroborate or hexafluorophosphate. The vacant coordination site in many transition metal compounds is stabilized by a coordinating solvent such as tetrahydrofuran. The alkali salt of an X-type ligand, such as a salen-type ligand, can also displace halide ligands (Cozzi, 2004). The abstraction of the halide is driven by the precipitation of the resulting alkali halide in an organic solvent, which is officially a transmetallation. The lattice energies of alkali halides are often very high. Sodium cyclopentadienide, for example, interacts with ferrous chloride to produce ferrocene equation (4.5) (Wilkinson, 1963):

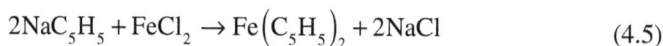

$$2NaC_5H_5 + FeCl_2 \rightarrow Fe\left(C_5H_5\right)_2 + 2NaCl \tag{4.5}$$

Although inorganic catalysis compounds can be synthesized and separated, they can also be made in situ by mixing the metal halide with the appropriate ligand. For palladium-catalyzed coupling processes, palladium chloride and triphenylphosphine are frequently substituted for bis(triphenylphosphine) palladium (II) chloride.

Although halides have long been recognized as suitable ALD precursors, fluorides, bromides, and iodides have received less attention. Fluorides and iodides have recently received increasing attention, but novel chloride precursors have also been introduced. BCl_3 is a novel precursor that was only recently introduced. In an ALD reaction, BCl_3 combines with ammonia to generate a BN film (George, 2003). In their flow modulation CVD technique, Shimogaki et al. (2001) reported on the first chloride adduct precursor. In order to deposit TaN films, they employed tantalum chloride thioether adduct and ammonia. The fact that this adduct is a liquid, unlike TaCl, is a significant advantage. Tungsten hexafluoride has been investigated as a tungsten metal and tungsten nitride film precursor. Iodides have been used as a metal precursor in the deposition of titanium nitride (Ritala et al., 1998) and titanium oxides (Aarik et al., 2020, Niemelä et al., 2017), zirconium (Liu et al., 2019, Hausmann and Gordon, 2003), and tantalum (Hsu et al., 2021, Kukli et al., 2000a). When H_2O_2 is used instead of water, the oxide production reaction is accelerated. At low temperatures, the reaction is self-limiting, but at higher temperatures, metal iodides disintegrate slightly. Studies of reaction mechanisms demonstrate that the situation is difficult in the case of TiO_2 because of the structural change (anatase-rutile) of the oxide, which alters the growth mechanism (Kukli et al., 2000b). On single crystal sapphire and MgO substrates, the TiI_4–H_2O_2 method has even supported epitaxial growth at low temperatures (<400°C) (Schuisky et al., 2000).

4.1.1.6 Nonmetal Precursors

Hydrides (H_2O, H_2S, NH_3, and AsH_3) are commonly used as nonmetal precursors because they have low volatility and thermal stability. Many ALD articles have shown that H_2O and H_2S have reactivity at moderate temperatures (<500°C) and the production of sufficient surface species for the metal precursor to be anchored, although NH_3 and AsH_3 are examples of precursors with less reactivity. The relevance of surface hydroxyl groups in the deposition of oxide coatings has been noted (Matero et al., 2000).

4.1.1.7 Atomic Layer Deposition Process

The ALD process is classified as thermal and plasma. They are here discussed in detail.

4.1.1.8 Thermal Atomic Layer Deposition

It uses heat to initiate the surface reaction. A net negative heat of reaction is required for the process. It happened spontaneously at different temperatures. Ritala et al. (1993) obtained 0.4 Å when they use thermal ALD deposition for TiO_2 in the temperature range of 150–600°C. Also, Yamada et al. (1997) obtained 2.2–2.5 Å at a temperature between 100°C and 160°C when ZnO was thermally deposited. Thermal processes are called thermal because they can be carried out without the usage of plasma or radicals. Binary metal oxides such as TiO_2, Al_2O_3, Ta_2O_5, ZrO_2, ZnO, and HfO_2 are the most used thermal ALD chemistry. Binary metal nitrides like TiN, W_2N, and TaN are also frequent thermal ALD systems. Sulfides such as CdS and ZnS and phosphides such as InP and GaP have also been thermally deposited.

4.1.1.9 Plasma-enhanced ALD

The conformal production of thin films of diverse materials with atomic-scale control is possible with a plasma-enhanced atomic layer deposition (PE-ALD) method. Plasma-enhanced ALD has the advantage of being able to deposit films at significantly lower temperatures than thermal ALD. This is because it creates particles that are highly reactive to the substrate surface, resulting in a lower reaction temperature and superior film quality. For the precursor and substrates, the processing conditions are more acceptable. A greater variety of thin-film materials can be made using it. It is able to deposit single-element that is impossible in thermal ALD (Kim, 2003). Plasma-enhanced ALD, for example, can generate Al_2O_3 films using TMA and O_2 plasma at temperatures as low as ambient temperature (Haiying et al., 2018, Heil et al., 2006). This is especially useful for thermally sensitive substrates like polymer, where a low deposition temperature is required (Langereis et al., 2006). It was discovered that plasma-enhanced ALD Al_2O_3 films had higher electrical characteristics than thermal ALD Al_2O_3 and provide good passivation of silicon substrates (Hoex et al., 2006). Plasma-enhanced ALD has mostly been used to deposit metal nitrides like TiN and TaN, which need a high temperature for thermal ALD (Chowdhury et al., 2021).

4.1.2 ALD Reactors

The detailed operations of an ALD reactor system are depicted in Figure 4.5. It is an adaptation of earlier schematic by Travis and Adomaitis (2013).

CV2 and CV5 control valves are typically open during all exposure and purge periods. CV1 is closed during the purge time prior to Exposure X to allow TMA to fill the ballast chamber; the TMA partial pressure in this chamber could potentially surpass the vapor pressure of liquid-phase TMA in the supply bottle. CV1 is opened during Exposure X, allowing TMA vapor to flow into the reactor and lowering the

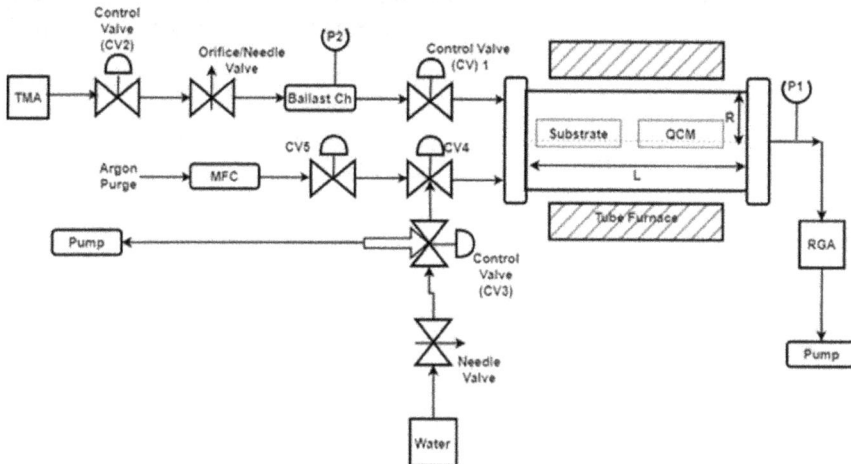

FIGURE 4.5 Schematic of ALD reactor system.

pressure in the ballast chamber. During Exposure X, a tiny flow of TMA will continue to flow via the orifice/needle valve. CV1 is closed at the end of Exposure X, and the pressure in the TMA ballast chamber is rebuilt. During all purge and exposure durations, Argon (Ar) purge gas flows continuously regardless of CV4 location. CV3 is open between the water source and the water purge pump, as well as to CV4, during both purge periods and during Exposure X; however, CV4 is closed in the direction of CV3, resulting in no water flow to the reactor. CV4 is turned to all-open during the water dose (Exposure Y), but CV3 is closed in the direction of the water purge pump, enabling Ar and water to flow to the reactor (XY). This arrangement was created to avoid condensation in the water delivery system and to increase the water dose's repeatability (Kimes and Maslar, 2011).

The flow-type ALD reactors used on industrial scale and frequently in laboratories allow for quick pulsing and purging. The pressures of flow-type reactors are 1–10 mbar, limiting the types of in situ characterization methods that can be used. Several ALD oxide techniques have been successfully investigated using quartz crystal microbalance (QCM) (Elam et al., 2002, Yoshida et al., 2022). A quadrupole mass spectrometer (QMS)-ALD reactor setup was recently demonstrated (Stevens et al., 2018, Nattermann et al., 2018), as well as a QMS–QCM–ALD combination (Nieminen et al., 2020). Another group of approaches that have been employed to characterize surface processes in flow-type ALD reactors is optical techniques (Kimes et al., 2012). The researchers used both surface photo absorption (SPA) and reflectance difference spectroscopy. ALD reactors are classified based on several parameters as shown in Figure 4.6.

ALD reactor is classified based on the presence of a carrier gas into two, viz. without carrier gas and with carrier gas as shown in Figures 4.7 and 4.8, respectively. A carrier gas was not used in the first few generations of ALD reactors. The two precursors feed the ALD chamber, which is coupled to a pump, as shown in Figure 4.7. The arrows show which way the water is flowing. To saturate the sample surface, valve 1 is first opened to allow precursor A to enter the ALD reactor. Then, by closing valve 1 and opening valve 3 to the pump reactor, precursor A is withdrawn. After closing valve 3, the same procedure is followed to expose the sample

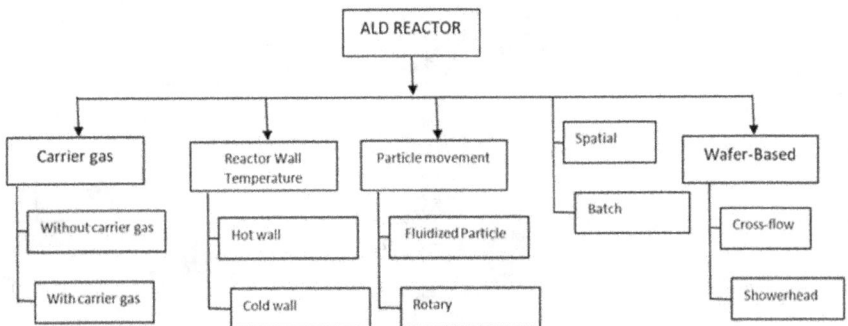

FIGURE 4.6 Classification of ALD reactor.

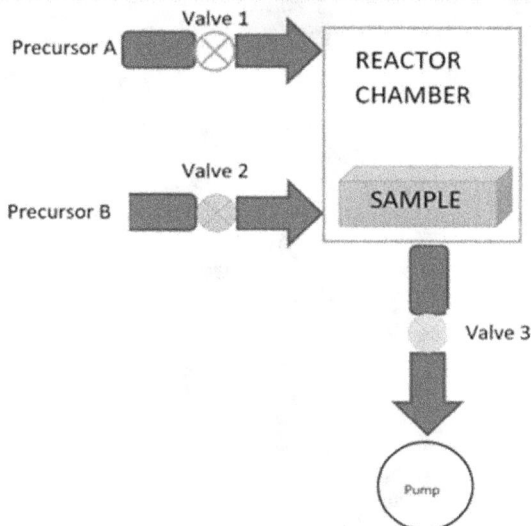

FIGURE 4.7 Schematic of the ALD reactor without the use of a carrier gas.

to precursor B via pressure B (regulated by valve 2). These exposures can employ precursors very efficiently due to the long residence durations of the precursors in the reactor. In the absence of a purge gas, however, the evacuation time for these ALD reactors is long.

In the ALD reactor with carrier gas, the carrier gas was then injected to shorten the ALD cycle time (Figure 4.8). The carrier gas flows continuously to the pump in a viscous flow. If the precursors have enough vapor pressure, they can be blended into the carrier gas stream. Alternatively, when they have lower vapor pressures, the carrier gas can flow over or through a solid or liquid precursor. The carrier gas allows for a shorter residence time in the reactor. Showerheads are frequently utilized to evenly spread the gas onto the sample (He, 2015).

The majority of ALD reactors today use an inert carrier gas in a viscous flow. For viscous flow reactors, the optimal pressure is roughly 133.3 Pa. Gas interdiffusion and entrainment are balanced at this optimum pressure. At 1 torr, precursor vapor diffusion in N_2 gas is sufficient to remove precursors and products from stagnant gas in an ALD reactor in a reasonable amount of time. The pressure of 1 torr is also sufficient for N_2 to function as a carrier gas. The gas mean free pathways are longer at pressures less than 1 torr, and entrainment of stationary gas is less effective. Then, to optimize reaction time and purging time for ALD reactors, synchronously modulated flow and draw (SMFD) was developed (Sneh, 2005). The synchronized modulation of the inert flowing gas between the reactor inlet and reactor outlet allows for high-speed gas flow switching, such as ALD cycle times of <1 s for Al_2O_3 ALD systems.

ALD reactor can also be classified based on the temperature of the reactor wall as hot and cold. The walls, gas, and substrates in "hot wall" reactors are all heated to the same temperature as the walls. Only the substrate is heated in "cold wall" reactors, with the walls remaining at room temperature or only slightly warmed.

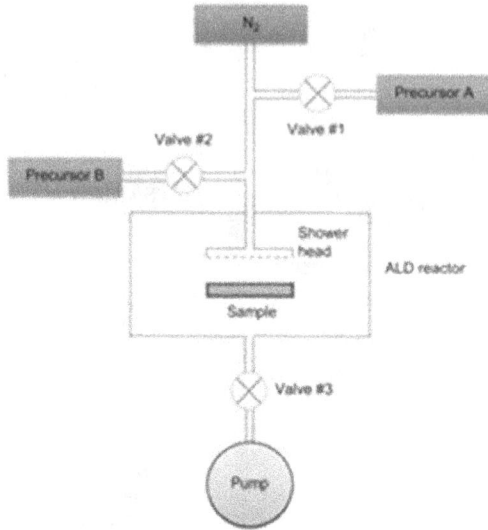

FIGURE 4.8   Schematic of the ALD reactor with the use of a carrier gas.

In comparison to conventional ALD reactors, the gas flow for wafer-based ALD reactors is more adaptable. The ALD reactor can also be classified based on the wafer-based gas flow to be showerhead and cross-flow. The cross-flow reactors, which feature parallel gas flows across the wafer surface, are used in several wafer-based ALD reactors. The "showerhead" reactor, on the other hand, uses a distributor plate to inject gas into the reactor perpendicular to the wafer surface. The gas then spreads out across the wafer's surface in a radial pattern. Plasma-enhanced ALD and ultrahigh-aspect-ratio deposition modules are designed to be fitted into the wafer-based ALD reactor to generate a variety of thin films on a variety of substrates. In a multiwafer atomic layer deposition (ALD) reactor, Pan et al. (2015b) analyzed two types of wafer arrangements: vertical and horizontal. The growth rate of an ALD-deposited alumina thin film is characterized and statistically compared. It has been discovered that the wafer layout has a substantial impact on the deposition process. Because of the increased collisions between precursor molecules and wafer surfaces in vertical multiwafer arrangement, it is found to be superior to horizontal multiwafer arrangement in terms of film deposition rate.

The gas particle movement has also been used to classify ALD reactor to be fluidization and rotary ALD reactor and schematically represented in Figure 4.6. The upward force of gas to float the particles equals the downward force of gravity in fluidization (Ali et al., 2013, Hakim et al., 2005). To prevent particle agglomeration, a rotary ALD reactor that tumbles the particles in a porous metal cylinder has also been used (Cavanagh et al., 2009). The rotary reactor, unlike the fluidized bed reactor, uses static exposures because of the continual movement of particles caused by the reactor's rotation. Longrie et al. (2014) investigated thermal and plasma-enhanced atomic layer deposition of TiN on particles agitated in a rotary reactor using tetrakis(dimethylamino)titanium (TDMAT) and NH_3. The researchers used a rotating ALD reactor with a quartz tube as the reactor vessel, which allows for both thermal and plasma-enhanced ALD on powders. All of the unique features of fluidized particle ALD systems are designed to reduce particle aggregation while creating core-shell structures. McCormick et al. (2007) used a rotary reactor to deposit atomic layers on massive amounts of nanoparticles. In order to overcome these obstacles, a new rotary reactor was created to maintain consistent particle agitation during static ALD reactant exposures. A cylindrical drum with porous metal walls was positioned inside a vacuum chamber in the design of this novel reactor. A magnetically connected rotational feedthrough rotated the porous cylindrical drum. The particles were agitated by a continual "avalanche" of particles created by rotating the cylindrical drum at a centrifugal force of less than 1 gravitational force. In addition, an inert N_2 gas pulse assisted in dislodging the particles from the porous walls and provided a quick way to remove reactants and products from the particle bed. Duan et al. (2015) combined both fluidized and rotary ALD reactors to coat nanoparticles.

A schematic of rotary was depicted by Cavanagh et al. (2009) and the fluidization done by Hakim and associates in 2005 (Hakim et al., 2005).

4.1.2.1 Batch ALD Reactors

These are reactors that can deposit many samples at the same time, as opposed to reactors that only treat a single wafer. Batch ALD reactors have a significantly larger chamber than standard ALD reactors, allowing them to coat numerous samples at once. The cost of chemicals and the time it takes to complete the operation are both drastically decreased. In batch reactors, the pulse time (and exposure time) for each precursor, as well as the purging time constants, are longer due to bigger reactor volumes and surface area. However, the average price is still lower.

The precursor gases are delivered into the reactor through a single-point injector at the top and evacuated through an exhaust at the bottom. Convection transports gases from the top to the bottom, while diffusion transports gases between adjacent wafers. In most batch reactors, the number of wafers is between 50 and 150. When thicker films are required, the overall (single-wafer) process time increases, making processing 50–100 wafers in a big batch appealing. Batch tools are especially appealing from an economic standpoint when huge batches of identical samples must be handled. This is usually the case with commodity products like DRAM and/or flash memory. Batch and single-wafer systems, for example, have throughputs of 16 and 7 wafers/h for a 10 nm HfO_2 ALD process, respectively. Batch-type systems typically have longer uptimes than single-wafer systems, resulting in cheaper total cost of ownership.

Granneman et al. (2007) did a detailed study of the characteristics, comparisons with single-wafer ALD, and examples of batch ALD. Dingemans et al. (2014) discussed the benefits and uses of batch ALD in a vertical furnace. It looks at new material and process advances, as well as throughput improvements, which are important considerations for future high-volume production applications. Fischer et al. (2008) studied Batch ALD of HfO_2 and ZrO_2 films using cyclopentadienyl precursors. Thermal ALD deposition of ZrO_2 and HfO_2 films was carried out utilizing bis-cyclopentadienyl precursors in the ASM A412(TM) 300 mm vertical furnace. The thermal stability and volatility of the precursor are crucial due to the extended precursor residence period and large surface area to be covered in a batch reactor. The thermal stability of the precursor was determined by depositing it on a high surface area of silica powder, and then NMR tests were used to determine the reactivity of the precursor. The first deposition studies were done in single-wafer reactors to see if batch processing could be scaled up. It has been established in experiments that the bis-cyclopentadienyl precursor family has sufficient thermal stability to generate ALD films that meet deposition criteria in batch reactors. Also, thermal atomic layer deposition (ALD) was used to study batch processing of aluminum nitride (AlN) at high temperatures of 500–550°C utilizing aluminum chloride ($AlCl_3$) and ammonia (NH_3) as metal and nitrogen precursors (Chen et al., 2019).

4.1.2.2 Spatial ALD Reactor

In this reactor, precursors are continually provided in separate locations and kept isolated by an inert gas region or zone. Figure 4.9 shows the schematic of the spatial ALD modified using Muñoz-Rojas et al. (2019) concept. The substrate is exposed to the locations holding the various precursors to grow the film. Because the purging phase is removed, the process becomes faster, compatible with high-throughput techniques like roll-to-roll (R2R), and considerably more adaptable, as well as easier and less expensive to scale up. Furthermore, it has the advantage of being able to be performed at ambient pressure and even in the open air (i.e., without the use of any deposition chamber) without impairing the deposition rate.

There is no pulse control in spatial ALD, as the precursor gases are continually supplied in different physical places (Illiberi et al., 2013). As a result, as long as the substrate is present, there are (at least) two zones where a half-reaction can occur. The period of substrate exposure for each zone must be long enough to generate a saturated monolayer. To complete one ALD cycle and generate one ALD monolayer, the substrate flows to both zones. To clear the unreacted precursors, the substrate must still pass through an inert gas zone between the two precursor gas zones. Alternatively, the substrate location can be fixed while the gas supply is relocated, or

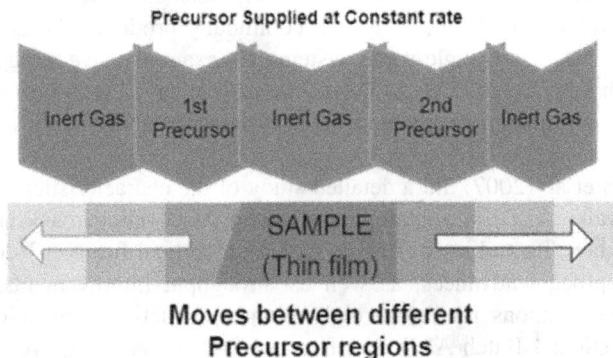

FIGURE 4.9 Schematic of spatial atomic layer deposition.

a combination of the two can be used. By repeating the necessary number of cycles, thicker films can be created.

Avoiding contact between two precursors, which involves decoupling the half-reaction zones, is the most important aspect of spatial ALD reactor design. Purge steps in between the precursor pulses are used in conventional ALD to alleviate this problem. A combination of physical barriers and continually flowing purge streams separate the precursor gases in spatial ALD. The total isolation of the reaction zones in spatial ALD ensures that precursors never contact, resulting in a CVD-like product. As a result, unlike traditional ALD, no deposition occurs on the reactor walls, only on the required substrates.

Atmospheric-pressure plasma-enhanced spatial ALD was used to generate high-quality ZrO_2 at lower temperatures than temporal ALD (Mione et al., 2017). PE-sALD delivers faster effective deposition rates and uses atmospheric-pressure plasma to trigger surface reactions at lower temperatures than temporal ALD. The researchers used spatial atmospheric plasma-enhanced ALD to produce ZrO_2 films utilizing tetrakis(ethylmethylamino)zirconium (TEMAZ) as a precursor and O_2 plasma as a coreactant at temperatures ranging from 150 to 250°C. With N and C levels as low as 0.4% and 1.5%, respectively, deposition rates as high as 0.17 nm/cycle were attained. With increasing deposition temperature, growth rate, film crystallinity, and impurity content in the films all improved. van den Bruele et al. (2015) used a unique combination of atmospheric process factors to study the formation of thin silver films: spatial ALD and an atmospheric pressure surface dielectric barrier discharge plasma source. Resistivity values as low as 18 cm and C- and F-levels below detection limits of energy dispersive X-ray analysis demonstrated that silver films were formed on top of Si substrates with good purity. The formation of silver films begins with the nucleation of islands, which then consolidate. The authors demonstrate that surface island shape is influenced by surface diffusion, which is influenced by temperature in the deposition temperature range of 100–120°C.

4.1.2.3   Other Reactor Configurations

Table 4.1 summarizes the various ALD types and key parameters associated with the precursors, temperature, and areas of application.

4.1.2.4   Method of Understanding ALD

ALD can be studied via two routes, viz. experimental and theoretical using numerical approach. The limitation of the experiment necessitated the use of numerical approach to design a pathway for the experimental actualization. The experimental optimizes variables such as concentration of precursor, purge time, temperature, and other variables successfully designed via the numerical approach. However, the numerical approach uses multiscale and mesoscopic scale. Atomic bond formation, species chemisorption/adsorption, chemical kinetics, and film deposition are all part of the multiscale process (Shaeri et al., 2014, Pan et al., 2015a). On a macroscopic level, it also comprises a reactor scale that encompasses material selections/interactions, geometrical effects, and fluid and energy transfer. The fact that neither the feature nor the reactor scales address the additional limits placed on the prediction capability

TABLE 4.1
The Various ALD Types and Key Parameters Associated with the Precursors, Temperature, and Areas of Application

ALD type	Viable reactants	Reactants	Temperature	Application
Plasma-enhanced ALD of metal oxides and nitrides	Al_2O_3, ZnO_x, SiO_2, $InOx$, HfO_2, TaN_x, SiN_x	$M(C_5H_5)_2$, $(CH_3C_5H_4)M(CH_3)_3$, $Cu(thd)_2$, $Pd(hfac)_2$, $Ni(acac)_2$, H_2	20–300°C	Semiconductor devices.
Metal ALD using thermal chemistry	Organometallics, metal fluorides, and catalytic metals	$M(C_5H_5)_2$, $(CH_3C_5H_4)M(CH_3)_3$, $Cu(thd)_2$, $Pd(hfac)_2$, $Ni(acac)_2$, H_2	175–400°C	MOS devices, conductive channels, catalytic surfaces.
Catalytic ALD	TiO_2, ZrO_2, $SnO2_2$ (metal oxides precursors)	$(Metal)Cl_4$, H_2O	> 32°C	Layers with a high k-dielectric constant, protective layers, antireflective layers, and so on.
Alumina ALD	Metal oxides and alumina precursors	$(Metal)Cl_4$, H_2O, $Ti(OiPr)_4$, $(Metal)(Et)_2$	30–300°C	Solar cell surface passivation includes dielectric layers, insulating layers, and others.
Plasma or radical-enhanced ALD for single element ALD materials	Pure metals (i.e., Ta, Ti, Si, Ge, Ru, Pt), metal nitrides (i.e., TiN, TaN, etc.)	Organometallics, bis(ethylcyclopentadienyl)ruthenium), NH_3, terrbutylimidotris(diethylamido) tantalum (TBTDET), MH_2Cl_2,	20–800°C	MOSFET and semiconductor devices, capacitors, DRAM structures.
ALD on polymers	Common polymers (polyethylene, PMMA, PP, PS, PVC, PVA, etc.)	H_2O, $Al(CH_3)_3$, $M(CH_3)_3$	25–100°C	Surface functionalization of polymers, composite construction, diffusion barriers, and so on.

of the deposition process across the substrates is a challenge (Merchant et al., 2000). The mesoscopic scale is used to remedy this. Researchers have attempted to compare results obtained from experimental with numerical approach. Jones III and Jones Jr (2014) developed a process to forecast the chemical species in a reactor as a function of time and space to produce an ideal ALD process. The actual and modeled results agreed well. Regardless of the pressure, the increased dissociation rate reduces the overall concentration in the chamber, according to the simulated result. Also, Cremers et al. (2019) discussed the current state of knowledge about the conformality of ALD techniques. They used an experimental and numerical approach to describe the basic concepts related to ALD conformality, including an overview of relevant gas transport regimes, definitions of exposure and sticking probability, and a distinction between different ALD growth types observed in high aspect ratio structures. To make comparisons easier, they introduced a geometry-independent equivalent aspect ratio (EAR). Muneshwar and Cadien (2016) proposed a model for enhancing the precursor usage in the ALD scaling process. The study used numerical growth approach and was validated with experiment.

4.1.3 ROLE OF ATOMIC LAYER DEPOSITION IN SUSTAINABLE DEVELOPMENT AND CLIMATE CHANGE

Sustainable development is key to continuity of life. A lot has been done to ensure man continues to live longer, healthier, and safe, with improved livelihood. This is achieved by reducing greenhouse gas emissions and ensuring replenishing of scarce commodities including energy. Climate change is changing the way things are done and the way humans live. ALD continues to play a vital role in this regard as it helps to ensure improved quality of deposition used in various applications including solar energy, sensor, and water desalination, among others. ALD is designed to reduce emission into the surrounding during the deposition process using fume chamber. ALD of high-quality thin films brings about improvement in the selected applications and new emerging applications.

The deposition process will ensure improvement in process that supports reduction in greenhouse gas emissions. ALD is now being used in water desalination and water splitting to improve the quality of water for man and animals. This results in improved livelihood. It is also used in sensor applications that help to detect different emissions hitherto not detected by human senses.

4.1.4 TRENDS IN ATOMIC LAYER DEPOSITION

The Fourth Industrial Revolution is the new wave of revolution shaping humanity. In the past, the introduction of a new industrial revolution birth new technologies and applications and also elimination of some technologies. ALD is a flexible deposition technique that is capable of playing key role in technologies that will survive and emerge with the 4IR. ALD will also play pivotal role in combating climate change as most of the applications are useful in climate actions. The precursors, reactors, and materials used may be modified and improved with time to accommodate any changes if need be.

4.2 CONCLUSION

This chapter explained the concept of atomic layer deposition using tools to make it easy to understand. The chemistry underlying atomic layer deposition, the characteristics, and the different precursors used were discussed. The various atomic layer deposition types and key parameters associated with the precursors, temperature, and areas of application were discussed. Atomic layer deposition reactor was examined with illustrative tools to help readers understand it. Atomic layer deposition is the deposition of choice for uniform, conformal, and quality thin film deposition for various applications including the fourth industrial application emerging technologies of self-cleaning and energy, among others. The chapter will help readers understand atomic layer deposition and possible areas of applications.

REFERENCES

Aarik, L., Arroval, T., Mändar, H., Rammula, R., & Aarik, J. (2020). Influence of oxygen precursors on atomic layer deposition of HfO_2 and hafnium-titanium oxide films: Comparison of O_3-and H_2O-based processes. *Applied Surface Science, 530*, 147229.

Ahvenniemi, E., Akbashev, A. R., Ali, S., Bechelany, M., Berdova, M., Boyadjiev, S., Cameron, D. C., Chen, R., Chubarov, M., & Cremers, V. (2017). Recommended reading list of early publications on atomic layer deposition—Outcome of the "Virtual Project on the History of ALD." *Journal of Vacuum Science & Technology A: Vacuum, Surfaces, and Films, 35*, 010801.

Ali, K. A., Jawad, A. H., & Ibrahim, S. I. (2013). Experimental investigation of minimum fluidization velocity in three phase inverse fluidized bed system. *Engineering and Technology Journal, 31*, 1194–1203.

Bacsa, J., Hanke, F., Hindley, S., Odedra, R., Darling, G. R., Jones, A. C., & Steiner, A. (2011). Rücktitelbild: The solid-state structures of dimethylzinc and diethylzinc (Angew. Chem. 49/2011). *Angewandte Chemie, 123*, 12008–12008.

Cavanagh, A. S., Wilson, C. A., Weimer, A. W., & George, S. M. (2009). Atomic layer deposition on gram quantities of multi-walled carbon nanotubes. *Nanotechnology, 20*, 255602.

Chen, Z., Zhu, Z., Härkönen, K., & Salmi, E. (2019). Batch processing of aluminum nitride by atomic layer deposition from $AlCl_3$ and NH_3. *Journal of Vacuum Science & Technology A: Vacuum, Surfaces, and Films, 37*, 020925.

Cho, W., Sung, K., An, K.-S., Sook Lee, S., Chung, T.-M., & Kim, Y. (2003). Atomic layer deposition of Al_2O_3 thin films using dimethylaluminum isopropoxide and water. *Journal of Vacuum Science & Technology A: Vacuum, Surfaces, and Films, 21*, 1366–1370.

Chowdhury, M. I., Sowa, M., Kozen, A. C., Krick, B. A., Haik, J., Babuska, T. F., & Strandwitz, N. C. (2021). Plasma enhanced atomic layer deposition of titanium nitride-molybdenum nitride solid solutions. *Journal of Vacuum Science & Technology A: Vacuum, Surfaces, and Films, 39*, 012407.

Cozzi, P. G. (2004). Metal-Salen Schiff base complexes in catalysis: Practical aspects. *Chemical Society Reviews, 33*, 410–421.

Cremers, V., Puurunen, R. L., & Dendooven, J. (2019). Conformality in atomic layer deposition: Current status overview of analysis and modelling. *Applied Physics Reviews, 6*, 021302.

Crutchley, R. (2013). CVD and ALD precursor design and application. *Coordination Chemistry Reviews, 257*, 3153.

Dingemans, G., Jongbloed, B., Knaepen, W., Pierreux, D., Jdira, L., & Terhorst, H. (2014). Merits of batch ALD. *ECS Transactions, 64*, 35.

Duan, C.-L., Liu, X., Shan, B., & Chen, R. (2015). Fluidized bed coupled rotary reactor for nanoparticles coating via atomic layer deposition. *Review of Scientific Instruments, 86*, 075101.

Efimova, A., Pfützner, L., & Schmidt, P. (2015). Thermal stability and decomposition mechanism of 1-ethyl-3-methylimidazolium halides. *Thermochimica Acta, 604*, 129–136.

Elam, J., Groner, M., & George, S. (2002). Viscous flow reactor with quartz crystal microbalance for thin film growth by atomic layer deposition. *Review of Scientific Instruments, 73*, 2981–2987.

Fischer, P. R., Pierreux, D., Rouault, O., Sirugue, J., Zagwijn, P. M., Tois, E., & Haukka, S. (2008). Batch atomic layer deposition of HfO_2 and ZrO_2 films using cyclopentadienyl precursors. *ECS Transactions, 16*, 135.

Fouache, J., & Lincot, D. (2000). Study of atomic layer epitaxy of zinc oxide by in-situ quartz crystal microgravimetry. *Applied Surface Science, 153*, 223–234.

George, S. M. (2003). *Fabrication and properties of nanolaminates using self-limiting surface chemistry techniques.* Colorado University at Boulder Department of Chemistry and Biochemistry.

George, S. M. (2010). Atomic layer deposition: An overview. *Chemical Reviews, 110*, 111–131.

Gordon, R. G. (2014). ALD precursors and reaction mechanisms. In C. Hwang (ed.), *Atomic Layer Deposition for Semiconductors* (pp. 15–46). Springer.

Goujon, F., Malfreyt, P., & Tildesley, D. J. (2014). The gas-liquid surface tension of argon: A reconciliation between experiment and simulation. *The Journal of Chemical Physics, 140*, 244710.

Granneman, E., Fischer, P., Pierreux, D., Terhorst, H., & Zagwijn, P. (2007). Batch ALD: Characteristics, comparison with single wafer ALD, and examples. *Surface and Coatings Technology, 201*, 8899–8907.

Gupta, B., Hossain, M. A., Riaz, A., Sharma, A., Zhang, D., Tan, H. H., Jagadish, C., Catchpole, K., Hoex, B., & Karuturi, S. (2022). Recent advances in materials design using atomic layer deposition for energy applications. *Advanced Functional Materials, 32*, 2109105.

Haiying, W., Hongge, G., Lijun, S., Xingcun, L., & Qiang, C. (2018). Study on deposition of Al_2O_3 films by plasma-assisted atomic layer with different plasma sources. *Plasma Science and Technology, 20*, 065508.

Hakim, L. F., George, S. M., & Weimer, A. W. (2005). Conformal nanocoating of zirconia nanoparticles by atomic layer deposition in a fluidized bed reactor. *Nanotechnology, 16*, S375.

Hausmann, D. M., & Gordon, R. G. (2003). Surface morphology and crystallinity control in the atomic layer deposition (ALD) of hafnium and zirconium oxide thin films. *Journal of Crystal Growth, 249*, 251–261.

Hausmann, D. M., Kim, E., Becker, J., & Gordon, R. G. (2002). Atomic layer deposition of hafnium and zirconium oxides using metal amide precursors. *Chemistry of Materials, 14*, 4350–4358.

He, W. (2015). ALD: Atomic layer deposition—precise and conformal coating for better performance. In: A. Nee (Ed.), *Handbook of manufacturing engineering and technology* (pp. 2959–2996). Springer.

He, J., Li, J., Han, Q., Si, C., Niu, G., Li, M., Wang, J., & Niu, J. (2019). Photoactive metal–organic framework for the reduction of aryl halides by the synergistic effect of consecutive photoinduced electron-transfer and hydrogen-atom-transfer processes. *ACS Applied Materials & Interfaces, 12*, 2199–2206.

Heil, S., Kudlacek, P., Langereis, E., Engeln, R., Van De Sanden, M. A., & Kessels, W. (2006). In situ reaction mechanism studies of plasma-assisted atomic layer deposition of Al_2O_3. *Applied Physics Letters, 89*, 131505.

Hoex, B., Heil, S., Langereis, E., Van De Sanden, M., & Kessels, W. (2006). Ultralow surface recombination of c-Si substrates passivated by plasma-assisted atomic layer deposited Al_2O_3. *Applied Physics Letters, 89*, 042112.

Hsu, C.-H., Chen, K.-T., Lin, L.-Y., Wu, W.-Y., Liang, L.-S., Gao, P., Qiu, Y., Zhang, X.-Y., Huang, P.-H., & Lien, S.-Y. (2021). Tantalum-doped TiO_2 prepared by atomic layer deposition and its application in perovskite solar cells. *Nanomaterials, 11*, 1504.

Illiberi, A., Scherpenborg, R., Poodt, P., & Roozeboom, F. (2013). Spatial atomic layer deposition of transparent conductive oxides. *ECS Transactions, 58*, 105.

Johnson, R. W., Hultqvist, A., & Bent, S. F. (2014). A brief review of atomic layer deposition: From fundamentals to applications. *Materials Today, 17*, 236–246.

Jones Iii, A.-A., & Jones JR, A. (2014). Numerical simulation and verification of gas transport during an atomic layer deposition process. *Materials science in Semiconductor Processing, 21*, 82–90.

Kanjolia, R. K., Anthis, J., Odedra, R., Williams, P., & Heys, P. (2008). Design and development of ALD precursors for microelectronics. *ECS Transactions, 16*, 79.

Kim, H. (2003). Atomic layer deposition of metal and nitride thin films: Current research efforts and applications for semiconductor device processing. *Journal of Vacuum Science & Technology B: Microelectronics and Nanometer Structures Processing, Measurement, and Phenomena, 21*, 2231–2261.

Kim, Y., Koo, J., Han, J., Choi, S., Jeon, H., & Park, C.-G. (2002). Characteristics of ZrO_2 gate dielectric deposited using Zr t–butoxide and Zr $(NEt_2)_4$ precursors by plasma enhanced atomic layer deposition method. *Journal of Applied Physics, 92*, 5443–5447.

Kimes, W. A., & Maslar, J. E. (June 2011). In situ water measurements as a diagnostic of flow dynamics in ALD reactors. *Proceedings of the ALD 2011* (pp. 26–29). Cambridge.

Kimes, W. A., Moore, E., & Maslar, J. (2012). Perpendicular-flow, single-wafer atomic layer deposition reactor chamber design for use with in situ diagnostics. *Review of Scientific Instruments, 83*, 083106.

Klejna, S., & Elliott, S. D. (2011). Understanding "clean-up" of III–V native oxides during atomic layer deposition using bulk first principles models. *Journal of Nanoscience and Nanotechnology, 11*, 8246–8250.

Knoops, H. C., Potts, S. E., Bol, A. A., & Kessels, W. (2015). Atomic layer deposition. In: P. Rudolph (Ed.), *Handbook of crystal growth*. Elsevier.

Koo, J., Kim, Y., & Jeon, H. (2002). ZrO_2 gate dielectric deposited by plasma-enhanced atomic layer deposition method. *Japanese Journal of Applied Physics, 41*, 3043.

Koponen, S. E., Gordon, P. G., & Barry, S. T. (2016). Principles of precursor design for vapour deposition methods. *Polyhedron, 108*, 59–66.

Kukli, K., Aarik, J., Aidla, A., Forsgren, K., Sundqvist, J., Hårsta, A., Uustare, T., Mändar, H., & Kiisler, A.-A. (2000a). Atomic layer deposition of tantalum oxide thin films from iodide precursor. *Chemistry of Materials, 13*, 122–128.

Kukli, K., Aidla, A., Aarik, J., Schuisky, M., Hårsta, A., Ritala, M., & Leskelä, M. (2000b). Real-time monitoring in atomic layer deposition of TiO_2 from TiI_4 and $H_2O–H_2O_2$. *Langmuir, 16*, 8122–8128.

Langereis, E., Creatore, M., Heil, S., Van De Sanden, M., & Kessels, W. (2006). Plasma-assisted atomic layer deposition of Al_2O_3 moisture permeation barriers on polymers. *Applied Physics Letters, 89*, 081915.

Larrabee, T. J. (2012). *A quantified dosing ALD reactor with in-situ diagnostics for surface chemistry studies*. The Pennsylvania State University.

Leal, D. A. A., Shaji, S., Avellaneda, D. A., Martínez, J. A. A., & Krishnan, B. (2020). In situ incorporation of laser ablated PbS nanoparticles in $CH_3NH_3PbI_3$ films by spin-dip coating and the subsequent effects on the planar junction $CdS/CH_3NH_3PbI_3$ solar cells. *Applied Surface Science, 508*, 144899.

Leskelä, M., & Ritala, M. (2002). Atomic layer deposition (ALD): From precursors to thin film structures. *Thin Solid Films, 409*(1), 138–146.

Linstorm, P. (1998). NIST chemistry webbook, NIST standard reference database number 69. *Journal of Physical and Chemical Reference Data Monographs, 9*, 1–1951.

Liu, J., Li, J., Wu, J., & Sun, J. (2019). Structure and dielectric property of high-k ZrO_2 films grown by atomic layer deposition using tetrakis(dimethylamido)zirconium and ozone. *Nanoscale Research Letters, 14*, 1–12.

Longrie, D., Deduytsche, D., Haemers, J., Smet, P. F., Driesen, K., & Detavernier, C. (2014). Thermal and plasma-enhanced atomic layer deposition of TiN using TDMAT and NH_3 on particles agitated in a rotary reactor. *ACS Applied Materials & Interfaces, 6*, 7316–7324.

Łukasiewicz, M., Wójcik-Głodowska, A., Guziewicz, E., Jakieła, R., Krajewski, T., Łusakowska, E., Paszkowicz, W., Minikayev, R., Kiecana, M., & Sawicki, M. (2008). ZnCoO films obtained at low temperature by atomic layer deposition using organic zinc and cobalt precursors. *Acta Physica Polonica A, 5*, 1235–1240.

Malik, M. A., & O'brien, P. (2008). Basic chemistry of CVD and ALD precursors. In: A. C. Jones, & M. L. Hitchman (Eds.), *Chemical Vapour Deposition* (pp. 207–271). The Royal Society of Chemistry.

Matero, R., Rahtu, A., Ritala, M., Leskelä, M., & Sajavaara, T. (2000). Effect of water dose on the atomic layer deposition rate of oxide thin films. *Thin Solid Films, 368*, 1–7.

Matero, R., Ritala, M., Leskelä, M., Jones, A. C., Williams, P. A., Bickley, J. F., Steiner, A., Leedham, T., & Davies, H. (2002). Atomic layer deposition of ZrO_2 thin films using a new alkoxide precursor. *Journal of Non-crystalline Solids, 303*, 24–28.

Mccormick, J., Cloutier, B., Weimer, A., & George, S. (2007). Rotary reactor for atomic layer deposition on large quantities of nanoparticles. *Journal of Vacuum Science & Technology A: Vacuum, Surfaces, and Films, 25*, 67–74.

Merchant, T. P., Gobbert, M. K., Cale, T. S., & Borucki, L. J. (2000). Multiple scale integrated modeling of deposition processes. *Thin Solid Films, 365*, 368–375.

Miikkulainen, V., Leskelä, M., Ritala, M., & Puurunen, R. L. (2013). Crystallinity of inorganic films grown by atomic layer deposition: Overview and general trends. *Journal of Applied Physics, 113*, 2.

Min, J. S., Park, J. S., Park, H. S., & Kanga, S. W. (2000). The mechanism of Si incorporation and the digital control of Si content during the metallorganic atomic layer deposition of Ti-Si-N thin films. *Journal of the Electrochemical Society, 147*, 3868.

Mione, M. A., Katsouras, I., Creyghton, Y., Van Boekel, W., Maas, J., Gelinck, G., Roozeboom, F., & Illiberi, A. (2017). Atmospheric pressure plasma enhanced spatial ALD of ZrO_2 for low-temperature, large-area applications. *ECS Journal of Solid State Science and Technology, 6*, N243.

Moshe, H., & Mastai, Y. (2013). Atomic layer deposition on self-assembled-monolayers (pp. 63–84). In: Y. Mastai (Ed.), *Materials Science-Advanced Topics*. Intech publishers.

Muneshwar, T., & Cadien, K. (2016). AxBAxB… pulsed atomic layer deposition: Numerical growth model and experiments. *Journal of Applied Physics, 119*, 085306.

Muñoz-Rojas, D., Huong Nguyen, V., Masse De La Huerta, C., Jiménez, C., & Bellet, D. (2019). Spatial atomic layer deposition. In: P. Mandracci (Ed.), *Chemical Vapor Deposition for Nanotechnology* (pp. 1–25). IntechOpen. https://doi.org/10.5772/int echopen.82439

Nattermann, L., Maßmeyer, O., Sterzer, E., Derpmann, V., Chung, H., Stolz, W., & Volz, K. (2018). An experimental approach for real time mass spectrometric CVD gas phase investigations. *Scientific Reports, 8*, 1–7.

Niemelä, J.-P., Marin, G., & Karppinen, M. (2017). Titanium dioxide thin films by atomic layer deposition: A review. *Semiconductor Science and Technology, 32*, 093005.

Nieminen, H.-E., Kaipio, M., & Ritala, M. (2020). In situ reaction mechanism study on atomic layer deposition of intermetallic Co_3Sn_2 thin films. *Chemistry of Materials, 32*, 8120–8128.

Niinistö, L., Nieminen, M., Päiväsaari, J., Niinistö, J., Putkonen, M., & Nieminen, M. (2004). Advanced electronic and optoelectronic materials by atomic layer deposition: An overview with special emphasis on recent progress in processing of high-k dielectrics and other oxide materials. *Physica Status Solidi (A), 201*, 1443–1452.

Oviroh, P. O., Akbarzadeh, R., Pan, D., Coetzee, R. A. M., & Jen, T.-C. (2019). New development of atomic layer deposition: Processes, methods and applications. *Science and Technology of Advanced Materials, 20*, 465–496.

Pan, D., Ma, L., Xie, Y., Jen, T. C., & Yuan, C. (2015a). On the physical and chemical details of alumina atomic layer deposition: A combined experimental and numerical approach. *Journal of Vacuum Science & Technology A: Vacuum, Surfaces, and Films, 33*, 021511.

Pan, D., Ma, L., Xie, Y., Wang, F., Jen, T.-C., & Yuan, C. (2015b). Experimental and numerical investigations into the transient multi-wafer batch atomic layer deposition process with vertical and horizontal wafer arrangements. *International Journal of Heat and Mass Transfer, 91*, 416–427.

Park, J.-S., Park, H.-S., & Kang, S.-W. (2001). Plasma-enhanced atomic layer deposition of Ta-N thin films. *Journal of the Electrochemical Society, 149*, C28.

Ponraj, J. S., Attolini, G., & Bosi, M. (2013). Review on atomic layer deposition and applications of oxide thin films. *Critical Reviews in Solid State and Materials Sciences, 38*, 203–233.

Puurunen, R. L. (2014). A short history of atomic layer deposition: Tuomo Suntola's atomic layer epitaxy. *Chemical Vapor Deposition, 20*, 332–344.

Ritala, M., Leskelä, M., Nykänen, E., Soininen, P., & Niinistö, L. (1993). Growth of titanium dioxide thin films by atomic layer epitaxy. *Thin Solid Films, 225*, 288–295.

Ritala, M., Leskelä, M., Rauhala, E., & Jokinen, J. (1998). Atomic layer epitaxy growth of TiN thin films from TiI_4 and NH_3. *Journal of the Electrochemical Society, 145*, 2914.

Saynatjoki, A. (2012). Atomic-layer-deposited thin films for silicon nanophotonics. *SPIE Newsroom, 10*, 004218.

Schuisky, M., Hårsta, A., Aidla, A., Kukli, K., Kiisler, A. A., & Aarik, J. (2000). Atomic layer chemical vapor deposition of TiO_2 low temperature epitaxy of rutile and anatase. *Journal of the Electrochemical Society, 147*, 3319.

Shaeri, M. R., Jen, T.-C., & Yuan, C. Y. (2014). Improving atomic layer deposition process through reactor scale simulation. *International Journal of Heat and Mass Transfer, 78*, 1243–1253.

Shahmohammadi, M., Mukherjee, R., Sukotjo, C., Diwekar, U. M., & Takoudis, C. G. (2022). Recent advances in theoretical development of thermal atomic layer deposition: A review. *Nanomaterials, 12*, 831.

Shon, H., Vigneswaran, S., Kim, I. S., Cho, J., Kim, G., Kim, J., & Kim, J.-H. (2007). Preparation of titanium dioxide (TiO_2) from sludge produced by titanium tetrachloride ($TiCl_4$) flocculation of wastewater. *Environmental Science & Technology, 41*, 1372–1377.

Sneh, O. (2005). *ALD apparatus and method*. US Patent Application.

Stevens, E. C., Mousa, M. B. M., & Parsons, G. N. (2018). Thermal atomic layer deposition of Sn metal using $SnCl_4$ and a vapor phase silyl dihydropyrazine reducing agent. *Journal of Vacuum Science & Technology A: Vacuum, Surfaces, and Films, 36*, 06A106.

Travis, C. D., & Adomaitis, R. A. (2013). Dynamic modeling for the design and cyclic operation of an atomic layer deposition (ALD) reactor. *Processes, 1*, 128–152.

Van Den Bruele, F. J., Smets, M., Illiberi, A., Creyghton, Y., Buskens, P., Roozeboom, F., & Poodt, P. (2015). Atmospheric pressure plasma enhanced spatial ALD of silver. *Journal of Vacuum Science & Technology A: Vacuum, Surfaces, and Films, 33*, 01A131.

Wilkinson, G. (1963). "Ferrocene." Organic syntheses. *Collective, 4*, 473.

Xie, Q., Jiang, Y.-L., Detavernier, C., Deduytsche, D., Van Meirhaeghe, R. L., Ru, G.-P., Li, B.-Z., & Qu, X.-P. (2007). Atomic layer deposition of TiO_2 from tetrakis-dimethyl-amido titanium or Ti isopropoxide precursors and H_2O. *Journal of Applied Physics, 102*, 083521.

Yamada, A., Sang, B., & Konagai, M. (1997). Atomic layer deposition of ZnO transparent conducting oxides. *Applied Surface Science, 112*, 216–222.

Yoshida, K., Nagata, I., Saito, K., Miura, M., Kanomata, K., Ahmmad, B., Kubota, S., & Hirose, F. (2022). Room-temperature atomic layer deposition of iron oxide using plasma excited humidified argon. *Journal of Vacuum Science & Technology A: Vacuum, Surfaces, and Films, 40*, 022408.

5 Demystifying 4IR and Emerging Modern Technology (Solar Energy)

5.1 ATOMIC LAYER DEPOSITION (ALD) IN FOURTH INDUSTRIAL REVOLUTION (4IR)

The wave of 4IR is sweeping across the globe, with many countries still lagging. Prior to the 4IR, there have been three different revolutions. The First Industrial Revolution (1IR) was characterized by decreased dependence on a beast of burden, giving way to the steam engine and its industrial and residential applications (Barham, 2013). This resulted in railway construction, steel, raw materials, and finished goods, which improved the quality of life of affected persons and led to urbanization (Kanji, 1990). The Second Industrial Revolution (2IR) took place from the 1880s to the 1950s. It was characterized by electricity's discovery, leading to increased production rates and improved communication and other facets of life (Mohajan, 2019). The discovery of the personal computer in the early 1950s and the internet around the 2000s paved the way for the Third Industrial Revolution (3IR) from the 1950s to 2000 (Zhong, 2017). The introduction of computing in the industry allowed for significant progress in terms of developing new technologies. This computing capacity made it possible for humans to run relatively complex sets of instructions and even facilitated humanity reaching the surface of the Moon. This development soon enhanced the quality of life of commercial firms with the introduction of personal computers by companies such as IBM and Apple in the 1980s (Popkin, 1999). Table 5.1 gives the critical features of the Industrial Revolution.

4IR is a term that was introduced in 2011 to describe marriage of information and communication and related technology into production and industry (Schuh et al., 2017). The focus is on industries and related sectors. It is a technology that aims to increase productivity, reduce downtime, and speed up production activities (Kagermann, 2014, Lasi et al., 2014). In some parts of the world, the Fourth Industrial Revolution (4IR) is sweeping across all sectors from companies to research centers (Hermann et al., 2016). In Europe and other developed continents, it has become a lullaby and is being discussed at every event and meeting (Drath and Horch, 2014). Countries in those continents have included it in their national key developmental initiative (Kagermann et al., 2013). The US government committed about 2 billion dollars to it in 2014 (Accelerating, 2014). The key features of the 4IR include

 DOI: 10.1201/9781003364481-5

TABLE 5.1
Critical Features of the Industrial Revolution

Industrial revolution	Period	Main features
First (machine age)	1780–1880	Mechanical production, energy transformation, and the start of globalization
Second (electricity age)	1880–1950	Mass production, electrical, energy transformation, conversion to clean energy, and electrification
Third (electronic age)	1950–2000	The advent of the internet, automation, conversion of
Fourth (age of the internet)	2000–2050	analog to digital (both mechanical and electronic devices), and introduction of information technology
		Introduction of social technology, new techno materials and cyber-physical systems, Internet of Things, additive manufacturing, industrial internet, and driverless cars

accelerating digitalization, artificial intelligence (AI), cloud computing, robotics, and 3D printing as shown in Figure 5.1. They have obvious and important implications for education, employment, and the future of work.

5.2   KEY TECHNOLOGIES OF THE 4IR

The key technologies that are characteristic of the 4IR are discussed below.

5.2.1   ARTIFICIAL INTELLIGENCE

It is the replication of human reasoning in machines. Machines replicate and act similarly to humans, especially in problem-solving, thinking, and learning (Ahmad et al., 2021). An application of AI in daily living is detecting and eliminating bank card fraud by identifying spending patterns. The technology can decode voices and handwriting among other traits that follow specific ways. It is already applied in medicine, construction, and big data, among others (Bakhtiyari et al., 2021; Bauer and Lizotte, 2021; Leenen and Meyer, 2021; Sharma et al., 2021).

5.2.2   INTERNET OF THINGS (IoT)

It links physical items using sensors and software to transmit data across the internet, mainly via Wi-Fi (Yahyaoui et al., 2021). Although not all IoT devices require the internet to function, they need to be interfaced with other devices. It is used in smart homes, connected appliances, biometric scanners, and innovative factory equipment (Ayvaz and Alpay, 2021; Knight et al., 2021). The purpose is for the device to feed data without manual human intervention. Examples include Google Home, Amazon Alexa, and baby monitors.

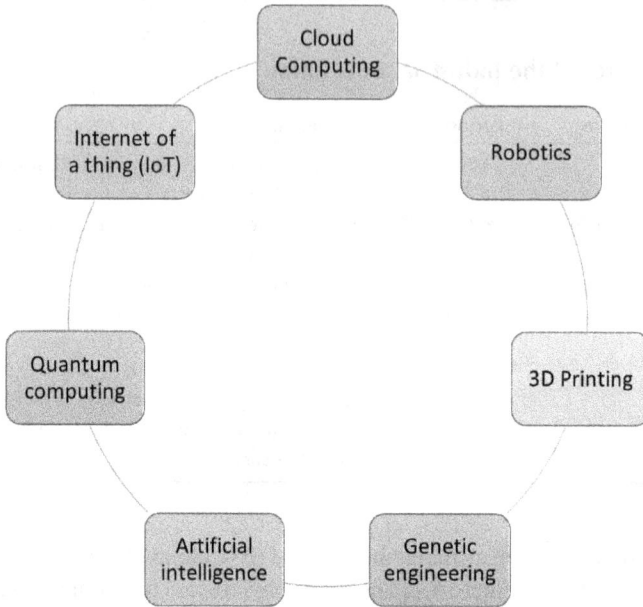

FIGURE 5.1　Key technologies of the Fourth Industrial Revolution.

5.2.3　3D PRINTING

It is also known as additive manufacturing and converts computer-aided objects into reality, especially three-dimensional objects (Li and Pumera, 2021). The materials are built layer by layer until the final thing is formed. The object to be printed is first constructed or modeled using computer-aided tools. Recently, several entities have been printed using 3D printers. It finds application in medicine for printing human organs, building, and other sectors (Goh et al., 2021).

5.2.4　CLOUD COMPUTING

It provides computer resources such as data storage, software, servers, and computing resources on-demand and made available over the internet (Houssein et al., 2021; Shahidinejad et al., 2021). However, quantum computing uses the principle of quantum theory for performing calculations (Jurcevic et al., 2021). It is different from conventional computers that use 1s or 0s, but quantum computing uses the object state before measurement (Emani et al., 2021). This allows quantum computing to perform exponentially compared to conventional computing.

5.2.5　ROBOTICS

They are machines built by combining technology, mathematics, engineering, and science (Vicentini, 2021). This machine is used for performing tasks hitherto tricky and sometimes dangerous for humans. Robots have been deployed for several applications, including surgery and manufacturing (Akinradewo et al., 2021; Boehm et al., 2021; Holland et al., 2021; Parmar et al., 2021; Tselegkaridis and Sapounidis, 2021; Vrontis et al., 2021).

The above-mentioned components of the 4IR have some similarities, with thin films and deposition being one such. Therefore, a key deposition technique will play a key role in actualizing the 4IR. Hence, ALD applications in some selected sectors that are vital in the era of 4IR are hereby discussed.

This study discussed some selected applications of ALD in emerging modern and conventional industrial and everyday technologies. These include self-healing, bio-medical, smart coatings, corrosion, energy storage and conversion (solar technology, lithium batteries, hydrogen), and machine learning/artificial intelligence, among others.

5.3 FOURTH INDUSTRIAL REVOLUTION IN AFRICA

The 4IR is gaining momentum across different parts of the world. However, the same cannot be said concerning Africa. Despite the claim that 4IR has a high penetration rate globally, some countries still lack some of the main features of the 2IR and 3IR. About 66.67% of the world population is behind the 2IR as they lack access to electricity (Ouedraogoa et al.). Similarly, 50% of the world population is behind the 3IR due to lack of access to the internet (Schwab, 2016). Africa and some parts of India and Asia account for the bulk of these figures (Dalla Longa and van der Zwaan, 2021). Although, some momentum is being gathered in parts of South Africa and Kenya. In these two countries, the emphasis of the 4IR is focused on education, entrepreneurship, the national economy, and manufacturing. Fwaya and Kesa (2018) examined the implication of 4IR on hotel businesses in South Africa and Kenya. The study argued that hotels in the two countries benefit from 4IR in revenue and exposure via social media usage. However, the lack of regulation of the 4IR features and unpredictable aspects of the 4IR poses a real challenge to the sector. Waghid et al. (2019) researched advancing cosmopolitan education in Africa about 4IR. The study believed that the best route of actualizing the 4IR is by implementing and tapping into technologies of previous IR, especially electricity and the internet. The study suggests that new skill sets such as problem-solving, social, processing, and cognitive skills should be included in Africa's school curriculum. The study opined that teaching and learning should be done to allow global acquisition of skills and competencies by African schools while developing these skillsets indigenously. Kayembe and Nel (2019) used unobtrusive methods to examine the challenges and windows of opportunities available to education in South Africa with 4IR. The study identified lack of adequate infrastructure, skill shortage, and funding as significant factors mitigating the success of 4IR. However, the 4IR offers an opportunity for global participation in collaborative partnerships and the digital economy.

Naudé (2017) did comprehensive work on the link between entrepreneurship, education, and 4IR. The paper suggested that consented efforts need to be given to entrepreneurship and education for Africa to benefit from the 4IR.

However, in line with Kayembe and Nel (2019) assertions, the implementation of 4IR with existing infrastructure in African countries needs to be tempered and observed. This is especially true in the case of industrial levels of energy generation and provision are concerned, specifically the fact that outside of countries like Nigeria, Ghana, and Sub-Saharan Africa, transmission and distribution infrastructure is sparsely distributed, often focusing entirely on capital cities. The more rural areas are often more dependent on decentralized, or local, generation and electricity, which is provided by fossil fuel. These include diesel for electricity and coal or wood for domestic heating and cooking.

Energy consumption is also limited by the availability of suitably sized off-takers, with most being relatively rudimentary, such as open-pit mining, small-scale materials processing, and logging. Therefore, despite a substantial influx of funds from IFCs (International Finance Corporation) and local financial institutions, the introduction of energy, more specifically renewable energy, in the African context needs to be carefully considered, as the production of power cannot be viewed on a stand-alone basis. However, the next section will aim to clarify what renewable energy is, how it functions in terms of generation, how it affects a transmission/distribution network in practice, and what is required.

The key features of the 4IR include the shift in population, shift in mass production, and rise of the sophisticated transportation system. Population shift will occur at the height of the 4IR. Also, there will be a shift in mass production by deploying advanced technologies such as robotics and 3D printing. A whole building will be completed within a short time using these technologies (Freire et al., 2021). There will also be a shift in transportation and power. A high-speed transportation system will be deployed, and goods and humans will move in a faster and large quantity. Electric cars will replace fossil-fueled cars (Balali and Stegen, 2021).

5.3.1 RENEWABLE ENERGY

Renewable energy is a cumulative term used to describe the creation of electron flow in a system without using substances/resources that are finite. The various types of renewable energy sources (Ellabban et al., 2014) are shown in Figure 5.2. Commonly, nonrenewable energy generation is often termed fossil fuel generation (Awodumi and Adewuyi, 2020). It encompasses technologies such as coal-fired power stations, gas turbine power stations, integrated gasification combined cycle, and other syngas power stations that operate using finite resources, such as solid fuel in the form of coal, derivatives of crude oil (petroleum products), or natural gas (found in pockets underground).

Nonrenewable technologies have been primary drivers in the 2IR and 3IR (Alaloul et al., 2020). Environmentalists, energy specialists, and laypersons have—and with good reason given the finite nature of the current fuel source and the impact on the environment caused by burning same—looked at implementing alternative sources of energy to lower dependence on nonrenewable energy as well as to be more

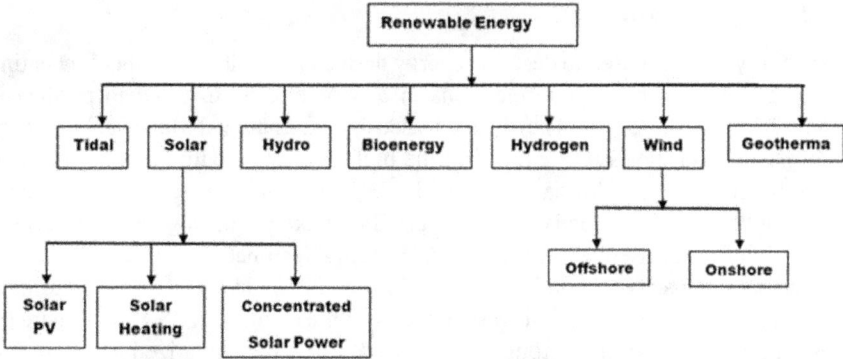

FIGURE 5.2 Classification of renewable energy sources.

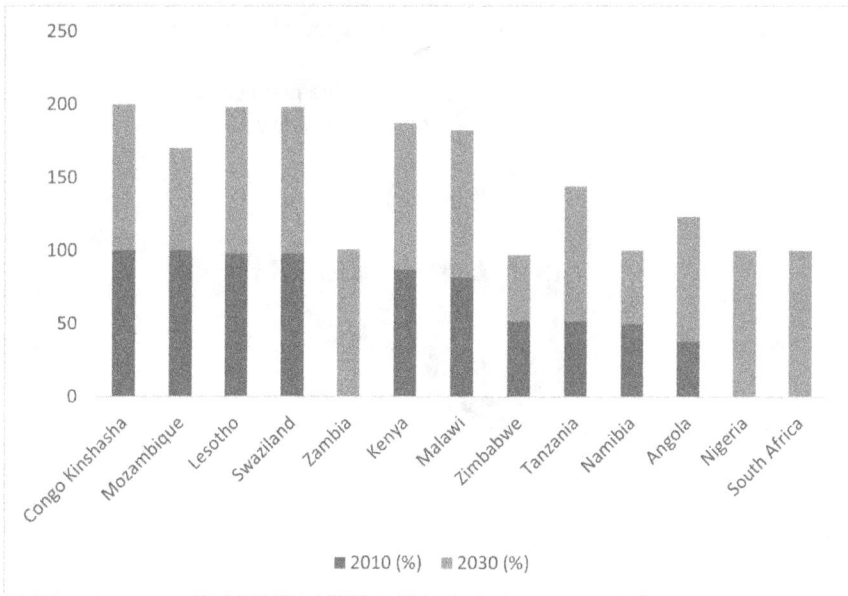

FIGURE 5.3 Renewable-based generation in total electricity generation.

environmentally friendly in existing under our current quality of life expectations. The following sections will describe a few standard renewable energy technologies currently implemented worldwide, noting their specific benefits and shortcomings in our daily lives. Figure 5.3 gives renewable energy installed capacity in some African countries with selected data from IRENA (Miketa and Merven, 2013) and Gebrehiwot and Van den Bossche (2015).

5.3.2 SOLAR ENERGY

Solar energy is a term used to describe energy derived from the Sun, either by heating or through photons that excite electrons in a photoelectric material to produce a current (Koli et al., 2020). The data released by International Renewable Energy Agency (IRENA) shows that Egypt had the highest growth, with 581 MW of solar energy in Africa. South Africa contributed 373 MW, making it second on the continent. South Africa has a total capacity of 2.5 GW, making it the highest market operational solar system. Kenya was third in 2018, with an estimated 55 MW added to the installed solar capacity. Namibia added 33 MW in 2018, taking installed solar capacity to 79 MW. The fifth highest was Ghana, with the addition of 25 MW to increase installed capacity to 64 MW (Kougias et al., 2018). This is summarized in Figure 5.4. The breakdown of South Africa's installed solar capacity shows that photovoltaic accounts for 2321 MW and concentrated solar power is 600 MW.

There are currently three technologies that use solar energy: solar photovoltaic implementations, concentrated solar photovoltaic implementations, and concentrated solar power—Central Tower/Heliostat Field and Parabolic Trough, as illustrated in Figure 5.5.

Solar photovoltaic (solar PV) power plants use the photoelectric effect on a large scale. The electrons energized by the electromagnetic waves do not escape the atoms

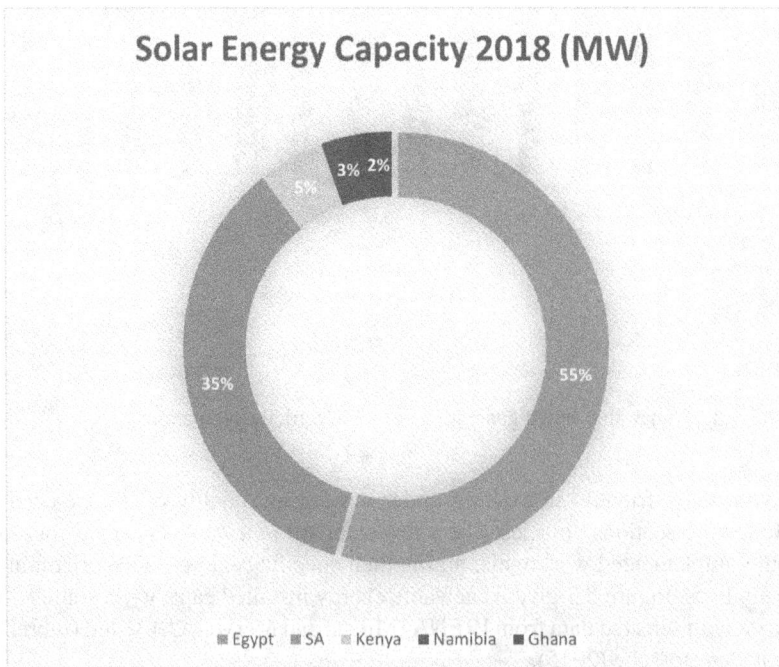

FIGURE 5.4   A chart of the 5 most installed solar energy capacities in 2018 in Africa.

```
                        ┌─────────────────────┐
                        │  Solar Technology   │
                        └─────────────────────┘
                                   │
         ┌─────────────────────────┼─────────────────────────┐
         ▼                         ▼                         ▼
┌─────────────────┐     ┌─────────────────────┐     ┌─────────────────┐
│ Solar Photovoltaic │   │ Concentrated solar  │     │  Solar Heating  │
│                 │     │       Power         │     │  and Cooling    │
└─────────────────┘     └─────────────────────┘     └─────────────────┘
```

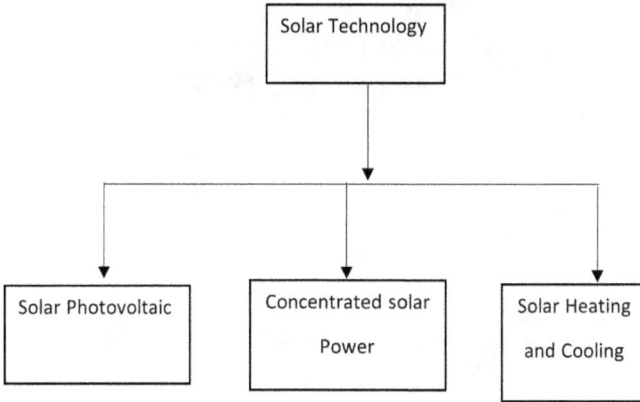

FIGURE 5.5　Types of solar energy technology.

in the material (most commonly silica). The valence electrons move freely to produce a direct current (DC) at a specific voltage. This electric power—produced by substantial numbers of solar cells arranged in panels and strings—is converted to MV (medium voltage—6.6 to 33 kV) alternating current (AC). The conversion to AC from DC aims to minimize heat losses from high currents in cables ($P_{loss} = I^2R$).

The MV AC is then distributed to a facility substation. It is transformed once more to obtain the electrical grid connection operating voltage (usually 132 kV, 220 kV,[1] 330 kV, or 400 kV) before being injected into the grid. Solar PV panels are sold based on their rated peak power output (output produced under STP conditions utilizing perpendicular irradiance at 1 kW/m², considered "Full Sun"). Therefore, one can consider a perpendicular placement concerning solar PV panels' incident sunlight optimal. However, given the Earth's rotational behavior on its axis and tilt to the Sun during the year, the solar PV panels' placement and performance are often optimized using software such as PVSyst. Solar PV power plants are also susceptible to some other issues. Soiling losses can reach up to 3%, while temperature losses can exceed 7%. The meteorological irradiance losses are also quite substantial, with deployed solar PV power plants only capturing between 18% and 25% of the available sunlight to convert into energy.

Taking all losses and irradiance losses into consideration, on any given day, you may only produce peak power for a combined time of 4–6 hours, depending on placement. The following Figure 5.6 demonstrates an example of the Sun's path throughout a given year in South Africa, read from left to right in terms of the time of day:

According to the information in Figure 5.6, there is little to no sunshine in the morning, while peak sun occurs around mid-day. This leads to irradiance losses in solar fields, meaning there are periods where the power output is significantly lower than the rated capacity.

To maintain outputs closer to the summer solstice, three solutions have been presented:

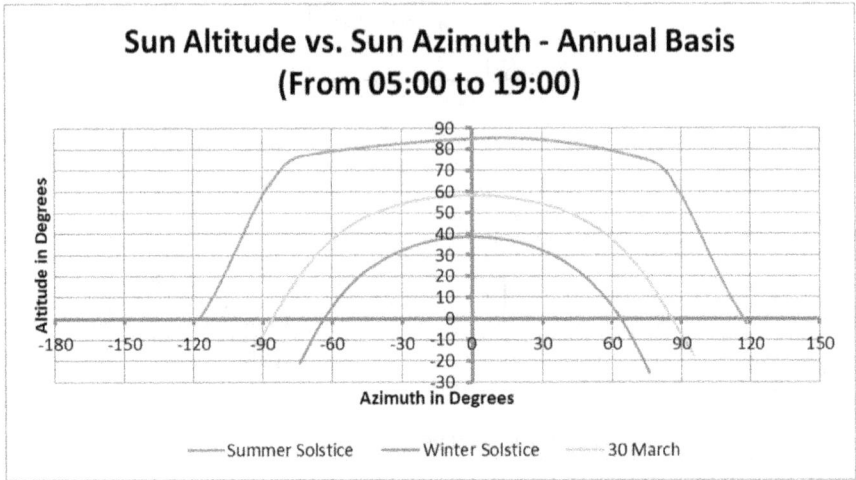

FIGURE 5.6   Sun Altitude vs Azimuth in South Africa over a given year.

Source: Adapted from http://andrewmarsh.com/apps/releases/sunpath2d.html

- Fixed frame installation facing 23° North
- Single-axis tracking from East to West daily
- Dual-axis tracking, which tracks East to West and North to South, based on data such as provided previously

Each of these solutions has its drawbacks, and decisions regarding which solution will be used usually come down to cost and ROI (return on investment)

Another issue facing solar PV installations is that of overcast weather. Should the plant experience cloud coverage (even if only partial), the plant output is significantly decreased.

Concentrated solar PV plants provide more energy output by using magnification to intensify the irradiance of a single solar PV cell. This technology, primarily due to the cost of magnification and the cost of increased spatial requirements, has fallen out of favor worldwide. Concentrated solar power, while being relatively costly compared to solar PV, is the technology most capable of consistently providing power and avoiding the pickup and drop-off in energy experienced by solar PV. This is entirely due to the operating mechanism behind concentrated solar power.

Concentrated solar power installations are primarily similar to the standard layout of a coal-fired power station. The exception is that the coal-fired boiler is replaced with a heat exchanger that operates on heat energy stored in molten salt. The heat in the molten salt comes from a secondary heat exchanger phase, where a heat transfer fluid (usually nonflammable oil) is heated directly by the Sun using either a central tower collector or tubes in parabolic trough mirrors.

In the case of the central tower, the facility is surrounded by concave mirrors connected to motors, allowing motion in three dimensions to track the Sun's position, called a heliostat field. This heliostat field uses incident sunlight reflection to a collector at the top of the central tower, where heat transfer occurs to the fluid, which is then pumped to the heat exchanger with the molten salt solution. The molten salt solution can then be stored and used to operate the power station at any given time and rate, with plants generally built with a specific amount of working time in mind, which affects the salt storage capacity and the heliostat field size. Parabolic trough installations work similarly, except that instead of having a central collector tower, the heat transfer fluid is pumped through piping systems at the focal point of the parabolic trough mirror. The key drawback of concentrated solar power is the plant's overall energy density. A significant field of mirrors, often on hectare footprint orders, must produce between 100 MW and 300 MW, which can be achieved and exceeded by nonrenewable energy power plants across a much smaller footprint. Another draw-back is that concentrated solar power installations must be located in hot places with minimal rainfall to ensure the thermal energy storage (TES) remains at adequate levels. Without consideration of other technologies for energy storage, concentrated solar power is the most capable of solar technologies to provide something close to baseload energy or the energy currently supplied by nonrenewable technology.

The continent of Africa has the potential to harness and benefit greatly from solar energy. The energy is clean, available to all, and capable of generating more than needed. However, there are a few impediments to the full implementation of the technology (Kabir et al., 2018). The initial cost of solar energy is not affordable to the average African populace. This is further compounded by the cost associated with importing solar equipment as there is a limited manufacturing outlet in Africa. Also, the issue of reliability related to weather fluctuation poses another impediment to solar technology. A global trade war poses another threat to the successful implementation of the technology in Africa.

5.4   SOLAR CELLS (DEPOSITION AND SELF-CLEANING)

Solar cells are devices whose current, resistance, and voltage are affected by the amount of sunlight falling on them. It is a device that generates electricity when sunlight falls upon it via chemical and physical phenomena. A solar cell must have three vital attributes; viz. solar cells must be able to absorb light resulting in generation of excitons or electron-hole pairs. Also, it can separate charge carriers, and finally solar cells should have the ability to extract carriers to external circuit.

5.4.1   THEORY OF SOLAR CELLS

A beam of sunlight containing photons hits the solar cells device and is absorbed by the solar cells material (semiconductor). This striking of the photons causes agitations of electron from orbital (molecular or atomic). This agitation is given off as heat or it is transported to the electrode. The potential established by this movement is countered by flow of current, resulting in electricity generation. The theory of solar cells is illustrated in Figure 5.7 as modified from STELR (Pentland, 2009). Several solar cells are staked together to generate sufficient electricity in form of direct current (DC). The DC is converted to useful alternating current (AC) for appliances by inverter. Transparent conducting film is the active part of the solar

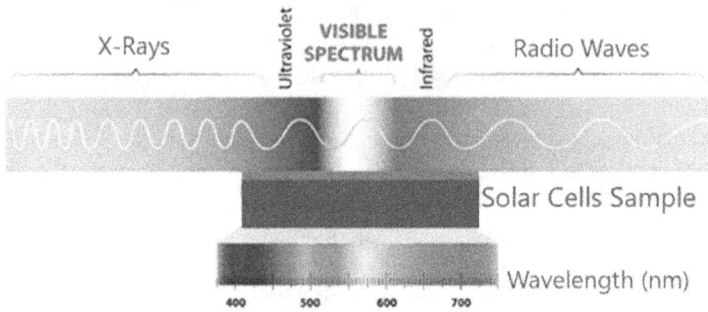

FIGURE 5.7 Theory of solar cells.

cells device used for collection of the charged carriers when sunlight beams on it. Figure 5.7 gives the theory of solar cells with the radio waves, visible spectrum, and X-rays classified.

The general principles that govern the operation of all solar cells involve photons ("packets" of energy that make up light) that strike a solar cell and are absorbed by the cell. Photons with sufficient energy trigger the cell to release electrons. If photons have insufficient energy, their energy is converted to heat energy. The liberated electrons flow around an electrical circuit in wires. The electrical current that results are in the form of a DC. This is a one-way current that only runs in one direction. More electrons will be emitted each second if the light is more intense (brighter light), and the electrical current will be larger. The cell's voltage is the same. Therefore, solar cells convert the packet of energy from the sunlight into electricity and heat.

5.4.2 Classifications of Solar Cells

Solar cells are arranged and oriented in one plane to form a solar panel. They are used for providing electricity to power electrical devices. Solar cells deposition is used to optimize the properties of the cells (chemical and physical properties). It aims to improve the optoelectronic and thermal ability of the material. Solar cells can be deposited using different deposition. The operation of a solar cell is determined by whether it is silicon based or not, such as an organic solar cell (made of polymers) or a dye-sensitized solar cell (also known as a Grätzel cell). The deposition of solar cells depends on the type of solar cells. First-generation solar cells are deposited mainly by plasma-enhanced chemical vapor deposition (PECVD) using mixture of silane and hydrogen. The hydrogen, about 5–14% hydrogen atoms, is used to passivate the inherent defect of the material. Solar cells can be classified based on several parameters. It is classified into two types based on the junction. A single junction and heterojunction solar cells. The mono junctions existed prior to the 1990s in which two types of solar cells existed, viz. the p-n junction and the p-i-n solar cells. The p-i-n solar cells are a device in which high electric field is generated around the i-layer via Fermi level difference existing between the p-doped and n-doped layers. The generated electric field causes electron-hole pair to generate electricity. The three

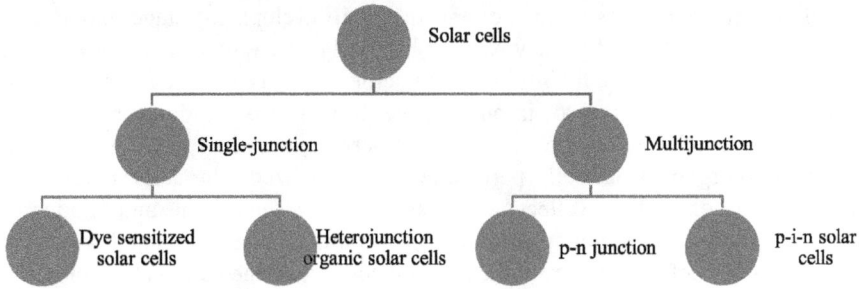

FIGURE 5.8   Classification of solar cells based on structure of the solar cells.

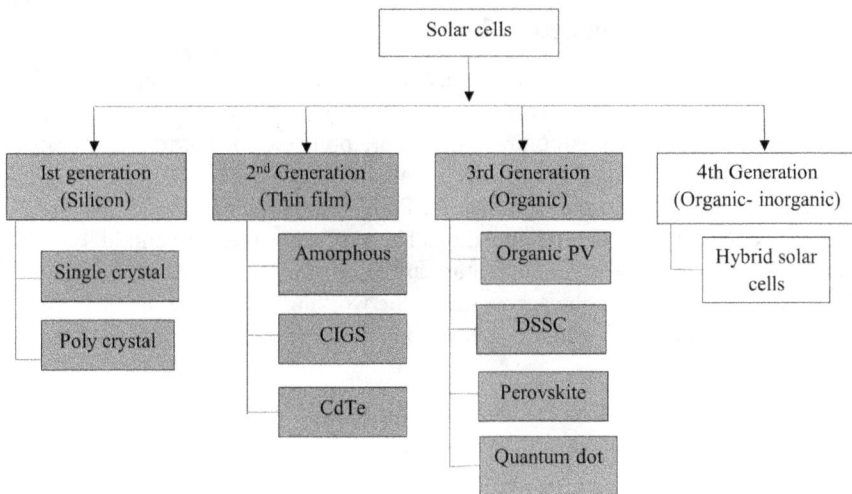

FIGURE 5.9   Classification of solar cells based on generation.

layers consist of two doped p and n layers and an intrinsic i-layer. The nanostructured heterojunctions include dye-sensitized solar cells (DSSC) and heterojunction organic solar cells. Figure 5.8 shows the classification of solar cells based on structure of the solar cells.

Also, solar cells can be classified based on the generation of the solar cells. Four generations of solar cells exist till today. This classification is based on the period of discovery of the solar cell technology and the solar cells material and is shown in Figure 5.9. The advantage of the first generation is the high carrier mobility, but it is expensive and there is loss of photon energy. This generation of solar cells has been commercialized on large scale into monocrystalline and polycrystalline solar panels. The second generation is less expensive compared with the first generation, but it has reduced efficiency. The second generation has been commercialized and used in stand-alone solar systems and utility-scale solar power stations. However, the third generation has affordable production cost and improved efficiency but has short-term stability.

The third generation is still in the research and development stage and not yet commercialized as their efficiency is either low or they have short absorber material.

Solar cell materials are mainly semiconductor materials designed to absorb sunlight on the Earth's surface or in outer space. Silicon was the first material to be used for solar cells. Over the years, researchers have discovered alternate solar cells such as organic solar cells, perovskite, dye-sensitized solar cells (DSSC), and quantum dot solar cells. Reflectance, charger carrier separation, and conductive and thermodynamic efficiency are some of the efficiencies associated with solar cells. The solar cells efficiency is these efficiencies combined and is the amount of sunlight converted into electricity. NREL gives the timeline of reported solar cells efficiencies.

5.4.3 PEROVSKITE SOLAR CELLS

The perovskite has low activation energy of 56.6–97.3 kJ/mol compared to 280–470 kJ/mol of silicon (Moore et al., 2015). Third-generation solar cells, especially perovskites, are deposited using low-temperature processes owing to the low activation energy and classified (Roy et al., 2020) as shown in Figure 5.10.

The single-step deposition technique for Perovskite is the most used owing to the inexpensive nature and ease of deposition. The low vapor pressure and high boiling point of the solvents at room temperature are the distinguishing factor of this technique. Figure 5.11a shows the schematic of the working of a single-step deposition. However, this technique produces Perovskites with nonuniform and poor surface coverage.

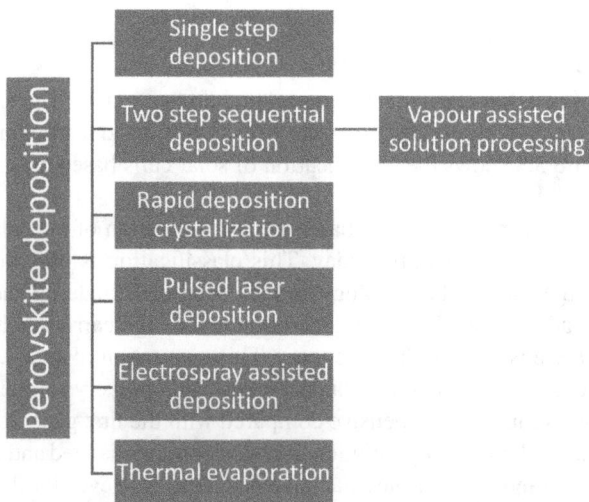

FIGURE 5.10 Classifications of Perovskite solar cells.

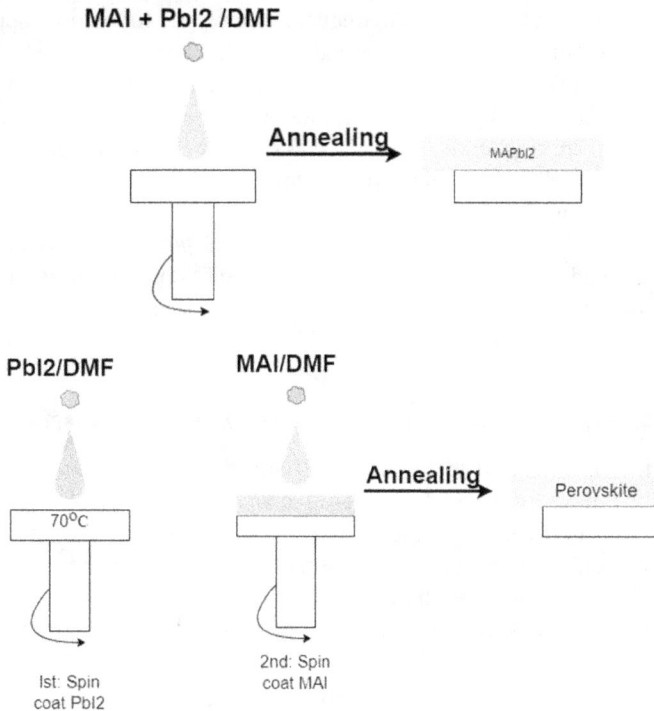

MAI + PbI2 /DMF

Annealing

MAPbI2

PbI2/DMF

MAI/DMF

Annealing

Perovskite

70°C

Ist: Spin
coat PbI2

2nd: Spin
coat MAI

FIGURE 5.11 Schematic representation of Perovskite: (a) single-step and (b) two-step deposition.

5.4.4 THIN FILMS SOLAR CELLS

The process of depositing one or more thin layers, or thin films (TF), of photovoltaic material on a substrate made of plastic, glass, or metal results in thin-film solar cells, which are second-generation solar cells. Thin-film solar cells come in a variety, and they are frequently utilized due to their effectiveness in generating power and their comparatively low cost. In comparison to monocrystalline and polycrystalline solar cell types, thin-film solar panels are less efficient and have lower power capabilities. Depending on the type of PV material used in the cells, the thin-film system's efficiency varies, but in general, they typically range from 7% to 18%.

Many different thin-film deposition methods, such as chemical vapor deposition, physical vapor deposition, and sputter deposition, result in the formation of thin films. Thin-film solar cells, one of the solar panel technologies, have the shortest life expectancy of 10–20 years. Thin-film panels provide the quickest payback time despite having a limited lifespan. In other words, it saves a lot of money on electricity and will pay for itself within 8 years.

This solar technology dates back to the 1970s, when Karl Bower invented the first solar home, called Solar One, and led the way in research on thin-film solar cells. Zhores Alferov and his students created the first Gallium Arsenide (GaAs) thin-film solar panel in 1970.

Three thin-film technologies that are frequently employed for outdoor applications include amorphous silicon (a-Si), copper indium gallium selenide (CIGS), and cadmium telluride (CdTe). Other thin-film technologies, such as organic, dye-sensitized, as well as quantum dot, copper zinc tin sulfide, nanocrystal, micromorph, and perovskite solar cells, are frequently categorized as emerging or third-generation photovoltaic cells. These technologies are still in an early stage of ongoing research or have limited commercial availability.

In 2013, crystalline silicon dominated 91% of the global deployment market, compared to around 9% for thin-film technology. CdTe controls more than half of the thin-film industry with a 5% share, leaving CIGS and amorphous silicon with a combined 2%.

5.4.5 INCREASING THIN FILM SOLAR CELLS EFFICIENCY AND ABSORPTION

The proportion of light that reaches the cell and the quantity that exits without being absorbed have both been increased and decreased using a number of methods. The most efficient approach is to reduce top contact coverage of the cell surface, which lessens the area that prevents light from entering the cell. Long wavelength light that has been partially absorbed can be obliquely linked with silicon and passed across the film multiple times to increase absorption. In order to maximize absorption by lowering the quantity of incident photons that are reflected away from the cell surface, numerous techniques have been developed. By altering the surface coating's refractive index, an additional antireflective coating can produce harmful interference inside the cell. Destructive interference removes the reflective wave, resulting in all incident light to enter the cell. Surface texturing is another approach.

5.5 SELF-CLEANING SOLAR TECHNOLOGY

Self-cleaning is the ability of a material to get rid of impurities by itself. Self-cleaning in solar is important as impurities reduce the efficiency of the solar panel. The cost of cleaning solar farms and the downtime for maintenance of solar farms also hinder productivity. Self-cleaning eliminates the cost and resources associated with robotic and/or manual cleaning of solar panels. Although self-cleaning existed in nature since time immemorial, Paz created the first instance of self-cleaning in 1995 by developing a coated glass with transparent titanium dioxide infused with nanoparticle that aid the glass to clean itself. A two-stage implementation cleaning process heralded the first commercial self-cleaning in 2001 by Pilkington Activ.

5.5.1 MECHANISM OF SELF-CLEANING AND EXAMPLES

Self-cleaning can be seen in nature and around us. The feet of geckos, lotus leaves, butterfly wings, shark skins, pitcher plants, and water striders are the most common form of self-cleaning in nature. Water is a common mechanism used for self-cleaning. It is classified as superhydrophobic and superhyrdrophilic, representing

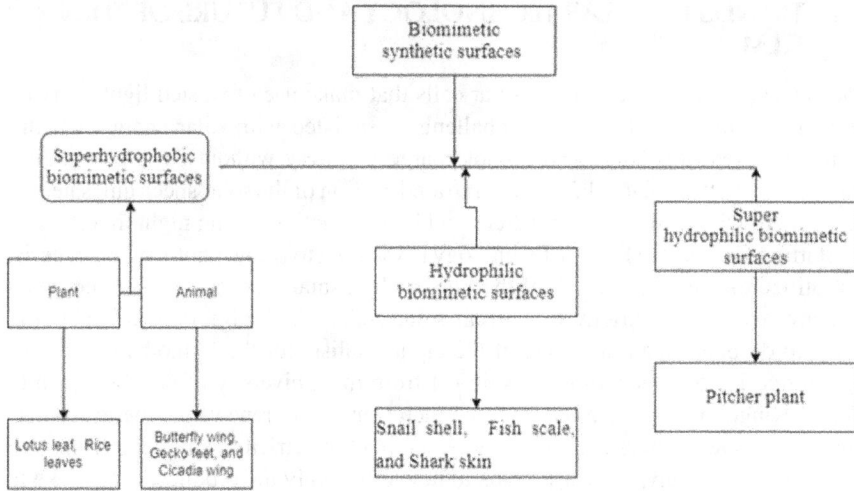

FIGURE 5.12   Classification of biomimetic synthetic surfaces.

repulsion and attraction, respectively. Another mechanism is electric curtain, photo-catalysis, and Joule heating for biomimetic synthetic surfaces are classified as shown in Figure 5.12.

Self-cleaning surfaces are at the heart of laboratory research for the optimization of some of our products. As we have seen in the application of solar panels, the addition of this coating would make electricity production more stable in the long term. But also to reduce considerably the maintenance of the panels, which was one of the factors of degradation of the solar panels. In the field of glazing, we have seen that the application of self-cleaning glass allows better visibility over time. This considerably reduces maintenance. In the same way as for glazing, self-cleaning building walls have reduced the maintenance of these surfaces. The maintenance of these surfaces, which are often difficult to access, is reduced. The technological improvement of self-cleaning surfaces has greatly improved the performance of these products. Unfortunately, the use of some products remains dangerous for health but also for the environment, which is why it is very important to remain very vigilant about the use of these products.

Nowadays, the development of self-cleaning transparent coatings based on the application of photovoltaic panels, strongly influenced by the study of wettability, has revolutionized the development of the self-cleaning property. This has led to an increased demand from industries. The use of mechanical and robotic cleaners for cleaning purposes seems impractical as it damages the surface of the glass, limits power generation during heavy rainfall, and requires expensive equipment.

MIT researchers inspired by the "lotus effect" have developed a way to eliminate glass reflections that also allows water droplets to bounce off its surface. The new surface is composed of nano textures through a special treatment to make the material self-cleaning.

5.6 TRENDS IN SOLAR TECHNOLOGY AND FUTURE OF THIN FILM INDUSTRY

The current trend is discovery of solar cells that make use of wasted light. This will result in elimination of the major challenge associated with solar energy, which is reliability. Present solar panels cannot charge a battery without day light sunlight. However, with this solar cell, only the infrared section of the solar spectrum is needed; hence there will be no need for battery and it can function during night time.

Starting Cambridge Photon Technology is working to create photovoltaic materials that utilize the entire spectrum of the Sun. Only a small portion of solar energy can be converted into electricity by current solar cells. Cambridge Photon Technology wants to do better than that. One of the eight finalists for the Spinoff Prize 2021 is Cambridge Photon Technology, a spin-off from the University of Cambridge in the United Kingdom. As the globe focuses increasingly on renewable energy sources, solar cell manufacturers are trying to get as much electricity as possible from their panels. Unfortunately, there are limits to how effectively the producers can make the products. Cambridge Photon Technology, situated in the United Kingdom, believes it has discovered a technique to greatly increase the amount of electricity that solar cells' photovoltaic material can generate.

The basic principle of solar cells operation is the same: light strikes the device, energizing the electrons within the cell and causing an electric current to flow. Silicon is the chosen photovoltaic (PV) material because it can efficiently capture and transform a significant part of incident sunlight into power. However, silicon performs best when exposed to red and near-infrared photons. Far infrared, microwave, and radio waves are longer-wavelength, lower-energy photons that are insufficiently energetic to start the current flowing. Green and blue photons with shorter wavelengths carry more energy than silicon can manage; the extra energy is lost as heat.

By transforming higher-energy photons into lower-energy ones that the solar cell can utilize, Cambridge Photon Technology claims to have discovered a technique to stop this waste. According to David Wilson, head of business development at the company, "We are attempting to deal with this dilemma of how you improve solar PV efficiency and bring down costs dramatically without throwing away the established silicon technology."

The Shockley-Queisser limit, a phenomenon, determines the maximum efficiency. The band gap, a feature of all PV materials, controls how much energy can fit into a single electron; for silicon, it is 1.1 electron volts. That is equivalent to photons in the near-infrared region of the electromagnetic spectrum. The entire visible light spectrum, which includes photons with energies higher than this band gap, can produce electrons, but the energy that exceeds the material's band gap escapes as heat. Due to this restriction, a traditional solar cell can only convert, at most, 29% of solar energy into electricity when used under perfect circumstances.

The novel method was created by University of Cambridge physicist Akshay Rao and his team. It is based on a phenomenon known as singlet exciton fission. Rao serves as the chief scientific officer for the start-up. When light strikes a PV material, it produces an exciton, an electrical connection between a negatively charged electron and a positively charged electron vacancy. However, if

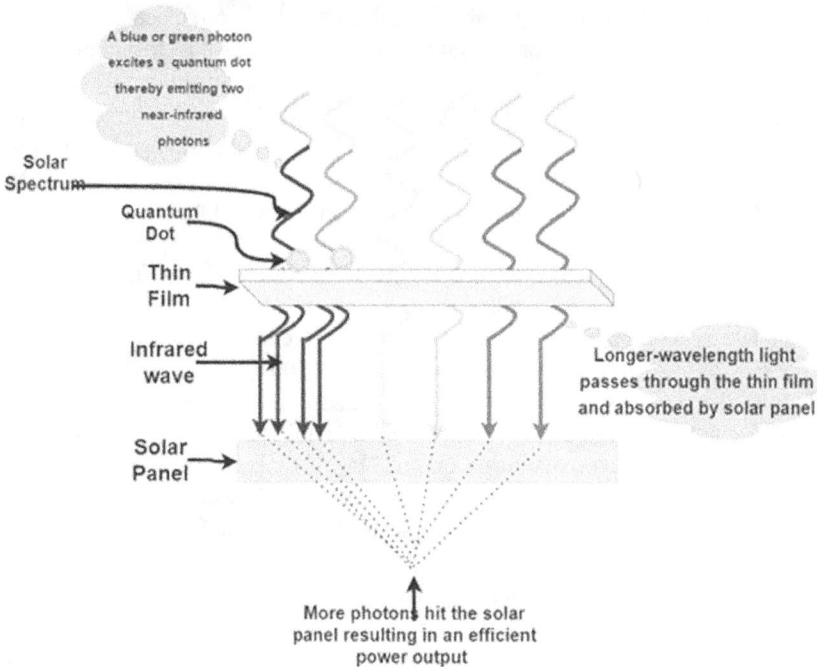

A blue or green photon
excites a quantum dot
thereby emitting two
near-infrared
photons

Solar
Spectrum

Quantum
Dot

Thin
Film

Infrared
wave

Longer-wavelength light
passes through the thin film
and absorbed by solar panel

Solar
Panel

More photons hit the solar
panel resulting in an efficient
power output

FIGURE 5.13   Harvesting wasted solar spectrum.

the substance is an organic polymer semiconductor, the photon can produce two lower-energy excitons, each of which can be turned into an electric current. Wilson explains that by increasing the photon flux that silicon receives in the part of the spectrum that it is good at turning into electricity, "you are maintaining the overall energy that comes in and out."

As shown in Figure 5.13, there is a color shift in the solar spectrum. Solar materials such as silicon and metal oxides convert energy from the sunlight into electricity using solar cells. The materials hitherto convert lights in the red spectrum (end of the solar spectrum). The bulk of the energy is in the blue and green (beginning of the solar spectrum) photons, and they are usually wasted by existing solar cells. Modern research uses polymer semiconductors to create quantum that absorbs these photons into useful wavelength to produce an efficient electricity. The research suggests that it may be able to harvest electricity in the dark (including at night).

Splitting photons is not a novel concept. According to Wilson, people have known for a long time that there is a way to bypass the Shockley-Queisser limit via the occurrence of singlet exciton fission in organic semiconductors. But it wasn't until 2014 when Rao and his colleagues at Cambridge University, working in physicist Richard Friend's group, came up with the first workable solution.

According to Claudio Marinelli, the company's CEO and an electrical engineer–turned-entrepreneur; the goal was always to try to commercialize this work. Rao consulted with a solar panel producer to better understand the needs of the industry and how his technology might be able to help. He then approached Marinelli and Wilson, two business experts, for assistance in developing a product that could be sold.

Rao created a photon-multiplier film composed of an organic polymer called pentacene and tiny, light-emitting inorganic particles called lead selenide quantum dots. Blue and green photons are absorbed by the polymer and changed into pairs of excitons. The quantum dots receive these excitons, which they absorb and release as lower-energy red or infrared photons. The light from the quantum dots shines onto the silicon when the film is placed on top of a silicon solar cell. The polymer coating does not stop the red and infrared wavelengths from the Sun from passing through and striking the silicon as they normally would. As a result, silicon is hit by more usable photons, which boosts the generation of electrical current.

Theoretically, according to Rao's calculations, this double-exciton approach might raise the conversion efficiency of solar cells to 35%. According to Wilson, the business has not even gone close to that point yet, but by the end of 2022, it hopes to have developed a prototype that turns around 31% of sunlight into electricity.

PV efficiency can also be raised using other methods. For instance, perovskites, a class of crystals that can absorb photons with shorter wavelengths, are used in tandem solar cells. The components can be used to construct silicon and solar cells, which can then be connected to create a hybrid device that generates more electricity. Wilson claims that the challenge with such a setup is that it may be difficult to get two devices to cooperate while generating differing currents. A new manufacturing procedure and equipment are needed to create solar cells made of a different substance, which could increase the cost. The goal of our entire strategy, according to Wilson, "has been to avoid these issues and to create a straightforward, non-toxic material with no electrical connections that add very little complexity to an existing design."

According to Christopher Bardeen, a University of California, Riverside scientist who is not connected to the company, the concept put forth by Cambridge Photon Technology sounds workable. He calls it a "promising technique" that offers tandem cells a straightforward substitute.

According to Wilson, the company's photon-multiplier film could readily be incorporated into current production methods. Manufacturers of solar panels could purchase a finished film to apply to their PV modules. Selling a precursor solution to the businesses that produce either the vinyl acetate layer that encases the silicon or the glass panels that protect the solar cells would be a simpler strategy. The finished product would be assembled by panel manufacturers using the already-treated parts. Wilson anticipates that a product will be ready for the market in around three years, regardless of the strategy.

About a dozen people work for Cambridge Photon Technology, which has raised £1 million ($1.4 million) in equity capital. It also has some research funding and access to Cambridge University's facilities and researchers to advance the technology. From the university, it has licensed four significant patents.

The company has created film and quantum dot prototypes to demonstrate that they are effective enough to use in a product. Still, it has not put all the components into a working solar cell with increased efficiency. According to Wilson, the potential reward might be enormous once it establishes its technique's viability. It's obvious that there is a pressing demand, he claims. And if the technology lives up to expectations, it will significantly fill that demand.

5.6.1 SOLAR CELLS SLICED AND DICED

Another approach to increasing solar cell efficiency is using the Gallium arsenide by peel-and-stamp technique. Better night-vision cameras, high-efficiency solar cells, and various other uses could result from a different process for creating light-sensitive semiconductors. A team led by the University of Illinois at Urbana-Champaign materials scientist John Rogers has proposed a technique that could be economical for manufacturing microchips made of the photoresponsive semiconductor gallium arsenide (Yoon et al., 2010). A transfer-printing technology is utilized to peel and print thin semiconductor layers onto glass or plastic, which addresses a long-standing difficulty in gallium arsenide manufacturing that could alter the solar-cell sector.

Modern semiconductor manufacturing relies heavily on silicon, which is utilized in everything from solar cells to digital cameras. However, scientists understand that there are better suitable materials for absorbing light. Since some semiconductors have a substantially higher light absorption capacity than silicon, they create better solar cells and infrared detectors. One of the most researched silicon substitutes is gallium arsenide. It is twice as efficient as silicon, producing energy from around 40% of incident solar radiation. Due to its efficiency, gallium arsenide is the preferred material for making spaceship solar cells. However, gallium arsenide is extremely expensive, the same as its finest uses. Rogers claims that part of this is the need for highly regulated chambers to be used to grow high-quality gallium arsenide wafers. The thick wafers are often cut into pieces after being created, but only their surfaces are utilized. According to Rogers, a lot of the expensive material is wasted. Rogers and his coworkers have now discovered a different approach. The researchers created a "pancake" comprising alternating layers of gallium arsenide and aluminum arsenide rather than generating a single layer of gallium arsenide. The team then removed the individual gallium arsenide layers using a precise chemical sequence and peeled them off using a silicon-based rubber stamp. They used more well-known methods to etch the thin slices into circuits after stamping the wafers onto a different surface, like glass or plastic. The group has developed infrared imaging devices, very small solar cells (500 micrometers broad), and several mobile phone parts. The start-up business Semprius, which plans to use the method to lower the cost of gallium arsenide electronics, has several coauthors on the research. According to Rogers, "We think this strategy can be cost-competitive with anything out there." Gallium arsenide has considerable promise, according to Rogers. His laboratory is currently focusing on creating solar cells that can produce electricity for less than $1 per watt, making them economically viable. We believe we can get it there, he continues, but nothing is truly proven until it is carried through.

As Roger reports, the fabrication of gallium arsenide solar cells using peel-and-stamp technology is highly intriguing, but costs and the production of huge cells remain difficult. Implementing computer-based technologies involves circuit drawings that utilize gallium arsenide as the ink.

5.7 CONCLUSION

The Fourth Industrial Revolution is changing the ways things are done and human lives, with most countries already implementing it in all sectors of their economy. Just like other previous industrial revolutions, it will make or mar most countries based on implementation. Electricity will play a crucial role in the successful implementation of 4IR. Solar energy is among the leading source of electricity generation. A thorough understanding of it and implementation of solar are therefore needed. This chapter helped in demystifying 4IR, solar energy, and solar cells. A snippet of the current trend of solar cells was also discussed. Harvesting of wasted sunlight from the solar spectrum will help reduce reliance on battery and more harvested electricity both during the day and night by solar panels.

NOTE

1 Not a commonly used transmission voltage in South Africa.

REFERENCES

Accelerating, U. (October 2014). *Advanced manufacturing: Report to the President*. Executive Office of the President, President's Council of Advisors on Science and Technology: website. 93 pp. www.whitehouse.gov/sites/default/files/microsites/ostp/PCAST/amp20_r eport_final. pdf.

Ahmad, Z., Rahim, S., Zubair, M., & Abdul-Ghafar, J. (2021). Artificial intelligence (AI) in medicine, current applications and future role with special emphasis on its potential and promise in pathology: Present and future impact, obstacles including costs and acceptance among pathologists, practical and philosophical considerations. A comprehensive review. *Diagnostic Pathology, 16*(1), 1–16.

Akinradewo, O., Oke, A., Aigbavboa, C., & Molau, M. (2021). Assessment of the Level of Awareness of Robotics and Construction Automation in South African *Collaboration and Integration in Construction, Engineering, Management and Technology* (pp. 129–132). Springer.

Alaloul, W. S., Liew, M., Zawawi, N. A. W. A., & Kennedy, I. B. (2020). Industrial Revolution 4.0 in the construction industry: Challenges and opportunities for stakeholders. *Ain Shams Engineering Journal, 11*(1), 225–230.

Awodumi, O. B., & Adewuyi, A. O. (2020). The role of non-renewable energy consumption in economic growth and carbon emission: Evidence from oil producing economies in Africa. *Energy Strategy Reviews, 27*, 100434.

Ayvaz, S., & Alpay, K. (2021). Predictive maintenance system for production lines in manufacturing: A machine learning approach using IoT data in real-time. *Expert Systems with Applications, 173*, 114598.

Bakhtiyari, A. N., Wang, Z., Wang, L., & Zheng, H. (2021). A review on applications of artificial intelligence in modeling and optimization of laser beam machining. *Optics & Laser Technology, 135*, 106721.

Balali, Y., & Stegen, S. (2021). Review of energy storage systems for vehicles based on technology, environmental impacts, and costs. *Renewable and Sustainable Energy Reviews, 135*, 110185.

Barham, L. (2013). *From hand to handle: The first industrial revolution*. Oxford University Press.

Bauer, G. R., & Lizotte, D. J. (2021). *Artificial intelligence, intersectionality, and the future of public health.* American Public Health Association.

Boehm, F., Graesslin, R., Theodoraki, M.-N., Schild, L., Greve, J., Hoffmann, T. K., & Schuler, P. J. (2021). Current advances in robotics for head and neck surgery—A systematic review. *Cancers, 13*(6), 1398.

Dalla Longa, F., & van der Zwaan, B. (2021). Heart of light: An assessment of enhanced electricity access in Africa. *Renewable and Sustainable Energy Reviews, 136*, 110399.

Drath, R., & Horch, A. (2014). Industrie 4.0: Hit or hype? [Industry forum]. *IEEE Industrial Electronics Magazine, 8*(2), 56–58.

Ellabban, O., Abu-Rub, H., & Blaabjerg, F. (2014). Renewable energy resources: Current status, future prospects and their enabling technology. *Renewable and Sustainable Energy Reviews, 39*, 748–764.

Emani, P. S., Warrell, J., Anticevic, A., Bekiranov, S., Gandal, M., McConnell, M. J., ... Bastiani, M. (2021). Quantum computing at the frontiers of biological sciences. *Nature Methods, 18*(7), 1–9.

Freire, T., Brun, F., Mateus, A., & Gaspar, F. (2021). 3D printing technology in the construction industry. In H. Rodrigues, F. Gaspar, P. Fernandes, & A. Mateus (eds.), *Sustainability and automation in smart constructions* (pp. 157–167). Springer.

Fwaya, E. V. O., & Kesa, H. (2018). The Fourth Industrial Revolution: Implications for hotels in South Africa and Kenya. *Tourism: An International Interdisciplinary Journal, 66*(3), 349–353.

Gebrehiwot, M., & Van den Bossche, A. (2015). *Driving electric vehicles: As green as the grid.* Paper presented at the AFRICON 2015.

Goh, G., Sing, S., & Yeong, W. (2021). A review on machine learning in 3D printing: Applications, potential, and challenges. *Artificial Intelligence Review, 54*(1), 63–94.

Hermann, M., Pentek, T., & Otto, B. (2016). *Design principles for industrie 4.0 scenarios.* Paper presented at the 2016 49th Hawaii international conference on system sciences (HICSS).

Holland, J., Kingston, L., McCarthy, C., Armstrong, E., O'Dwyer, P., Merz, F., & McConnell, K. (2021). Service robots in the healthcare sector. *Robotics, 10*, 47.

Houssein, E. H., Gad, A. G., Wazery, Y. M., & Suganthan, P. N. (2021). Task Scheduling in cloud computing based on meta-heuristics: Review, taxonomy, open challenges, and future trends. *Swarm and Evolutionary Computation, 62*, 100841.

Jurcevic, P., Javadi-Abhari, A., Bishop, L. S., Lauer, I., Borgorin, D., Brink, M., ... Kanazawa, N. (2021). Demonstration of quantum volume 64 on a superconducting quantum computing system. *Quantum Science and Technology, 6*(2), 025020.

Kabir, E., Kumar, P., Kumar, S., Adelodun, A. A., & Kim, K.-H. (2018). Solar energy: Potential and future prospects. *Renewable and Sustainable Energy Reviews, 82*, 894–900.

Kagermann, H. (2014). Industrie 4.0 und Smart Services. In: W. Brenner, & T. Hess (Eds.), *Wirtschaftsinformatik in Wissenschaft und Praxis* (pp. 243–248). Springer Gabler.

Kagermann, H., Wahlster, W., & Helbig, J. (2013). *Recommendations for implementing the strategic initiative Industrie 4.0: Final report of the Industrie 4.0 Working Group.* Forschungsunion.

Kanji, G. K. (1990). Total quality management: The second industrial revolution. *Total quality management, 1*(1), 3–12.

Kayembe, C., & Nel, D. (2019). Challenges and opportunities for education in the Fourth Industrial Revolution. *African Journal of Public Affairs, 11*(3), 79–94.

Knight, P., Bird, C., Sinclair, A., Higham, J., & Plater, A. (2021). Testing an "IoT" Tide Gauge Network for Coastal Monitoring. *IoT, 2*(1), 17–32.

Koli, P., Dayma, Y., Pareek, R. K., & Jonwal, M. (2020). Use of Congo red dye-formaldehyde as a new sensitizer-reductant couple for enhanced simultaneous solar energy conversion and storage by photogalvanic cells at the low and artificial sun intensity. *Scientific reports, 10*(1), 1–10.

Kougias, I., SS, N. S., Monforti, F., Banja, M., Bódis, K., & Moner-Girona, M. (2018). *Water-Energy-Food Nexus Interactions Assessment: Renewable energy sources to support water access and quality in West Africa.* European Commission, Luxembourg.

Lasi, H., Fettke, P., Kemper, H.-G., Feld, T., & Hoffmann, M. (2014). Industry 4.0. *Business & Information Systems Engineering, 6*(4), 239–242.

Leenen, L., & Meyer, T. (2021). Artificial intelligence and big data analytics in support of cyber defense. In M. Khosrow-Pour (ed.), *Research Anthology on Artificial Intelligence Applications in Security* (pp. 1738–1753). IGI Global.

Li, J., & Pumera, M. (2021). *3D printing of functional microrobots.* Chemical Society Reviews.

Miketa, A., & Merven, B. (2013). *Southern African power pool: Planning and prospects for renewable energy.* Abu Dhabi, IRENA. International Renewable Energy Agency. www.irena.org/SAPP.

Mohajan, H. (2019). The second industrial revolution has brought modern social and economic developments. *Journal of Social Sciences and Humanities, 6*(1), 1–14.

Moore, D. T., Sai, H., Tan, K. W., Smilgies, D.-M., Zhang, W., Snaith, H. J., … Estroff, L. A. (2015). Crystallization kinetics of organic–inorganic trihalide perovskites and the role of the lead anion in crystal growth. *Journal of the American Chemical Society, 137*(6), 2350–2358.

Naudé, W. (2017). *Entrepreneurship, education and the Fourth Industrial Revolution in Africa.* IZA Discussion Papers, No. 10855, Institute of Labor Economics (IZA), Bonn, pp. 1–22.

Ouedraogoa, I., Jiya, A. N., & Diandac, I. Access to electricity and health capital accumulation in Sub-Saharan Africa. In *Proceedings of the 2021 International Development Economics Conference*, Bordeaux, France (vol. 30).

Parmar, H., Khan, T., Tucci, F., Umer, R., & Carlone, P. (2022). Advanced robotics and additive manufacturing of composites: Towards a new era in Industry 4.0. *Materials and Manufacturing Processes, 37*(5), 483–517.

Pentland, P. (2009). STELR program. *Chemistry in Australia, 76*(4), 11–13.

Popkin, B. M. (1999). Urbanization, lifestyle changes and the nutrition transition. *World Development, 27*(11), 1905–1916.

Roy, P., Sinha, N. K., Tiwari, S., & Khare, A. (2020). A review on perovskite solar cells: Evolution of architecture, fabrication techniques, commercialization issues and status. *Solar Energy, 198*, 665–688.

Schuh, G., Anderl, R., Gausemeier, J., ten Hompel, M., & Wahlster, W. (2017). Industrie 4.0 maturity index. In: G. Schuh, R. Anderl, R. Dumitrescu, A. Krüger, & M. ten Hompel (Eds.), *Managing the digital transformation of companies.* Herbert Utz.

Schwab, K. (2016). *The Fourth Industrial Revolution.* World Economic Forum. www.weforum.org/about/the-fourth-industrial-revolution-by-klaus-schwab

Shahidinejad, A., Ghobaei-Arani, M., & Masdari, M. (2021). Resource provisioning using workload clustering in cloud computing environment: A hybrid approach. *Cluster Computing, 24*(1), 319–342.

Sharma, S., Ahmed, S., Naseem, M., Alnumay, W. S., Singh, S., & Cho, G. H. (2021). A survey on applications of artificial intelligence for pre-parametric project cost and soil shear-strength estimation in construction and geotechnical engineering. *Sensors, 21*(2), 463.

Tselegkaridis, S., & Sapounidis, T. (2021). Simulators in educational robotics: A review. *Education Sciences, 11*(1), 11.

Vicentini, F. (2021). Collaborative robotics: A survey. *Journal of Mechanical Design, 143*(4), 040802.

Vrontis, D., Christofi, M., Pereira, V., Tarba, S., Makrides, A., & Trichina, E. (2021). Artificial intelligence, robotics, advanced technologies and human resource management: A systematic review. *International Journal of Human Resource Management, 33*(6), 1237–1266.

Waghid, Y., Waghid, Z., & Waghid, F. (2019). The Fourth Industrial Revolution reconsidered: On advancing cosmopolitan education. *South African Journal of Higher Education, 33*(6), 1–9.

Yahyaoui, A., Abdellatif, T., Yangui, S., & Attia, R. (2021). READ-IoT: Reliable event and anomaly detection framework for the internet of things. *IEEE Access, 9*, 24168–24186.

Yoon, J., Jo, S., Chun, I. S., Jung, I., Kim, H.-S., Meitl, M., ... Paik, U. (2010). GaAs photovoltaics and optoelectronics using releasable multilayer epitaxial assemblies. *Nature, 465*(7296), 329–333.

Zhong, Q.-C. (2017). Synchronized and democratized smart grids to underpin the third industrial revolution. *IFAC-PapersOnLine, 50*(1), 3592–3597.

6 Demystifying 3D Printing Technology and the Application in Emerging 4IR Technologies

6.1 INTRODUCTION

This study discusses in detail the concept of 3D printing in the era of the fourth industrial revolution (4IR) with emphasis on method, materials, applications in relation to 4IR, challenges, and opportunities of the technology. It aims to demystify the concept of 3D printing while canvassing for the implementation in daily life, especially in some key human challenges such as medicine and housing with the poor at heart. This is intended to shed light on a technology that can deliver rapid implementation of the benefits of 4IR. Also, it will open windows of opportunities that exist for the actualization of the fourth industrial revolution. The study used top-down approach focusing on 4IR and streamlining it to 3D printing. The main 3D methods such as fused deposition modeling, powder bed fusion, stereolithography (SLA), direct energy deposition, and laminated object manufacturing were discussed in the context of their application and 4IR. Recent advances and trending applications of 3D printing are also examined in relation to materials and related parameters.

Material, process, and design strategy selections are made in a nonlinear and integrated manner that necessitates thorough consideration and understanding of their relationship to an application. Materials, processes, and design methods for 3D printing are discussed in the context of 4IR applications, with a critical evaluation of how each of these choice variables influences applications and each other.

Different authors have done a review of 3D printing in selected applications. The application of 3D printing in medicine seems to have a lot of reviews (Yan et al., 2018, Tack et al., 2016, Dawood et al., 2015, Gopinathan and Noh, 2018, Prasad and Smyth, 2016), alongside ceramic (Chen et al., 2019) and composite (Wang et al., 2017). There are also some reviews focusing on the materials such as metal (Buchanan and Gardner, 2019, Wei et al., 2020, Ninpetch et al., 2020, Duda and Raghavan, 2018, Attarilar et al., 2021) and polymer (Singh et al., 2020, Zhou et al., 2020). However, very few discussed 3D printing in its entirety. Some did an executive summary of the 3D printing technology without a detailed discussion of the technology (Shahrubudin

DOI: 10.1201/9781003364481-6

et al., 2019, Choi and Kim, 2015). However, Ngo et al. (2018) did a review of 3D printing with an emphasis on the material and application. This study gives a comprehensive discussion of 3D printing, technology, materials, and recent trending applications and materials. It also discussed the fourth industrial revolution and in relation to 3D printing.

Industrial revolution is the deployment of new techniques and processes for manufacturing and related activities. It ushers development, transformation, and inequalities among countries. Mhlanga (2022) referred to 4IR as a confluence of technology that blurs the borders between the physical, digital, and biological domains. Countries that are prepared and able to tap into the technology of every new industrial revolution become better than countries that did not latch onto the revolution. The first industrial revolution introduced the steam engine (Stearns, 2020). This was followed by the second industrial revolution characterized by mass production powered by electricity (the age of science) (Hyun Park et al., 2017). The harbinger of Industry 3.0 is internet and electricity, known as rapid technology (Greenwood, 1997, Blinder, 2006). This saw a wide margin between developed and developing countries (Janicke and Jacob, 2013). Most developing countries were left to wallow in underdevelopment and dependence on other economies. Such countries (such as Africa), despite having the population and some of the raw materials, lagged due to the absence of the stable electricity and energy needed for powering batch systems and other mass production technology. The last decade has introduced the current industrial revolution called Industry 4.0 (Devezas and Sarygulov, 2017). Some countries like China have deployed and are tapping into the various features of this technology. In some parts of the world, the fourth industrial revolution (4IR) is sweeping across all sectors, from companies to research centers (Hermann et al., 2016). Although countries like South Africa seem to champion the deployment of the technologies for Industry 4.0, other countries in the continent of Africa continue to lag (Sutherland, 2020, Butler-Adam, 2018, Ayentimi and Burgess, 2019). Key features of the fourth industrial revolution (4IR) include accelerating digitalization, artificial intelligence (AI), cloud computing, robotics, and 3D printing, which have obvious and important implications for education, employment, and the future of work (Fwaya and Kesa, 2018). This is especially true for African countries and the rest of the developing countries (Cilliers, 2018).

The procedure of 3D printing is like how ink or toner is fused to paper in a printer (hence the term printing) (Sathish et al., 2018). Although, at each location in the horizontal cross-section where solid material is sought, it is actually the solidifying or binding of a liquid or powder. In the case of 3D printing, the layering process is repeated hundreds or thousands of times until the entire item is complete in every vertical dimension. 3D printing is used to create plastic or metal prototypes fast during the design of new parts, but it can also be used to create finished things for sale to clients (Günther et al., 2014). A 3D printing system can be contained in a cabinet the size of a big kitchen stove or less. Fused deposition modeling (FDM) is the most prevalent method of 3D printing that primarily uses polymer filaments (Gordelier et al., 2019). Inkjet printing, contour crafting, stereolithography, direct energy deposition (DED), and laminated object manufacturing (LOM) are the main methods of additive manufacturing (Yuan et al., 2019). Muhammad et al. (2021) did a systematic and bibliometric analysis of articles published on 3D printing starting from 1999 to 2019.

The limited materials accessible for 3D printing, on the other hand, make it difficult to apply this technology in a variety of industries. As a result, there is a need to produce appropriate materials for 3D printing. To improve the mechanical qualities of 3D-printed objects, more research is required. This review is arranged using the top-down approach. 4IR is discussed and closely followed by 3D printing. Methods and materials applications of 3D printing in relation to key 4IR are thereafter discussed. It is hoped that this study will generate the necessary interest and investment.

6.2   FOURTH INDUSTRIAL REVOLUTION

It is no longer new that 4IR has come to stay. This is evident with the technologies associated with 4IR being part of our daily life. Internet of things (IoT), machine learning, robotics, and other technologies are now being used industrially and domestically. Phones are now fitted with tools that communicate and run "errands," and robots are now being employed at home for cleaning and in industries for performing complicated and risky tasks. Despite this wide acceptance and penetration of 4IR, some countries still lag in using them. Some countries in developing world still battle with implementation of Second and Third Industrial Revolution.

There were about three other industrial revolution before 4IR. The first industrial revolution (1IR) was characterized by decreased dependence on a beast of burden, giving way to the steam engine and its industrial and residential applications (Barham, 2013). This resulted in railway construction, steel, raw materials, and finished goods, which improved the quality of life of affected persons and led to urbanization (Kanji, 1990). The second industrial revolution (2IR) took place from the 1880s to the 1950s. It was characterized by electricity's discovery, leading to increased production rates and improved communication and other facets of life (Mohajan, 2019). The discovery of the personal computer in the early 1950s and the internet around the 2000s paved the way for the third industrial revolution (3IR) from the 1950s to 2000 (Zhong, 2017). The introduction of computing in the industry allowed for significant progress in terms of developing new technologies. This computing capacity made it possible for humans to run relatively complex sets of instructions and even facilitated humanity reaching the surface of the Moon. This development soon enhanced the quality of life of commercial firms with the introduction of personal computers by companies such as IBM and Apple in the 1980s (Popkin, 1999).

4IR is a term that was introduced in 2011 to describe the marriage of information and communication and related technology with production and industry (Schuh et al., 2017). The focus is on industries and related sectors. It is a technology that aims to increase productivity, reduce downtime, and speed up production activities (Lasi et al., 2014, Kagermann, 2014). In some parts of the world, the fourth industrial revolution (4IR) is sweeping across all sectors, from companies to research centers (Hermann et al., 2016). In Europe and other developed continents, it has become a lullaby and is being discussed at every event and meeting (Drath and Horch, 2014). Countries in those continents have included it in their national key developmental

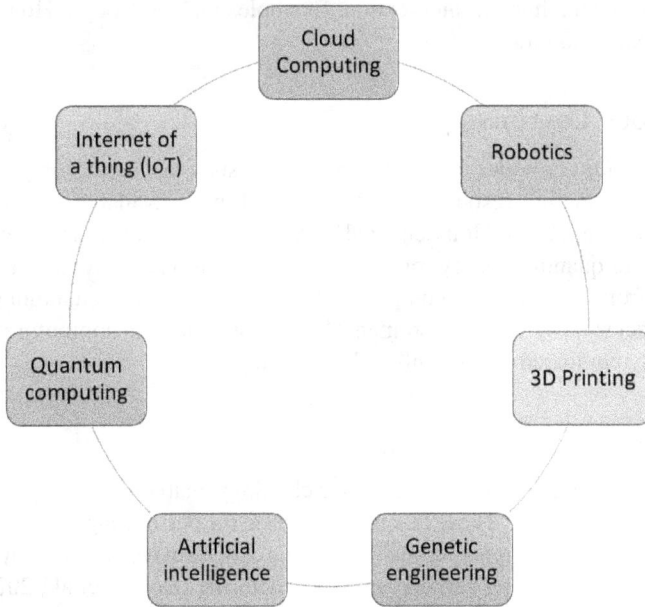

FIGURE 6.1   Key technologies of the fourth industrial revolution.

initiative (Kagermann et al., 2013). The US government committed about US$2 billion to it in 2014. The key features of the 4IR include accelerating digitalization, artificial intelligence (AI), cloud computing, robotics, and 3D printing as shown in Figure 6.1. They have obvious and important implications for education, employment, and the future of work.

6.2.1   ARTIFICIAL INTELLIGENCE

It is the replication of human reasoning in machines. Machines replicate and act similarly to humans, especially in problem-solving, thinking, and learning (Ahmad et al., 2021). An application of AI in daily living is detecting and eliminating bank card fraud by identifying spending patterns. The technology can decode voices, and handwriting, among other traits that follow specific ways. It is already applied in medicine, construction, and big data, among others (Sharma et al., 2021, Bakhtiyari et al., 2021, Bauer and Lizotte, 2021, Leenen and Meyer, 2021).

6.2.2   INTERNET OF THINGS (IOT)

IoT links physical items using sensors and software to transmit data across the internet, mainly via Wi-Fi (Yahyaoui et al., 2021). Although not all IoT devices require the internet to function, they need to be interfaced with other devices. It is used in smart homes, connected appliances, biometric scanners, and innovative factory equipment (Knight et al., 2021, Ayvaz and Alpay, 2021). The purpose is for the device to feed data without manual human intervention. Examples include Google Home, Amazon Alexa, and baby monitors.

6.2.3 CLOUD COMPUTING

Cloud computing provides computer resources such as data storage, software, servers, and computing resources on-demand and made available over the internet (Shahidinejad et al., 2021, Houssein et al., 2021). However, quantum computing uses the principle of quantum theory for performing calculations (Jurcevic et al., 2021). It is different from conventional computers that use 1s or 0s, but quantum computing uses the object state before measurement. This allows quantum computing to perform exponentially compared to conventional computing.

6.2.4 ROBOTICS

Robotics are machines built by combining technology, mathematics, engineering, and science (Vicentini, 2021). These machines are used for performing tasks that are hitherto tricky and sometimes dangerous for humans. Robots have been deployed for several applications, including surgery and manufacturing (Boehm et al., 2021, Vrontis et al., 2021, Tselegkaridis and Sapounidis, 2021, Parmar et al., 2021, Holland et al., 2021, Akinradewo et al., 2021).

6.3 3D PRINTING

It is also known as additive manufacturing. A technology that transforms computer-aided models into reality, especially 3D objects (Li and Pumera, 2021). The materials are built layer by layer until the final desired object is formed. The object to be printed is first constructed or modeled using computer-aided tools. Recently, several entities have been printed using 3D printers. It finds application in medicine for printing human organs, building, and other sectors (Goh et al., 2021). This study discusses this technology in detail.

3D printing is a relatively new method of manufacturing and processing. Conventional printers use ink to print artwork and text on 2D paper, but 3D printers use heat, light, and laser to convert raw materials into thin layers. These materials include metals, ceramics, and polymers. The layers are then joined together, much like a house, to form the final product. Some common 3D printing technologies include FFF, SLA, and SLM. The most widely used technology is the FFF. MakerBot was the first 3D printing company to release a model that costs less than £1,000. Maplin, a wall-street company, recently advertised a 3D printer for £700.

Recent advancements have lowered the cost of 3D printers, allowing them to be used in more places such as schools, residences, libraries, and laboratories. Due to its speedy and cost-effective prototyping capability, 3D printing was first widely employed by architects and designers to create aesthetic and functional prototypes. The usage of 3D printing has reduced the additional costs associated with the development of a product. However, 3D printing has just recently become widely used in a variety of industries, from prototypes to finished items. As a result of the high costs of developing custom-tailored items for end-users, product customization has been a difficulty for manufacturers. Addictive manufacturing has the ability to 3D print small batches of customized products for a reasonable price. With application in a wide range of medical implants using CT-imaged tissue replicas 3D printing has recently gained popularity in the healthcare business (Stansbury and Idacavage, 2016). WinSun successfully mass-produced a group of reasonably inexpensive houses in China (US$4,800 per unit) in less than a day (Wu et al., 2016). The increased acceptance of 3D printing over traditional methods can be linked to a few benefits, including the manufacture of complicated geometry with high precision, maximum material savings, design freedom, and personalization. Metals, polymers, ceramics, and concrete are just a few of the materials used in 3D printing today. The main polymers used in 3D printing composites are polylactic acid (PLA) and acrylonitrile butadiene styrene (ABS).

The number of 3D-printable materials is also increasing. Innovations in the energy, aerospace, and defense sectors will be enabled by refractory superalloys. More durable polymers are being developed today that are anticipated to pass the FAA's flame, smoke, and toxicity tests, resulting in lower maintenance costs for relevant sectors. The future of 3D printing remains bright, with new research and breakthroughs in additive manufacturing on the rise.

Despite being a ground-breaking approach for unique products and niche applications, 3D printing requires more research to compete with older methods in mass manufacturing of everyday goods due to its greater cost and slower pace. Nonetheless, AM has progressed tremendously in recent years. Increased investment, research, and development around the world will result in a rapid shift from traditional manufacturing processes to 3D printing in the not-too-distant future.

6.3.1 DIFFERENCE BETWEEN COMPUTER NUMERICAL CONTROLLED (CNC) AND ADDICTIVE MANUFACTURING (AM)

This section compares CNC machining versus additive manufacturing on a variety of issues. The goal isn't to persuade people to choose one technology over another but to figure out how they may be used at different stages of the product development process or for different sorts of products. Computer numerical controlled machining technique shares some DNA with addictive manufacturing. CNC stands for computer numerical control, and it is also a computer-based manufacturing method. CNC is distinguished by the fact that it is essentially a subtractive rather than an additive process, requiring a material block that is at least as large as the component to be manufactured (Gibson et al., 2021).

6.4 HISTORY OF 3D PRINTING

3D printing is a technology that has existed for 80 years in theory; it looks new but is 40 years old in practice. The timeline for 3D printing is shown in Figure 6.2. The first patent for a liquid metal recorder was issued in the 1970s, but the concept is far older. In 1945, a futuristic short story by Murray Leinster called "Things Pass By" described the process of:

> [feeding] magnetronic plastics— they make houses and ships of nowadays— into this moving arm. It scans drawings using photo-cells and produces drawings in the air. However, plastic comes out of the drawing arm's end and hardens as it goes.

What was science fiction in Leinster's day became a reality very quickly. 3D printing has been proposed as a faster and more efficient technique of producing items since 1945 and in practice—however primitively—since 1971.

Between 1971 and 1999, the first 3D printer is developed. This is split into two eras, viz. the pre-1990s and the 1990s. The pre-1990s saw the discovery of 3D printing technology. In the 1990s, the main 3D printer manufacturers and CAD tools emerged. Inkjet technology, a means of "drawing" a drop of material from a nozzle using electronics, was pioneered by the Teletype Corporation in the 1960s. It led to the development of a device that could print up to 120 characters per second, paving the path for consumer desktop printing. Teletype later experimented using melted wax, as stated in a 1971 patent by Johannes F. Gottwald, whose concept was to output a liquified metal item that solidified into a shape dictated by the inkjet's movement upon each successive layer. This gadget was the liquid metal recorder, which hypothesized that "printing" could extend beyond ink and was the foundation of fast prototyping.

The first recorded variations of 3D printing may be traced back to the early 1980s in Japan. Hideo Kodama was looking for a technique to design a quick prototyping system in 1981. He devised a layer-by-layer manufacturing method based on a photosensitive resin that was polymerized by UV radiation. Charles Hull pioneered stereolithography, a technology that allows designers to generate 3D models using digital data, which is subsequently utilized to make a physical thing, in 1984, making 3D printing history. Chuck Hull was the first person to build a 3D printer from the ground up. His technique sends spatial data from a digital file to the extruder of a 3D printer to build up the object one layer at a time, based on his patent for curing photopolymers using radiation, particles, a chemical reaction, or lasers. 3D printing was still a new technology at the time. As the product set, it tended to warp if it was constructed of popular polymers. Because 3D printing equipment cost hundreds of thousands of dollars at the time, they were exclusively used in large production plants, far out of reach of ordinary people.

Figure 6.2 gives a summary of the timeline of 3D printing technology. In a nutshell, Dr. Kodama of Japan was the first to file a patent. In 1980, rapid prototyping

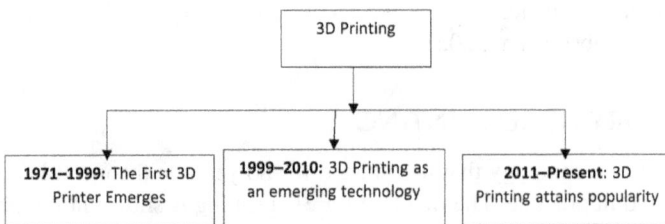

FIGURE 6.2 Timeline of 3D printing technology.

was introduced. In 1984, French engineers abandoned stereolithography. Charles Hull began studying stereolithography in 1986, and the first SLA-1 machine was built in 1988. DTM Inc. built the first SLS machine, which was later purchased by the firm called 3D Systems in 1988. In this era the first EOS Stereos system was released in 1990. In 1992, Stratasys received the FDM patent. Solidscape began operations in 1993. In 1995, MIT granted Z Corporation an exclusive license.

The next era was from 1999 to 2010. Despite public fears that the millennium bug would cause computer systems to crash and a digital Armageddon, 3D printing did show considerable promise in a variety of businesses. This era coincided with when bioengineering was also making significant progress. Scientists from the Wake Forest Institute for Regenerative Medicine in Winston–Salem, North Carolina, used additive manufacturing to print a human urinary bladder. Thereafter, they coated the organ with cells from the patient to ensure that the bladder was not rejected by the body. This unlocks several breakthroughs in the usage of 3D printing, especially in medicine. A small kidney, a complicated prosthetic leg, and the first bioengineered blood arteries created from donated human cells were built by scientists, technicians, and clinicians using 3D printing.

During this era, RepRap and MakerBot 3D printers were made and gained prominence, facilitated by the need to reduce cost to meet consumer needs. Adrian Bowyer's RepRap Project initiated an open-source endeavor in 2005 to produce a 3D printer that could self-assemble—or at least print most of its own parts. The first practical application of RepRap's principle was the 1.0 Darwin machine, which gave anyone the ability to build whatever they wanted. Around the same time, Kickstarter was launched, giving home 3D printing a tremendous boost. Manufacturing was rapidly becoming more democratic. For the MakerBot 3D printer, Objet (now Stratasys) brought commercial 3D printing to the desktop in 2006, allowing customers to send ideas to the device and have them printed in a variety of materials with varied qualities. Marketplaces and virtual swap meet for trading, sharing, and acquiring designs popped up all over the place, igniting a massive surge of enthusiasm. MakerBot made cofounder Bre Pettis a celebrity and gave 3D printing the same cachet as previous rising technologies like social networking, e-commerce, and even the Web itself when it launched open-source DIY kits to design and print just about anything in 2009. In summary, in 1999, medical advances were made possible by engineered organs. In the year 2000, a working kidney was 3D-printed. The SLM technology was also introduced by MCP Technologies (a well-known vacuum casting OEM). RepRap 3D printer built a 3D printer in 2004. Spectrum Z510 was introduced by Z Corp. in 2005. It was the world's first high-definition color 3D printer. An open-source project (RepRap) was started in 2006. He created the first 3D-printed prosthetic limb in 2008. FDM patents are released into the public domain, and Sculpteo was founded in 2009. Figure 6.3 shows a 3D printer manufactured with a 3D printer. This is capable of printing other components.

The current era of 3D printing started around 2011 to date. 3D printing is now a well-established technique. Consumer interest in industrial platforms expanded. Some believe additive manufacturing will eventually replace traditional CNC and milling manufacturing, while a Lux Research report from 2021 predicts that 3D printing would be valued at $51 billion by 2030.

FIGURE 6.3   A 3D printer manufactured with 3D printer.

This era saw more utilization of 3D printing to solve real-world human challenges as opposed to it being used for producing plastic toys. It was used for printing meals and extruding various materials simultaneously. The process became faster and less expensive and was also utilized in bioprinting human tissue. It heralds the beginnings of printing organs tailored to patients' needs and making objects out of silver or gold. The spectrum of materials available for 3D printing has expanded tremendously.

The applications are as diverse as the ideas of innovators and engineers. Scientists at the University of Southampton flew the world's first 3D-printed unmanned aircraft. Also, makers of a 3D-printed car achieved up to 200 mpg with a hybrid gas/electric engine. A start-up specializing in ecological living structures developed a robot-made habitat suitable for living on Mars. In summary of this era, Urbee was the world's first 3D-printed prototype automobile, unveiled in 2010. In 2011, Cornell University began developing a 3D food printer. In 2012, the first prosthetic jaw was printed and installed. The use of the term "3D printing" in Obama's 2013 State of the Union address expanded its popularity. In 2015, Carbon 3D released their innovative CLIP 3D printing machine, which is extremely quick. In 2016, Daniel Kelly's lab said that it will be able to 3D-manufacture bone. In 2018, the first family moved into a 3D-printed home.

So, what does the future hold for 3D printing? The global additive manufacturing industry is predicted to rise 17% yearly through 2023, according to Statista, as the technology's uses expand and metal additive becomes more viable. Between 2020 and 2026, the market for additive manufacturing products and services is predicted to nearly triple. Figure 6.4 gives the pictorial representation of the items printed using 3D printer.

Consumer products (eyewear, footwear, design, furniture) and industrial products are examples of 3D-printed parts (manufacturing tools, prototypes, functional end-use parts). It also includes dental goods, prosthetics, architectural scale models and maquettes, fossil reconstruction, recreating ancient artifacts, forensic pathology evidence reconstruction, and cinematic props.

FIGURE 6.4 Some 3D-printed items.

6.4.1 Process of 3D Printing

Subtractive manufacturing, which involves cutting or hollowing out a piece of metal or plastic with a milling machine, is the reverse of 3D printing. When compared to traditional production methods, 3D printing allows you to create complicated shapes with less material. The process of 3D printing requires a model, a 3D printer, and the materials to be used. The major components and diagram of a 3D printer are shown in Figure 6.5.

6.4.2 Components of 3D Printer

The extruder, print bed, nozzle, and filament used by the printer are the four main components of a 3D printer. Other components are the moving parts and touch screen. The printing bed is the platform on which models are printed, and it is frequently heated to assist layers to adhere to each other tightly. Figure 6.6 gives a summary of the key components of a 3D printer.

    i.   Extruder: Extruder is the heart of a 3D printer, melting and stretching filaments to construct the model. The extruder is the mechanism that feeds filament from the spool into the hotend, where it is melted and extruded onto the print bed through the nozzle. The extruder can be found in one of two places: either fastened to the printer frame down by the spool of filament or immediately

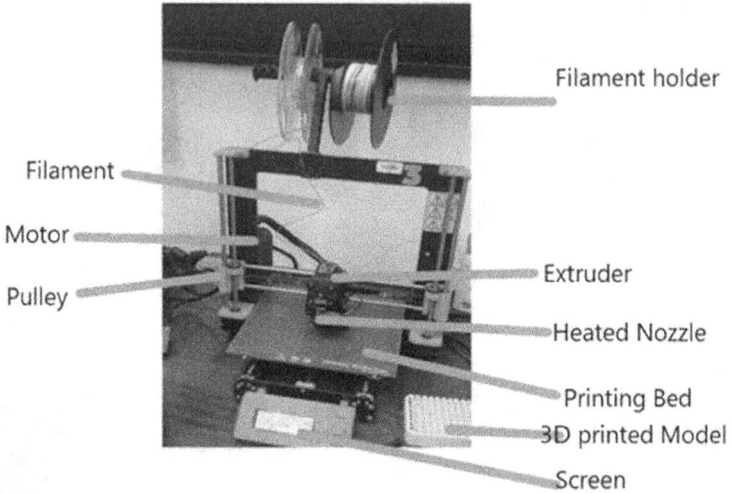

FIGURE 6.5 A 3D printer showing key components.

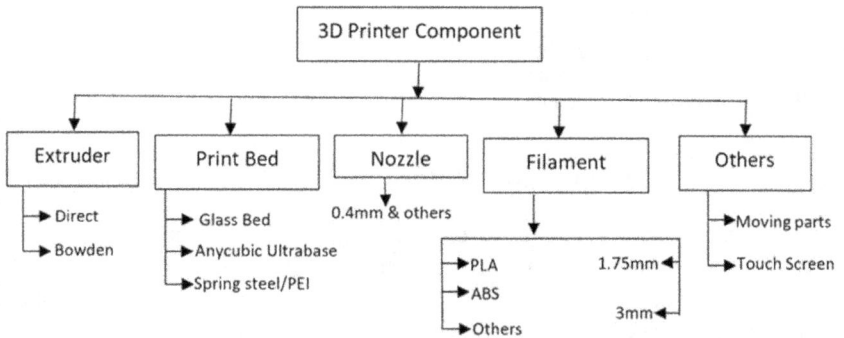

FIGURE 6.6 Key components of 3D printer.

above the hotend. It's a direct drive extruder if the extruder motor is situated directly above the hotend. This type of extruder pulls filament off the spool and feeds it into the hotend, where it is melted and extruded through the nozzle. The Bowden extruder, an alternate approach, is constructed with the motor situated near the spool.

ii. Nozzle: The melted filament that was previously pushed/pulled into the hotend will be deposited by the 3D printer nozzle. Most 3D printers come with a 0.4 mm brass nozzle, which has become the industry standard. The 0.4 mm diameter provides an excellent blend of precision and speed. Brass has excellent thermal conductivity at a low cost. The standard 0.4 mm brass nozzle should suffice for most printing unless otherwise. When the need arises, you can upgrade your nozzle. When it comes to nozzle diameters, keep in mind that if you require more detail and don't mind longer print times, a thinner-diameter nozzle is the way to go. A bigger diameter nozzle can improve the quantity of filament you can put down, which can help you reduce print times if you don't mind higher tolerances.

When it comes to nozzle materials, brass, stainless steel, and hardened steel are widely used. Brass has the highest thermal conductivity of the three, allowing it to heat up more quickly. Brass is also less expensive. This is balanced by the fact that it is the weakest and so most vulnerable to abrasive filaments and external impacts. While the damage may not be visible immediately away, it can accumulate over time and impair the quality of your prints. As you progress from carbon steel to stainless steel and finally toughened steel, the heat conductivity decreases and the price rises. However, the increased durability will be beneficial to you.

iii. Filaments: Filament is a thermoplastic that melts when heated, allowing it to be laid down layer by layer during printing. PLA and ABS filament are the most common filament materials, and they are made from thermoplastics. PLA and ABS are the most typically discussed filaments because they are widely available and well suited to their respective niches. PLA is tougher and stiffer than ABS; however, due to its poor heat resistance, it is primarily used by amateurs. ABS is a superior material for prototyping since it is weaker and less stiff. Figure 6.7 shows a 3D printer workstation with different colors of filament shown.

FIGURE 6.7 A 3D printer workstation with different filaments and other tools.

Polylactic acid (PLA) is an easier material to print with since it melts at a lower temperature and reduces warping. This improves bed adherence and enables for a finer printing detail. A thermoplastic filament made of acrylonitrile butadiene styrene is called ABS. It is slightly more durable in general and is better suitable for instances where the printed object must withstand high temperatures. ABS also can be smoothed out with acetone. PLA is perfect for 3D printing that requires a high level of aesthetics. It is easier to print with because of the lower printing temperature and hence it is more suited for pieces with fine details. ABS is best used in applications that require a high level of strength, ductility, machinability, and thermal stability. ABS is more prone to warping than other materials. Although it is a bit more technical than PLA, polyethylene terephthalate glycol (PET-G) is also considered a beginner-friendly material. However, businesses such as engineering and manufacturing benefit greatly from it. It is a suitable material for functional prototypes since it can resist greater temperatures and has a chemical composition that's ideal for these applications.

Polycarbonate/acrylonitrile butadiene styrene (PC-ABS) refers to the reinforcement of polycarbonate (PC) into an acrylonitrile butadiene styrene (ABS) base material to improve mechanical qualities and give increased strength in 3D-printed constructions and products. Some critical attributes of PC-ABS composites, such as impact strength, flexural modulus, and hardness, are relatively higher than pure polycarbonate and ABS materials.

Also, the printer type determines the filament diameter required. Each filament size has its own set of advantages and disadvantages. A 1.75 mm filament is the most prevalent size these days because it offers more manufacturers and pricing alternatives. Prices on Amazon differ in the range of US$2 to US$3 per kilogram. The biggest difference between the two filament sizes, aside from pricing and material availability, is what size printing they choose. For improved control, a smaller nozzle with a thinner 1.75 mm filament can be utilized to achieve finer details and longer print times. A bigger nozzle that uses a 3 mm filament to extrude more filament quicker is required for faster print times and for producing larger items where thicker layer lines are acceptable.

iv. Print Bed: The print bed is the flat surface on which the nozzle will deposit the plastic layers. Glass, Ultrabase, and Polyether Imide (PEI) are the most popular print bed materials found in printers. A heated bed is a standard feature on several printers. While a heated bed can help with PLA filament bed adherence, it is an optional feature. However, for printing involving ABS or other types of filaments, extra investment in a heated bed is required. A glass bed is probably the most popular print bed because of its availability, with a few nice qualities that make it ideal for a print bed. The biggest disadvantage of a glass bed is that printed things, particularly larger ones, tend to stick to the surface. Basic home products like painter's tape, glue sticks, or hairspray can be used to solve it. Most Anycubic's latest printers come with the Anycubic Ultrabase as the default bed. The Ultrabase is a glass bed with a unique surface coating to aid surface adhesion used to eliminate the flaw of glass beds. When the bed is heated, the surface adheres to the deposited plastic quite well. It should be rather easy to release the parts once the print is finished and the bed has cooled. The primary disadvantage of this bed is that the coating might wear down with time, causing the print bed to behave like a regular glass bed. This print bed, which is available from a variety of manufacturers these days, features a flat magnetic base on which a thin sheet of steel with a PEI coating can be attached. This allows printing onto the steal and then removing it after the print is complete. By bending the steel, the printed part can be easily removed. The problem with this form of bed is that it has a higher initial cost, and larger beds have a higher likelihood of warping over time. This may result in printing troubles, necessitating the purchase of a replacement PEI sheet.

v. Others: Moving parts are the components that move along three axes, namely X, Y, and Z. Forward and backward motions are controlled by the X- and Y-axes, while vertical movements are controlled by the Z-axis. Touchscreen uses the built-in RaiseTouch on the LED touch screen, and users can operate the printer and complete various settings.

A 3D printer, slicing software, and a filament are the three key items needed to start the 3D printing process. Computer-aided design (CAD) and slicing software are two essential pieces of software for 3D printing. Typically, any CAD system capable of creating a working model can be used. A customized model requires a CAD. The CAD software must be able to export an STL file. The second half of the equation is the slicing program. This program converts the STL file into a format that the printer can understand. The G-code specifies how and where the printer's axis should move, as well as how much material should be deposited. An SD card or Wi-Fi is used to send the G-code to the printer.

6.4.3 3D PRINTING PROCEDURE

There may be more or fewer steps in the printing procedure depending on the type of print. However, in general, 3D printing entails the following: modeling, slicing, printing, and postprocessing as shown in Figure 6.8.

i. Modeling: A digital model of the printed product is required. The object to be printed will be transformed into a digital model that can be printed on a 3D printer using modeling. There are about three ways of creating models for printing, viz. using 3D CAD, scanning it with a 3D scanner, and downloading from other users. A 3D printable file can be created using almost any 3D modeling software. Blender, Creo, SketchUp, Mastercam, AutoCAD, SolidWorks, Maya, Photoshop, Tinkercad, and other 3D modeling applications are used

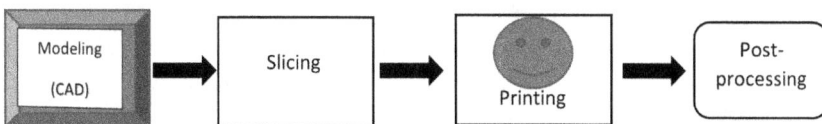

FIGURE 6.8 Schematic of 3D printing steps.

to make the designs. 3D scanning is another method for creating a three-dimensional digital file. 3D scanning is a technique that analyzes a real-world object and instantaneously makes a digital counterpart. It is closely related to 3D printing. Industry professionals frequently employ 3D scanning for reverse engineering activities. After an existing object has been digitized, it can be modified before printing. This procedure necessitates the use of a 3D scanner. Finally, 3D print–ready files can be downloaded or purchased from companies such as Thingiverse, YouMagine, GrabCAD, and MyMiniFactory Shapeways. Before transferring 3D files to the printer, they must meet several design standards. Because the design is for the actual world, when designing for additive manufacturing (3D Printing). These are suitable scale size, minimum wall thickness, and manifold/water-tightness.

ii. Slicing (STL file): Rapid prototyping, 3D printing, and computer-aided manufacturing all employ STL files (CAM). The next stage after creating the model is conversion to STL files. STL, or STereoLithography, was named after the first ever 3D printing method and is the most prevalent 3D printing file type. STL stands for "Standard Triangle Language" and "Standard Tessellation Language," among other things. The file extension that can be used is STL. The data that represents the layout/surface of a three-dimensional object is stored in this file format as triangular mesh (polygons). The other formats are: OBJ and.3MF. File formats such as X3D, WRL, DAE, and PLY provide the final printout in color. By default, not every STL or OBJ file is 3D printable. The file formats must meet specific requirements, such as a maximum polygon count, water-tightness, physical size, and minimum wall thickness, among others. This is the process of converting a 3D file into instructions that the 3D printer can understand. Slicing is the process of dividing or chopping a 3D model into hundreds or thousands of horizontal levels and instructing the machine step by step on what to do. After the files have been sliced, a new file format called G-code with the extension .gcode is created. The G-code programming language is the most extensively used numerical code programming language in computer-aided manufacturing to manage automated machine tools such as 3D printers and CNCs (computer numerical controls). In a nutshell, G-code is the machine's language and avenue of communicating with it. An individual 3D printer owner needs to configure this process by adjusting the setting. However, this setting is adjusted by a service provider when they handle the printing.

iii. The Printing: The printing machines are made up of a lot of moving and sophisticated elements, and they need to be properly maintained and calibrated to create good prints. After the printing has started, most 3D printers do not need to be monitored. The machine will follow the automated G-code instructions, so there should be no problems during the printing process provided there are no software errors or exhaustion of the raw material. The operation can be completed in a matter of minutes or over several hours, depending on the size of your object, your printer, and the materials used.

iv. Postprocessing: There may be extra postprocessing procedures after printing, such as painting, brushing off powder, and so on, depending on the quality

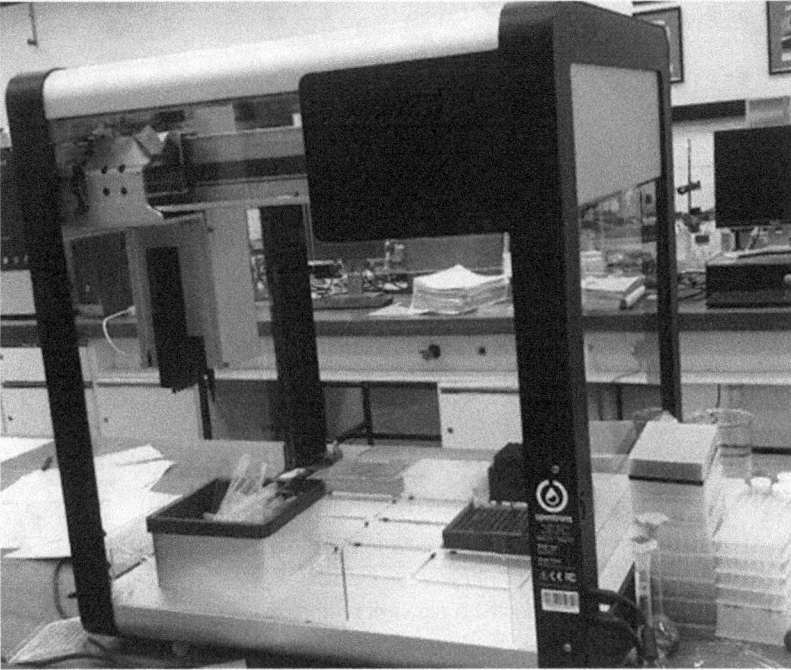

FIGURE 6.9   A laser cutter used for 3D printing postprocessing.

and finish of the final product or the material you used. The final stage of 3D printing is postprocessing. The phases in 3D printing postprocessing are as follows (not all of them must be completed); the support after printing should be removed (if the model contains any). The filament will remain on the model's surface. Sandpaper can be used to smooth out the model. Coloring of the model and addition of extra details is used to improve the quality and finishing. Also, to make the model surface smoother and brighter, a special coating or other techniques is used to polish it. Welding may be used for a multipart model or a huge model to subdivide it into several sections and then assemble the parts to produce a complete model when it is printed. Figure 6.9 shows a laser cutter used for cutting and finishing of 3D-printed component and Figure 6.10 gives a finished 3D-printed model with postprocessing.

The summary of the 3D printing process is represented in Figure 6.11. In some cases, postprocessing may be optional with some 3D printers; hence it is not completely part of the 3D printing cycle.

6.5   ADVANTAGES AND DISADVANTAGES OF 3D PRINTING

Although it is an emerging manufacturing technology, 3D printing will transcend traditional manufacturing processes and become a new option for future production

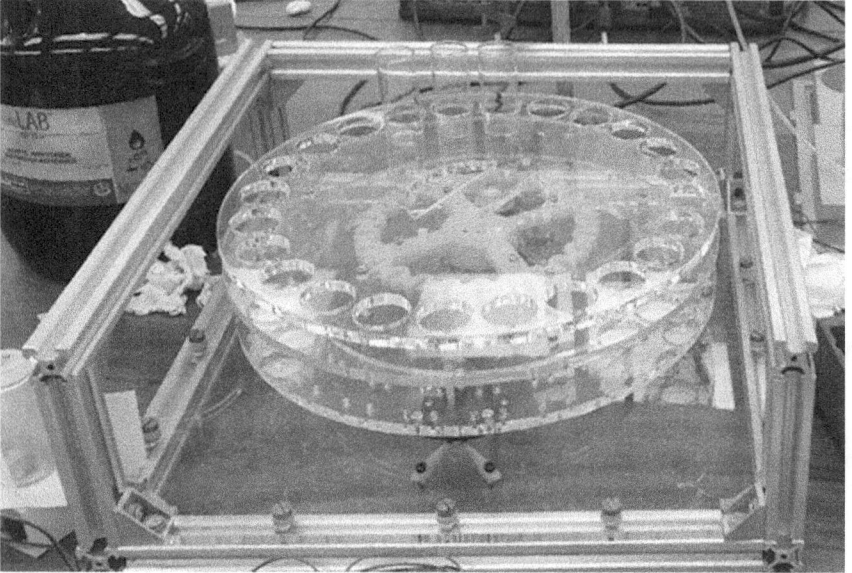

FIGURE 6.10 A finished 3D printed component postprocessed using the laser cutter.

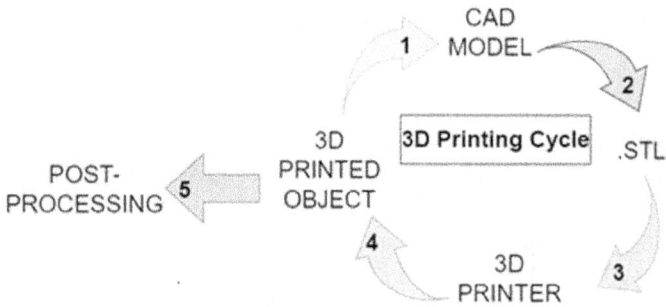

FIGURE 6.11 Schematic of 3D printing cycle.

due to its unique additive manufacturing method. The technology helps in shortening the production cycle. 3D printing drastically reduces the time it takes to create a product. The company can generate product prototypes more quickly utilizing a 3D printer, allowing clients to provide comprehensive and immediate feedback. The use of 3D printers for prototyping and small-scale manufacturing will continue to grow along with the ability to manufacture difficult parts. 3D printing has strong forming skills and is not limited by complex curved surfaces or exact structures; it can be used to make extremely complex parts. Thirdly, 3D printers have an unlimited design and production space provided a 3D model is developed. Combining different strands opens up more possibilities.

The technology of 3D printing also has demerits. Currently, the size of items made with 3D printers is limited. Because 3D printing is an additive method (layer by layer), the materials that can be used are limited: ceramics, resin, plastics, and so on.

6.6 CHALLENGES AND OPPORTUNITIES OF 3D PRINTING

6.6.1 OPPORTUNITIES OF 3D PRINTING

3D printing could contribute to the formation of manufacturing facilities in small towns and stimulate industrial development outside of major cities, due to the well-established software sector that is characterized by the deployment of previous industrial revolution and increased connection induced by "Digitalization." Switching to 3D printing technology, which is both cost-effective and efficient, can benefit traditional small and medium businesses. Manufacturing in the aviation, health, construction, and automotive industries could benefit from the technology. It can shorten production times while also improving product strength, weight, and environmental effect.

6.6.2 CHALLENGES OF 3D PRINTING

One disadvantage is the lack of domestic 3D printer manufacturers. While some attempts have been made to build 3D printers in the United States, they are not of industrial quality, and industries rely heavily on imports. Import costs are high. There is a lack of clarity on the import of 3D printers, which are subject to a customs duty of 30–40%, in addition to the shipping costs. The high expense of importing industrial-grade 3D printers is prohibitive for developing countries' medium and small-scale enterprises. 3D printing reduces the need for assembly labor, and it has hazardous consequences for the employment situation in emerging countries. As far as employment is concerned, manufacturing jobs will become outdated, resulting in a detrimental impact on developing economies.

6.6.3 INTERNATIONAL BEST PRACTICES

Although 3D printing appears to be shockingly simple, there are moral, business, and technical challenges that must be solved before technology becomes commonplace. The "Additive Manufacturing Industry Promotion Plan 2015–2016" was China's first national plan for 3D printing. Later, a new additive manufacturing Action Plan (2017–2020) was announced for the country's continued development of the technology. The Plan focuses on bolstering research and development, as well as advancing 3D printing applications and acceptance in the workplace. Printing a duplicate of a copyrighted action figure is not the same as going to a toy store and shoplifting the miniature. Because the law has not yet been tested in relation to 3D printing, there may be "a bit of a grey area" when replicating a component. Producing an exact copy of an action figure is a copyright violation; however, creating a spare part for a broken figure "might" be a copyright violation if the original creator's business strategy included fixing action figures as part of their business model. There are other issues related to morality, among others. Copyright infringements are a source of concern. Counterfeit printing of copyrighted or patented products is a source of concern. Anyone who has a blueprint will be able to effortlessly counterfeit things. Concerns about discouraging or restricting individuals from 3D printing potentially dangerous goods have led to the production of unsafe items. There are also concerns about cyber security. According to studies, a 3D printer connected to the internet is vulnerable to cyberattacks. In healthcare, there are ethical issues to consider.

6.7 CLASSIFICATION OF 3D PRINTING

Different approaches are used for classifying 3D printers based on the material (filament and resin) and based on the method of 3D printing (VAT photopolymerization, and extrusion 3D printers) as shown in Figure 6.12.

The American Society for Testing and Materials (ASTM) has created a set of standards that divides additive manufacturing into seven categories: vat photopolymerization [stereolithography (SLA), digital light processing (DLP), continuous liquid interface production (CLIP)], material extrusion [fused deposition modeling (FDM) and fused filament fabrication (FFF)], powder bed fusion [Multi Jet Fusion (MJF), selective laser sintering (SLS), and direct metal laser sintering (DMLS)], binder jetting, material jetting, sheet lamination, and direct energy. Table 6.1 is the summary of different types of 3D printing technologies.

In the consumer sector, there are now two main 3D printing technologies. Fused deposition modeling (FDM) and vat polymerization are the two technologies and are shown in Figures 5a and 5b, respectively. Stereolithography (SLA), masked stereolithography (MSLA), and digital light processing (DLP) make up vat polymerization.

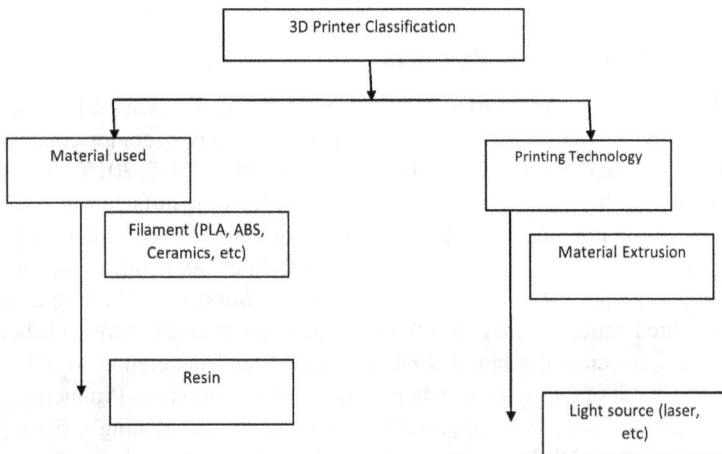

FIGURE 6.12 Schematic depiction of 3D printing classifications.

TABLE 6.1
Summary of Different Types of 3D Printing Technologies

Parameter	Fused deposition modeling	Stereolithography	Selective laser sintering	Selective laser melting
Abbreviation	FDM	SLA	SLS	SLM
Operation principle	Extrusion of melted filament	UV curing	Laser sintering	Laser melting
Material printed	Thermoplastic polymer in the form of string (filament), i.e., PLA, ABS	Resins/ photocurable liquid materials	Powdered sinterable polymers (i.e., polyamides, TPU, TPE)	Various metal alloys
Advantages	Low cost Fast printing time	High print resolution High process automatization	No support structure quality prototyping movable parts	printouts durability
Disadvantages	Need for support structures thermal shrinkage of filament	Narrow material variety high maintenance costs	Long printing time	High cost
Applications	Fast prototyping education low volume production	Complex internal geometry prototypes dental models	Education functional prototypes medical models prototyping moveable parts	Automotive and aviation industry functional parts
Layer thickness	0.1–0.3 mm	0.05–0.15 mm	0.060–0.15 mm	0.02–0.1 mm
Printing without support structures	No	Not always necessary	Yes	Not always necessary
Printing objects with movable parts	Not always achievable (lower precision)	No	Yes	No

6.7.1   3D Printing Main Methods

To fulfill the demand for printing complicated structures at fine resolutions, additive manufacturing (AM) methods have been created. Rapid prototyping, the ability to print massive structures, the reduction of printing flaws, and the improvement of mechanical characteristics are only a few of the important aspects that have fueled the growth of AM technology. Fused deposition modeling (FDM) is the most prevalent method of 3D printing that primarily uses polymer filaments. Inkjet printing, contour crafting, stereolithography, direct energy deposition (DED), and laminated object manufacturing (LOM) are the main methods of additive manufacturing of powders by selective laser sintering (SLS), selective laser melting (SLM), or liquid binding in 3D printing (3DP). These methods are briefly described, along with their applications and appropriate materials for each approach, as well as their pros and drawbacks.

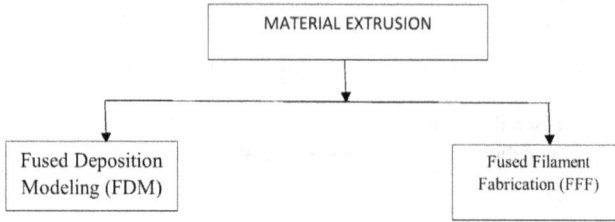

FIGURE 6.13   Classification of 3D printing based on material extrusion.

A) Material extrusion: This is classified as fused deposition modeling and fused filament fabrication, as shown in Figure 6.13.

1) Fused Deposition Modeling 3D Printer

The process of depositing layers of melted thermoplastic one on top of another as they cool into a final item is referred to as FDM. It is a subset of material extrusion 3D printing as shown in Figure 6.14. These printers are the most widely available 3D printers for customers right now. Furthermore, FDM printers are less expensive than their vat polymerization competitors. There's also the benefit of having a wide range of filaments to pick from, allowing customers to compare strengths, weaknesses, and pricing points.

FDM 3D printing is a horizontally and vertically oriented method in which an extrusion nozzle moves across a build platform. The procedure uses thermoplastic material that reaches melting point and is then forced out in layers to make a 3D object. FDM works by converting digital design files into physical dimensions by uploading them to the machine. It is a 3D printing technique that uses PLA, ABS, and other thermoplastic filaments that are heated and extruded by an extrusion head, then piled layer by layer under computer control to create a formed 3D model. With higher precision and cheaper cost, it is the most frequent and widely used 3D printing technology. FDM is an additive manufacturing technique in which molten plastic filament is precisely extruded to build a product. Parts are rigid, especially when compared to SLS, making them ideal for projects requiring rigidity. Some fine features may not resolve well because of the filament's thickness of 0.010 in. (0.254 mm). Certain features in FDM require support material (soluble or break-away), which should be considered while designing components to allow support structures to be minimized, reducing postprocessing, and making parts easily detachable.

In comparison to other 3D printing methods, FDM is relatively cost-effective, but it also has the lowest resolution. As a result, it's a less practical alternative for pieces with fine details. The FDM system is made up of a platform extrusion nozzle and a control system that allows for quick and easy 3D printing. As the object is produced layer by layer in a Z-direction, the thermoplastic filament is heated and deposited in X- and Y-coordinates. FDM printers produce objects from the bottom up, and if the object has overhanging elements that are more than 45 degrees, support structures are usually required. Individuals and small enterprises may build parts fast

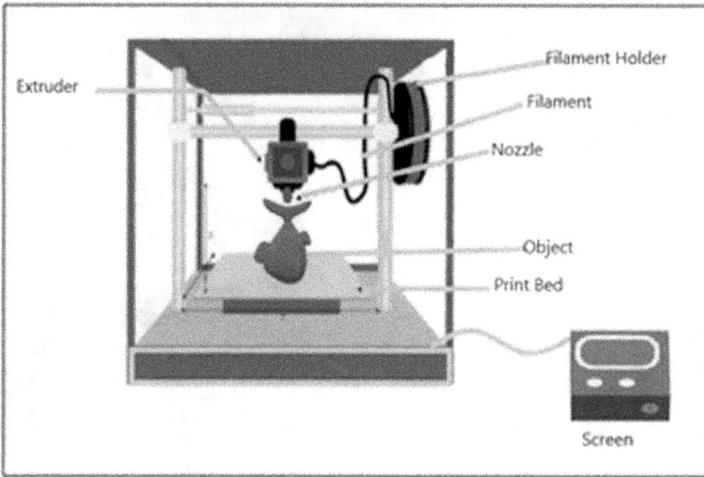

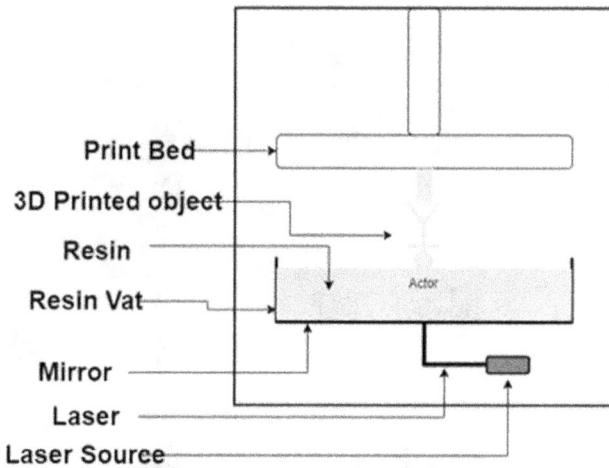

FIGURE 6.14 Schematic of classes of 3D printer: (a) fused deposition modeling; (b) vat polymerization.

and efficiently with this type of 3D printing because it is the most cost-effective. Printers have progressed from large, expensive equipment to smaller, faster, and less expensive machines that have become essential components in engineering and design departments. Figure 6.15 shows the schematic depiction of FDM, a snapshot of an FDM 3D printer with some printed parts, and a user with an FDM-printed frog.

6.7.1.1 Operation of FDM 3D Printer

To use an FDM printer, insert a spool of thermoplastic filament into the machine. The printer feeds the filament via an extrusion head and nozzle once the nozzle reaches the correct temperature. Several process parameters can be adjusted in most

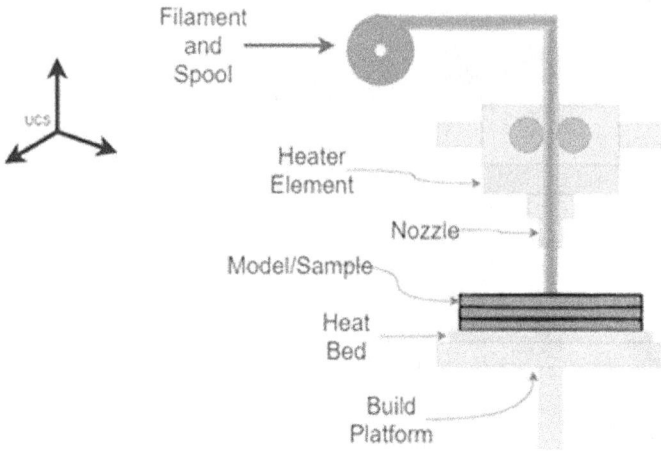

FIGURE 6.15   (a) A schematic depiction of FDM; (b) a snapshot of an FDM 3D printer with some printed parts; (c) a user with an FDM printed frog.

FDM systems. The temperatures of the nozzle and build platform, build speed, layer height, and cooling fan speed are all factors to consider. These processes are generally adjusted for designers by additive manufacturing operators. However, build size and layer height are crucial factors to consider. A home 3D printer's typical build size is $200 \times 200 \times 200$ mm, but industrial machines can exceed $1{,}000 \times 1{,}000 \times 1{,}000$ mm.

FIGURE 6.15   (Continued)

Although, a large model printed by a desktop machine needs to be fragmentized into smaller sections and reassembled. Layer heights in FDM typically range from 50 to 400 microns. Shorter layers generate smoother parts and more correctly capture curved geometries, while taller layers allow the building of parts more rapidly and at a cheaper cost.

6.7.1.2   Characteristics of FDM Print

Although the extrusion processes and part quality of different FDM 3D printers differ, there are some basic qualities common to the FDM printing process. Warping, layer adhesion, support structure, infill, and shell thickness are all factors to consider.

    i.   Layer adhesion

In FDM, the adherence of deposited layers of a part is crucial. When molten thermoplastic is extruded via the nozzle of an FDM machine, it presses against the previously printed layer. This layer remelts because of the high temperature and pressure, allowing it to bind with the prior layer. The molten material's shape deforms to an oval when it presses on the previously printed layer. This means that no matter what layer height is employed, FDM items will always have a wavy surface and that minor details like small holes or threads may require postprocessing. Figure 6.16 gives the schematic illustration of reality and theoretical extrusion profile for FDM printing.

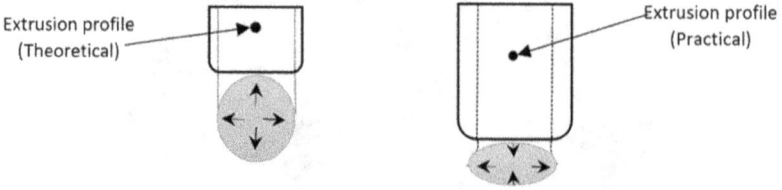

FIGURE 6.16   Extrusion profile for FDM printing.

ii. Warping

One of the most typical FDM flaws is warping. The dimensions of extruded material shrink as it cools during solidification because the printed item cools at different rates; the dimensions of separate sections change at different rates. Internal tensions develop up because of differential cooling, pulling the underlying layer higher and causing it to distort.

There are numerous methods for avoiding warping. One technique is to keep a tight eye on your FDM system's temperature, particularly the build platform and chamber. Another one is by increasing the adhesion between the part and the build platform.

6.7.1.3   Stages of FDM Printing

Preprocessing, production, and postprocessing are the three processes of FDM part manufacturing. In the preprocessing, layers are created by slicing a 3D CAD file with printer software. The software then turns the data into machine code for each slice, which sets the machine's tool paths. In the printing, extruder heads extrude melted plastic filament layer by layer along the tool path until the part is complete. And finally for the postprocessing, dissolving the support material in water or breaking it off are two options for removing it. After that, additional bespoke finishes like tapping, inlays, and sanding are performed. The way parts are oriented in FDM has a huge impact on their overall strength and look, especially for delicate and concentric features. When layers are printed parallel to the XY-axis, concentric features are best resolved. Many fine details (such as tabs) are also enhanced when printed parallel to the XY-axis. The ideal orientation for a part to print is determined by designing it so that fragile and concentric features grow in the same direction. Vertical tabs are extremely weak and may break off between layers when printing the part shown from the bottom up. Concentric features, on the other hand, are more refined and do not have layer stepping. Tabs are significantly stronger when printing the identical item on its side, yet concentric features have significant stepping.

6.7.1.4   Finishing and Postprocessing

Due to its layer-by-layer process of extruding plastic filament, FDM tends to produce a coarser and clearly layered surface texture when compared to other production processes. For shallow-angled and curved surfaces, this layer "stepping" is most visible. This stepping can be reduced by designing parts so that curves and angled surfaces grow parallel to the build surface, or by increasing angles or curvatures. ABS is a robust, cost-effective thermoplastic that comes in a variety of colors. Ultem is the most durable and heat-resistant FDM plastic available. ASA is UV-resistant, cost-effective, long-lasting plastic in a variety of colors. Polycarbonate (PC) is a common industrial thermoplastic with excellent tensile and flexural strength. Nylon is a high-impact material that is commonly utilized in traditional manufacturing. PPSF is a plastic that is sterilizable, robust, and high performing. Depending on the material, there are presently 16 distinct colors accessible in FDM, allowing for tremendous customization of any part. Parts made of ABS, PC, Ultem, and PPSF materials can be sanitized to make them last longer. FDM parts are noted for their strength and rigidity without jeopardizing the structural integrity of the item. Tapping, inserts, and pins are all possibilities.

2) Inkjet Printing and Contour Crafting

Contour crafting is a 3D printing technique that is used to create architectural structures. Contour crafting is an additive method that uses computer control to take advantage of troweling's better surface-forming capabilities to build smooth, accurate planar and free-form surfaces out of extruded materials (Davtalab et al., 2018). Contour carving necessitates the use of a large construction 3D printer capable of printing any form of infrastructure layer by layer. Most of the time, these construction 3D printers use a crane or a robotic arm to produce these large-scale projects. Quick-setting materials, such as concrete or sand, are employed in the 3D printing materials used for contour carving. The government of Dubai, United Arab Emirates, recently announced that by 2025, 25% of the new Emirati buildings will be 3D-printed using contour crafting technology (Akhnoukh, 2021). A whole house can be completed in a few hours and in 20 hours, contour crafting is anticipated to construct a 200 m^2 house (Khoshnevis and Hwang, 2006, Ndlovu et al., 2020). Although, the current lifespan is around 10 years, with a cost of US$5,000 (Bardwell, 2023).

This technology is gaining much attention as it is promising technology especially for the construction industry. It is capable of reducing the construction time and other bottlenecks associated with it (Khorramshahi and Mokhtari, 2017, Rouhana et al., 2014). It is capable of printing buildings in a matter of days (Fernandes and Feitosa, 2015). It is projected to be the future of sustainable construction even in space (Hager et al., 2016).

There has been an effort to understudy contour crafting (CC) 3D printing from different perspectives. Kazemian et al. (2019) did study the performance testing of Portland cement used as material for contour crafting printing. The study examined the shrinkage of 3D-printed concrete, the structural performance and longevity of 3D-printed structures, the robustness of printing concrete, and the printing method. Similarly, a framework was developed to study the performance of fresh mixture used in construction scale in contour crafting 3D printing (Kazemian et al., 2018). The study examined the 3D-printed concrete shrinkage, as well as structural durability and quality of products. Also, a study proposed four ways of monitoring real-time contour crafting 3D printing operations (Kazemian and Khoshnevis, 2021). The role that contour crafting can play in the field of civil engineering was studied (Fernandes and Feitosa, 2015). The study outlines how contour crafting technology might transform the civil engineering industry, as well as how it can alter the perceptions of the construction process. Because of the benefits it can offer to the environment and the construction process itself, it has been dubbed a forerunner of a new era in the construction business. The study gathered and categorized concepts to serve as a useful resource for future research. Features of contour crafting technology, concrete mixture for contour crafting finality, and a comparison of traditional and contour crafting construction methods are among the topics covered. The properties and processing of contour crafting 3D printing were discussed by Tay et al. (2016). The study discussed the process of 3D concrete printing, contour crafting, and D-shape alongside materials that have been used for construction. Attempts have also been made to compare the viability of conventional housing construction with contour-crafting 3D-printed houses (Jagoda, 2020).

6.7.2   HISTORY OF CONTOUR CRAFTING 3D PRINTING

Dr. Behrokh Khoshnevis of the University of Southern California's Information Sciences Institute studied this printing technology (Kazemian et al., 2017). The contour crafting technique was invented by Dr. Behrokh Khoshnevis, who worked on a machine that could build houses. In the summer of 2008, Caterpillar Incorporated provided funding to support Viterbi project research (Khoshnevis, 2008). Singularity University graduate students founded the ACASA project to commercialize contour crafting in 2009, with Khoshnevis as the CTO. Khoshnevis stated in 2010 that his method could build a complete home in a single day and that its electrically powered crane would produce very little waste (Torabi et al., 2014). According to the Science Channel's Discoveries the Weekly show in 2005, contour crafts might drastically minimize environmental effects by reducing 3–7 tons of material waste and exhaust pollution from construction vehicles during conventional home construction (Khoshnevis et al., 2006). NASA has been examining Contour Crafting since 2010 for use in the construction of bases on Mars and the Moon (Leach et al., 2012). There have been efforts at using contour crafting in space applications by NASA (Khoshnevis et al., 2013, Yuan et al., 2016, Khoshnevis and Bekey, 2003).

6.7.3   OPERATION OF CONTOUR CRAFTING 3D PRINTING

The machine is made up of two metal tracks and gantry frames, as well as a robotic extruding mechanism. The machine receives input from 3D CAD software that may be changed even during execution, and it pours concrete layers around the perimeter of the construction using a nozzle. By the time the nozzle circumnavigates the entire building, special hardeners and fibers are used in the concrete mortar mix to make each layer hard enough to carry the next one. In addition to the required reinforcement, it may install all the plumbing, electrical, and air conditioning conduits. To avoid impeding the CC machine's movement, all of these are put in little portions embedded through layers of concrete that are being poured by it. Furthermore, the gantry frames may be adjusted to travel along the height of the structure, building one story at a time, and multinozzle contour crafting machines can be employed to build large buildings faster (Khoshnevis et al., 2012).

6.7.4 Merit of Contour Crafting 3D Printing

Contour crafting allows for the creation of concrete shapes that were previously impossible to produce with the traditional pouring method; as a result, dwellings may be built more efficiently and environmentally friendly by adjusting to changing weather conditions. Unlike traditional plaster handwork and sculpting, it can create a wide range of surface shapes and complex structures with only a few types of troweling tools (Khoshnevis, 1999). Due to the absence of surface discontinuities, it is also a rapid prototyping method that produces a superior surface quality that is incredibly smooth and accurate. Because the layers are thicker, the production of a component is much faster than other prototyping methods. Thermosets, thermoplastics, metals, cement, clay, concrete, and other materials are all treated with contour crafting. It is a sustainable and viable technology for digital and fast construction in the era of Industry 4.0 (Craveiroa et al., 2019).

6.7.5 Challenges of Contour Crafting 3D Printing

The mobility of the CC machine would be one of the first obstacles it would face in the future. Because of its size and weight, the machine must be connected to another equipment that can move it. Second, there are environmental challenges. How will these machines overcome environmental challenges such as uneven ground, weather effects, and so on? Some studies have come up with innovative solutions to this problem, such as a machine suspended by cables orientated by a Cartesian system (Bosscher et al., 2007). Third, due to its recent discovery, the cost of developing this technology remains an impediment. The costs of maintaining contour crafting equipment are likewise quite high. Perhaps CC progress might be accelerated if sponsors provided greater support. The construction sector will also face a significant problem in adapting to the move from traditional construction to the CC technique. One of the drawbacks of CC technology is that it is incompatible with traditional design (very complex reinforcement systems). Furthermore, it is critical to establish management guidelines for this type of building. What will be done about it? What method will be used to replace the materials? These are some of the issues that will need to be addressed in the future.

6.7.6 Inkjet 3D Printing

One of the most common technologies for additive manufacturing of ceramics is inkjet printing. It's utilized to print intricate and advanced ceramic structures for purposes like tissue engineering scaffolds. A solid ceramic suspension, such as zirconium oxide powder in water, is pumped and deposited in the form of droplets onto the substrate via the injection nozzle in this approach (Dou et al., 2011). The droplets then create a closed pattern that solidifies to a strength that allows succeeding layers of printed materials to be held in place. This process is quick and efficient, giving you more options when it comes to designing and printing complicated structures. Wax-based inks and liquid suspensions are the two primary types of ceramic inks. To solidify wax-based inks, they are melted and applied on a cold substrate. Liquid suspensions, on the other hand, are cemented by liquid evaporation. The quality of inkjet-printed parts is determined by characteristics such as ceramic particle size distribution, ink viscosity, and solid content, as well as extrusion rate, nozzle size, and printing speed (Travitzky et al., 2014). The main disadvantages of this approach are its lack of workability, coarse resolution, and lack of adhesion between layers. The basic method of additive manufacture of massive architectural structures is contour crafting, which is analogous to inkjet printing. Using larger nozzles and high pressure, this approach can extrude concrete paste or soil. Construction on the moon has been prototyped using contour crafting (Khoshnevis, 2004).

6.7.7   Vat Polymerization 3D Printer

A bed of photosensitive resin is hardened by a laser or digital projector in the vat polymerization process. These printers can print with significantly finer precision than FDM printers. In comparison to the FDM printer's precision of motors and nozzle size, their resolution is only limited by the laser or pixels in the projector. Another advantage is the way the layers are created. Instead of the visible layer lines observed in FDM printers, resin prints offer finer sides. Because separated layers are less of a concern, this results in stronger 3D-printed items. Postprocessing is needed on print to complete the finishing. In an alcohol bath, extra resin is usually removed. The curing process must subsequently be completed, which is commonly done using a UV light. Vat polymerization printers have numerous advantages. Unfortunately, their higher cost, additional processing, and potential risks with the resin and its fumes make them difficult to suggest as a first printer. A container filled with photopolymer resin is used in a 3D printer that uses the vat photopolymerization process. A UV light source is used to solidify the resin as shown in Figure 6.17.

The ASTM classification of vat photopolymerization is shown in Figure 6.18.

3)   Stereolithography (SLA)

SLA is the oldest additive printing technique. Stereolithography is part of the vat photopolymerization family of additive manufacturing processes. These machines all work on the same principle, curing liquid resin into rigid plastic with the use of a light source such as a laser or projector. The arrangement of the fundamental components, such as the light source, the construction platform, and the resin tank, is the main physical difference. A layer of photosensitive liquid resin is exposed to a UV-laser beam in this type of 3D printing. The sculpture is developed one layer at a time while the resin cures in the desired pattern as seen in Figure 6.19. This is a modified schematic from MakeShaper.

DLP printing is like SLA printing, with the exception that it cures resin using a digital projector and light. SLA 3D printing has seen major transformations.

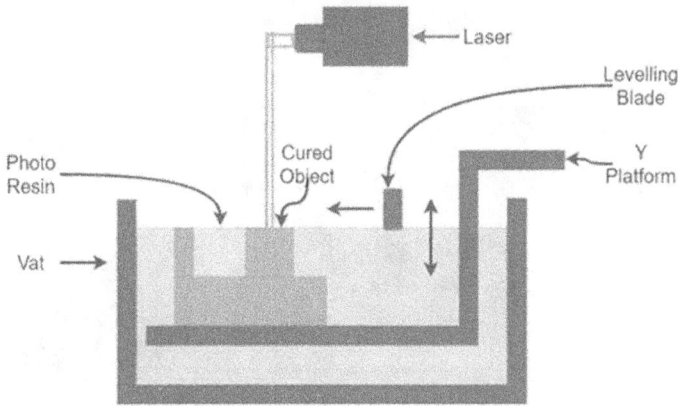

FIGURE 6.17   Schematic of vat photopolymerization.

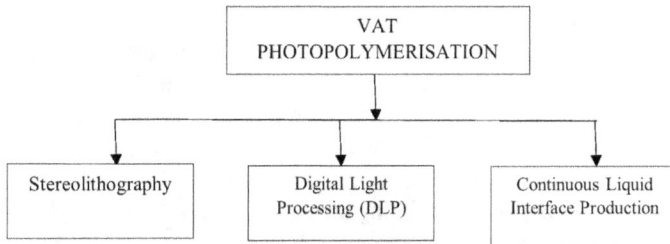

FIGURE 6.18   Schematic classification of vat photopolymerization process.

SLA machines have traditionally been monolithic and expensive, needing expert technicians and expensive service contracts. Small-format desktop printers now provide industrial-quality outputs at significantly lower costs and with unrivaled adaptability. Because of its capacity to generate high-accuracy, isotropic, waterproof parts in a variety of sophisticated materials with precise details and a smooth surface finish, SLA 3D printing has become extremely popular.

6.7.8   History of SLA 3D Printing

When Japanese researcher Dr. Hideo Kodama created the present layered technique of stereolithography in the early 1980s (Schmidleithner and Kalaskar, 2018), he used ultraviolet light to cure photosensitive polymers (Su and Al'Aref, 2018). Charles (Chuck) W. Hull invented the term stereolithography and patented the technology in 1986, founding 3D Systems to commercialize it (Hull, 2015). Hull described the approach as "printing" tiny layers of a UV-curable substance in successive layers to create 3D objects. SLA 3D printing, on the other hand, was not the first 3D printing method to become popular. With patents expiring at the end of the 2000s, the emergence of small-format desktop 3D printing extended access to additive manufacturing,

FIGURE 6.19 Schematic of stereolithography.

with fused deposition modeling (FDM) finding traction on desktop platforms initially. While this low-cost extrusion-based technology sparked the first wave of widespread adoption and awareness of 3D printing, FDM machines fell short of meeting a wide range of professional requirements, including repeatable, high-precision results, biocompatible materials in the dental industry, and the ability to create fine features for industries like jewelry and millifluidics.

SLA stands out among all 3D printing technologies as it has the highest resolution and precision, the finest details, and the nicest surface finishes, but SLA's main advantage is its versatility. SLA resin formulations with a wide variety of optical, mechanical, and thermal qualities that match those of standard, engineering, and industrial thermoplastics have been developed by material makers.

SLA printing has the disadvantage of requiring the object to be rinsed with a solvent after printing and, in some cases, baking in a UV lamp to complete the curing process.

6.7.9 Why SLA Is Preferred over Others

SLA is preferred over others due to the isotropy, water-tightness, precision and accuracy, material versatility, and fine features and surface finishing.

i. Isotropy: Because of the layer-to-layer variances caused by the print process, anisotropic print techniques like fused deposition modeling (FDM) are well known. This anisotropy either limits the usage of FDM for specific applications or necessitates more part geometry revisions to compensate for it. SLA printing, on the other hand, produces highly isotropic pieces. The integration of material chemistry with the print process allows for fine control of a variety of parameters that contribute to part isotropy. Resin components establish covalent bonds during printing, yet the part remains in a semireacted "green state" layer by layer. The resin contains polymerizable groups in the green state that can form bonds across layers, giving the part isotropy and water-tightness after final cure. There is no distinction between the X, Y, and Z planes at the molecular level. Parts having predictable mechanical performance, such as jigs and fixtures, end-use parts, and functional prototypes, are the result. Whether creating geometries with solid features or internal channels, SLA-printed parts are continuous.

ii   Water-tightness: Whether creating geometries with solid features or internal channels, SLA-printed parts are continuous. This water-tightness is critical for engineering and manufacturing applications that require precise control and predictability of air or fluid movement. SLA's water-tightness is used by engineers and designers to tackle air and fluid flow problems in automotive applications and biomedical research and to test part designs for consumer products such as kitchen appliances.

iii. Fine features and smooth surface finish: SLA printing is the gold standard for achieving a clean surface quality, with results that are equivalent to traditional production methods like machining, injection molding, and extrusion. Because parts can be easily sanded, polished, and painted, this surface quality is perfect for applications that demand a faultless finish. It also helps save postprocessing time. Leading corporations, such as Gillette, use SLA 3D printing to develop end-use consumer products, such as 3D-printed razors. In their Razor Maker platform, they have razor handles.

The Z-axis layer height is widely used to determine a 3D printer's resolution. On SLA 3D printers, this can be set between 25 and 300 microns, with a trade-off between speed and quality. FDM and SLS printers, on the other hand, typically produce Z-axis layers at 100 to 300 microns. An item produced at 100 microns on an FDM or SLS printer, on the other hand, does not look the same as a part generated at 100 microns on an SLA printer. Because the outermost perimeter walls of SLA prints are straight, and the newly produced layer interacts with the prior layer, smoothing out the staircase effect, the prints have a nicer surface quality directly out of the printer. FDM prints have visible layers, but SLS printers have a rough appearance. With an 85-micron laser spot size on the Form 3+, the smallest achievable detail is likewise significantly finer on SLA, compared to 350 microns on industrial SLS printers and 250- to 800-micron nozzles on FDM machines.

iv   Material versatility: Materials can be soft or rigid, highly packed with secondary materials like glass and ceramic, or imbued with mechanical qualities like high heat deflection temperature or impact resistance with SLA resins. Materials for prototype range from industry-specific materials, such as dentures, to materials that closely resemble final materials and are engineered to resist lengthy testing and perform under duress. In certain circumstances, it is because of this combination of versatility and functionality that businesses decide to implement SLA in-house. It's not long after one application is answered by a certain functional material that new possibilities are discovered, and the printer becomes a tool for harnessing the diverse capabilities of various materials.

6.7.10 THE NEXT PHASE OF SLA

Low-force stereolithography (LFS) is the next step in SLA 3D printing, and it meets the market's requirement for scalable, dependable, industrial-quality 3D printing. Using a flexible tank and linear illumination, this new version of SLA substantially decreases the stresses put on objects during the print process, resulting in remarkable surface quality and print accuracy. Lower print forces allow for light-touch support structures that pull away easily, and the approach opens a slew of new options for advanced, production-ready materials in the future.

4) Digital Light Processing

DLP, or digital light processing, is a printing technique that uses light and photosensitive polymers. While it is extremely like SLA, the light source is the main distinction. Other light sources, like arc lamps, are used in DLP. In comparison to other 3D printing methods, DLP is rather speedy.

5) Continuous Liquid Interface Production (CLIP)

Created by Carbon, it is one of the fastest vat photopolymerization methods. Digital light synthesis technology lies at the heart of the CLIP process. A sequence of UV pictures is projected by light from a proprietary high-performance LED light engine, exposing a cross-section of the 3D-printed object and allowing the UV-curable resin to partially cure in a finely regulated manner. The oxygen-permeable window creates a thin liquid contact of uncured resin between the window and the printed portion known as the dead zone when oxygen travels through it. The dead zone is only five microns thick. The presence of oxygen inside the dead zone prevents light from curing the resin closest to the window, allowing liquid to flow continuously beneath the printed object. The UV-projected light induces a cascade-like cure of the portion just above the dead zone.

End-use qualities with real-world applications cannot be achieved simply by printing using Carbon's technology. After the light has molded the item, a second programmable curing phase bakes the 3D-printed part in a thermal bath or oven to produce the appropriate mechanical qualities. The mechanical properties of a material are established by a subsequent chemical process that causes the substance to strengthen, resulting in the desired final qualities.

Carbon's technique allows components to be printed that are comparable to injection molded parts. Digital light synthesis creates completely isotropic parts with constant and predictable mechanical properties. Figure 6.20 gives the various resin 3D printing technologies.

C) Material Jetting

Similar to how a standard inkjet paper printer works, material is applied in droplets through a small diameter nozzle in this method, but it is deposited layer by layer to a build platform and then hardened by UV radiation as shown in Figure 6.21.

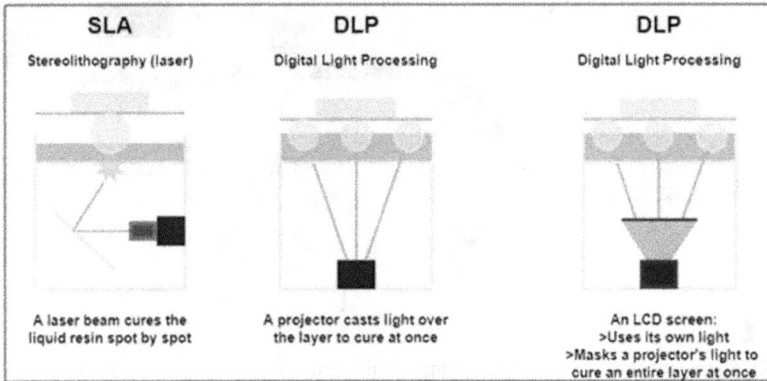

SLA	DLP	DLP
Stereolithography (laser)	Digital Light Processing	Digital Light Processing
A laser beam cures the liquid resin spot by spot	A projector casts light over the layer to cure at once	An LCD screen: >Uses its own light >Masks a projector's light to cure an entire layer at once

FIGURE 6.20   Resin 3D printing technologies.

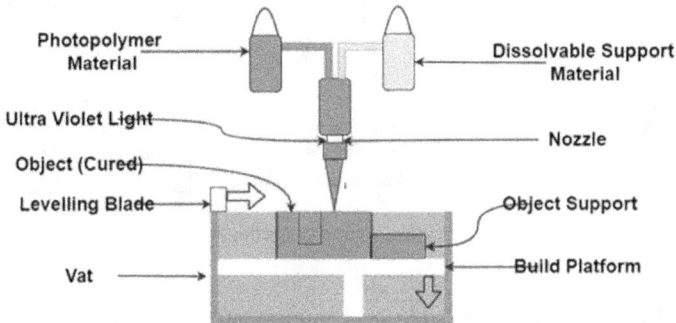

FIGURE 6.21   Schematic of material jetting.

D) Binder Jetting 3D Printing Technique

Binder jetting uses two types of materials: a powder base material and a liquid binder. Powder is dispersed in equal layers in the build chamber, and binder is sprayed by jet nozzles that "bond" the powder particles into the desired shape as shown in Figure 6.22.

The residual powder is cleaned when the print is completed, and it is frequently reused to print the following object. The Massachusetts Institute of Technology was the first to create this technology in 1993.

E) Powder Bed Fusion

Powder bed fusion (PBF) is an additive manufacturing technology that operates on the same basic idea as milling because objects are made by adding material rather than deleting it. The powdered material is dispersed on the bed and then melted selectively with a laser beam. The construction plate then falls to the ground, and the process begins again. During 3D printing, the parts are supported by the surrounding

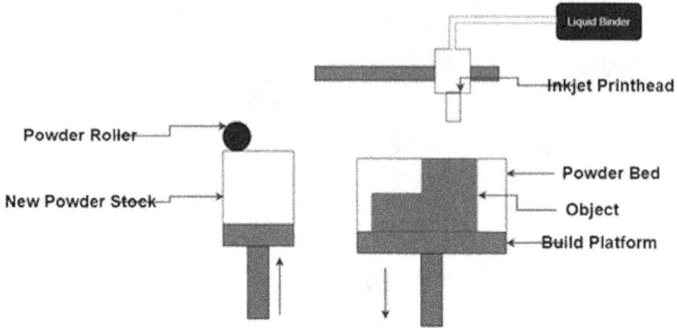

FIGURE 6.22   Schematic of binder jetting 3D printer.

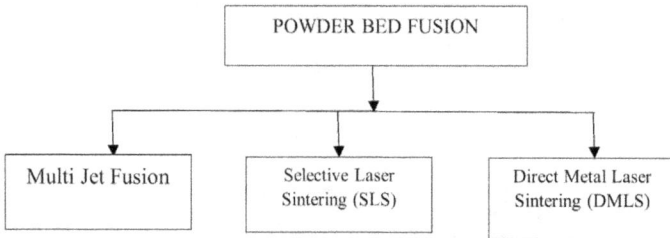

FIGURE 6.23   Schematic of powder bed fusion classification.

powder and do not require any additional support structures. Figure 6.23 shows the different types of powder bed fusion 3D printing technology.

Although machine prices are falling, a single laser powder-bed fusion system with a typical build volume (10 by 10 by 12 inches) still costs between $400,000 and $800,000.

F) Multi Jet fusion

Hewlett Packard invented Multi Jet Fusion technology, which uses a sweeping arm to deposit a coating of powder and then a second arm with inkjets to selectively apply a binder agent over the material. To achieve accurate dimensions and smooth surfaces, the inkjets additionally drop a detailing agent around the binder. Finally, a burst of thermal energy is applied to the layer, causing the agents to respond.

G) Selective Laser Sintering

It's an additive manufacturing technique that involves sintering microscopic particles of polymer powder into a solid structure based on a 3D model using a high-powered laser as shown in Figure 6.24.

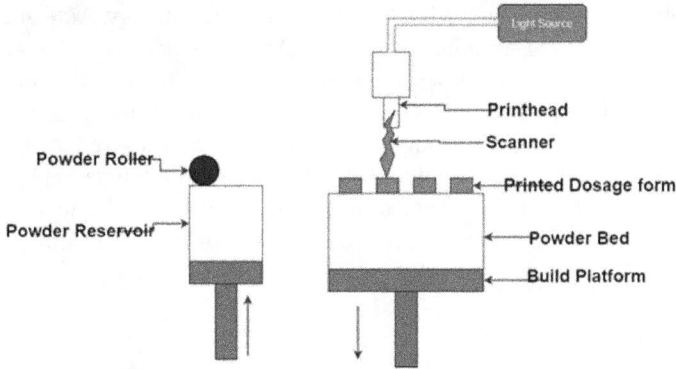

FIGURE 6.24   Schematic of selective laser sintering.

H)   Direct Metal Laser Sintering

DMLS is similar to SLS, but instead of water, it employs metal powder. All unused powder is left alone and serves as a support framework for the sculpture. Powder that has not been used can be reused for the next print. DMLS has evolved into a laser melting method as the laser power has risen. On our metal technologies overview page, you can learn more about that and other metal technologies.

I)   Directed Energy Deposition

The metal sector and quick production applications are the most common applications for this technique. A nozzle distributes metal powder or wire on a surface, and an energy source (laser, electron beam, or plasma arc) melts it, generating a solid item. The 3D printing apparatus is commonly attached to a multiaxis robotic arm. When it comes to producing midsize metal parts, a study indicated that DED is ten times faster and five times less expensive than powder bed fusion (PBF). The study compared the two processes for fabricating a 150 mm diameter, 200-mm-tall Inconel metal item.

J)   Sheet Lamination 3D Printing

Material in sheets is held together using external force in sheet lamination. Sheets can be made of metal, paper, or polymer. Metal sheets are fused together in layers using ultrasonic welding, then CNC machined into the desired shape. Paper sheets can also be used, but they must be adhered to the surface with adhesive glue and cut into shape using precise blades.

6.8   CONCLUSION

This chapter discussed a vital tool shaping every aspect of the economy known as additive manufacturing (3D printing). It lays a background for a better understanding of 3D printing by newbies and advanced users of the technology. A detailed explanation of the concept of 3D printing, history of 3D printing, process of 3D printing, and key components of 3D printers were discussed. Also the 3D printing procedure and advantages and disadvantages of 3D printing alongside the challenges and opportunities of 3D printing and classification of 3D printing were all covered. It lays the foundation for the materials and manufacturers of 3D printers discussed in Chapter 7. It will surely trigger the curiosity and learning sense of the reader due to the wide array of illustrative tools including pictures, charts, and vivid classification techniques employed.

REFERENCES

Ahmad, Z., Rahim, S., Zubair, M., & Abdul-Ghafar, J. (2021). Artificial intelligence (AI) in medicine, current applications and future role with special emphasis on its potential and promise in pathology: Present and future impact, obstacles including costs and acceptance among pathologists, practical and philosophical considerations. A comprehensive review. *Diagnostic Pathology, 16*, 1–16.

Akhnoukh, A. K. (2021). Advantages of contour crafting in construction applications. *Recent Patents on Engineering, 15*, 294–300.

Akinradewo, O., Oke, A., Aigbavboa, C., & Molau, M. (2021). Assessment of the level of awareness of robotics and construction automation in South African. In S. M. Ahmed, P. Hampton, S. Azhar, A. D. Saul (eds.), *Collaboration and Integration in Construction, Engineering, Management and Technology*. Springer.

Attarilar, S., Ebrahimi, M., Djavanroodi, F., Fu, Y., Wang, L., & Yang, J. (2021). 3D Printing technologies in metallic implants: A thematic review on the techniques and procedures. *International Journal of Bioprinting, 7*(7), 21–46.

Ayentimi, D. T., & Burgess, J. (2019). Is the fourth industrial revolution relevant to sub-Sahara Africa? *Technology Analysis & Strategic Management, 31*, 641–652.

Ayvaz, S., & Alpay, K. (2021). Predictive maintenance system for production lines in manufacturing: A machine learning approach using IoT data in real-time. *Expert Systems with Applications, 173*, 114598.

Bakhtiyari, A. N., Wang, Z., Wang, L., & Zheng, H. (2021). A review on applications of artificial intelligence in modeling and optimization of laser beam machining. *Optics & Laser Technology, 135*, 106721.

Bardwell, T. (2023). *The Best 3D Printers Under $5000 For Every Use Case (2023)*. Retrieved from www.3dsourced.com/3d-printers/best-3d-printers-under-5000/

Barham, L. (2013). *From hand to handle: The first industrial revolution*. Oxford University Press.

Bauer, G. R., & Lizotte, D. J. (2021). *Artificial intelligence, intersectionality, and the future of public health*. American Public Health Association.

Blinder, A. S. (2006). Education for the third industrial revolution. *American Prospect, 163*, 44–46.

Boehm, F., Graesslin, R., Theodoraki, M.-N., Schild, L., Greve, J., Hoffmann, T. K., & Schuler, P. J. (2021). Current advances in robotics for head and neck surgery—A systematic review. *Cancers, 13*, 1398.

Bosscher, P., Williams II, R.L., Bryson, L.S., & Castro-Lacouture, D. (2007). Cable-suspended robotic contour crafting system. *Automation in Construction, 17*(1), 45–55.

Buchanan, C., & Gardner, L. (2019). Metal 3D printing in construction: A review of methods, research, applications, opportunities and challenges. *Engineering Structures, 180*, 332–348.

Butler-Adam, J. (2018). The fourth industrial revolution and education. *South African Journal of Science, 114*, 1–1.

Chen, Z., Li, Z., Li, J., Liu, C., Lao, C., Fu, Y., Liu, C., Li, Y., Wang, P., & He, Y. (2019). 3D printing of ceramics: A review. *Journal of the European Ceramic Society, 39*, 661–687.

Choi, J.-W., & Kim, H.-C. (2015). 3D printing technologies-a review. *Journal of the Korean Society of Manufacturing Process Engineers, 14*, 1–8.

Cilliers, J. (2018). *Made in Africa-Manufacturing and the Fourth Industrial Revolution.* Institute for Security Studies.

Craveiroa, F., Duartec, J. P., Bartoloa, H., & Bartolod, P. J. (2019). Additive manufacturing as an enabling technology for digital construction: A perspective on Construction 4.0. *Sustainable Development, 4*(6), 251–267.

Davtalab, O., Kazemian, A., & Khoshnevis, B. (2018). Perspectives on a BIM-integrated software platform for robotic construction through Contour Crafting. *Automation in Construction, 89*, 13–23.

Dawood, A., Marti, B. M., Sauret-Jackson, V., & Darwood, A. (2015). 3D printing in dentistry. *British Dental Journal, 219*, 521–529.

Devezas, T., & Sarygulov, A. (2017). *Industry 4.0.* Springer.

Dou, R., Wang, T., Guo, Y., & Derby, B. (2011). Ink-jet printing of zirconia: coffee staining and line stability. *Journal of the American Ceramic Society, 94*, 3787–3792.

Drath, R., & Horch, A. (2014). Industrie 4.0: Hit or hype?[industry forum]. *IEEE Industrial Electronics Magazine, 8*, 56–58.

Duda, T., & Raghavan, L. V. (2018). 3D metal printing technology: The need to re-invent design practice. *Ai & Society, 33*, 241–252.

Fernandes, G., & Feitosa, L. (2015). Impact of contour crafting on civil engineering. *International Journal of Engineering Research & Technology, 4*, 628–632.

Fwaya, E. V. O., & Kesa, H. (2018). The fourth industrial revolution: Implications for hotels in South Africa and Kenya. *Tourism: An International Interdisciplinary Journal, 66*, 349–353.

Gibson, I., Rosen, D. W., Stucker, B., Khorasani, M., Rosen, D., Stucker, B., & Khorasani, M. (2021). *Additive manufacturing technologies.* Springer.

Goh, G., Sing, S., & Yeong, W. (2021). A review on machine learning in 3D printing: Applications, potential, and challenges. *Artificial Intelligence Review, 54*, 63–94.

Gopinathan, J., & Noh, I. (2018). Recent trends in bioinks for 3D printing. *Biomaterials Research, 22*, 1–15.

Gordelier, T. J., Thies, P. R., Turner, L., & Johanning, L. (2019). Optimising the FDM additive manufacturing process to achieve maximum tensile strength: a state-of-the-art review. *Rapid Prototyping Journal, 25*(6), 953–971.

Greenwood, J. (1997). *The third industrial revolution: Technology, productivity, and income inequality.* American Enterprise Institute.

Günther, D., Heymel, B., Günther, J. F., & Ederer, I. (2014). Continuous 3D-printing for additive manufacturing. *Rapid Prototyping Journal, 20*(4), 320–327.

Hager, I., Golonka, A., & Putanowicz, R. (2016). 3D printing of buildings and building components as the future of sustainable construction? *Procedia Engineering, 151*, 292–299.

Holland, J., Kingston, L., McCarthy, C., Armstrong, E., O'dwyer, P., Merz, F., & McConnell, K. (2021). Service robots in the healthcare sector. Robotics 2021, 10, 47. s Note: MDPI stays neutral with regard to jurisdictional claims in published … .

Houssein, E. H., Gad, A. G., Wazery, Y. M., & Suganthan, P. N. (2021). Task scheduling in cloud computing based on meta-heuristics: Review, taxonomy, open challenges, and future trends. *Swarm and Evolutionary Computation, 62*, 100841.

Hull, C. W. (2015). The birth of 3D printing. *Research-Technology Management, 58*, 25–30.

Hyun Park, S., Seon Shin, W., Hyun Park, Y., & Lee, Y. (2017). Building a new culture for quality management in the era of the Fourth Industrial Revolution. *Total Quality Management & Business Excellence, 28*, 934–945.

Jagoda, J. A. (2020). *An analysis of the viability of 3D-printed construction as an alternative to conventional construction methods in the expeditionary environment.* Air Force Institute of Technology Wright-Patterson AFB OH Wright-Patterson. https://scholar.afit.edu/etd/3240/

Janicke, M., & Jacob, K. (2013). A third industrial revolution. In B. Siebenhüner, M. Arnold, K. Eisenack, & K. H. Jacob (eds.), *Long-term governance for social-ecological change* (pp. 47–71). Routledge.

Jurcevic, P., Javadi-Abhari, A., Bishop, L. S., Lauer, I., Borgorin, D., Brink, M., Capelluto, L., Gunluk, O., Itoko, T., & Kanazawa, N. (2021). Demonstration of quantum volume 64 on a superconducting quantum computing system. *Quantum Science and Technology, 6*(2), 025020.

Kagermann, H. (2014). Industrie 4.0 und Smart Services. In W. Brenner & T. Hess (eds.), *Wirtschaftsinformatik in Wissenschaft und Praxis* (pp. 223–249). Springer Gabler.

Kagermann, H., Wahlster, W., & Helbig, J. (2013). *Recommendations for implementing the strategic initiative Industrie 4.0: Final report of the Industrie 4.0 Working Group.* Forschungsunion, Berlin, Germany.

Kanji, G. K. (1990). Total quality management: The second industrial revolution. *Total Quality Management, 1*, 3–12.

Kazemian, A., & Khoshnevis, B. (2021). Real-time extrusion quality monitoring techniques for construction 3D printing. *Construction and Building Materials, 303*, 124520.

Kazemian, A., Yuan, X., Cochran, E., & Khoshnevis, B. (2017). Cementitious materials for construction-scale 3D printing: Laboratory testing of fresh printing mixture. *Construction and Building Materials, 145*, 639–647.

Kazemian, A., Yuan, X., Meier, R., & Khoshnevis, B. (2018). A framework for performance-based testing of fresh mixtures for construction-scale 3D printing. *RILEM International Conference on Concrete and Digital Fabrication, 2018.* Springer, 39–52.

Kazemian, A., Yuan, X., Meier, R., & Khoshnevis, B. (2019). Performance-based testing of Portland cement concrete for construction-scale 3D printing. In J. G. Sanjayan, A. Nazari, & B. Nematollahi (eds.), *3D concrete printing technology.* Elsevier.

Khorramshahi, M. R., & Mokhtari, A. (2017). Automatic construction by contour crafting technology. *Emerging Science Journal, 1*, 28–33.

Khoshnevis, B. (1999). Contour crafting-state of development. *Solid Freeform Fabrication Proceedings, 1999.* 743–750.

Khoshnevis, B. (2004). Automated construction by contour crafting—related robotics and information technologies. *Automation in Construction, 13*, 5–19.

Khoshnevis, B. (2008). Caterpillar Inc. funds Viterbi'Print-a-House'construction technology. *Contour Crafting.*

Khoshnevis, B., & Bekey, G. (2003). Automated construction using contour crafting—Applications on earth and beyond. *Nist Special Publication Sp,* 489–494.

Khoshnevis, B., Carlson, A., Leach, N., & Thangavelu, M. (2012). Contour crafting simulation plan for lunar settlement infrastructure buildup. *Earth and Space 2012: Engineering, Science, Construction, and Operations in Challenging Environments.* 1458–1467.

Khoshnevis, B., & Hwang, D. (2006). Contour crafting. In A. Kamrani & E. A. Nasr (eds.), *Rapid prototyping* (pp. 221–251). Springer.

Khoshnevis, B., Hwang, D., Yao, K.-T., & Yeh, Z. (2006). Mega-scale fabrication by contour crafting. *International Journal of Industrial and Systems Engineering, 1*, 301–320.

Khoshnevis, B., Thangavelu, M., Yuan, X., & Zhang, J. (2013). Advances in contour crafting technology for extraterrestrial settlement infrastructure buildup. *AIAA SPACE 2013 Conference and Exposition, 2013*. 5438.

Knight, P., Bird, C., Sinclair, A., Higham, J., & Plater, A. (2021). Testing an "IoT" tide gauge network for coastal monitoring. *IoT, 2*, 17–32.

Lasi, H., Fettke, P., Kemper, H.-G., Feld, T., & Hoffmann, M. (2014). Industry 4.0. *Business & Information Systems Engineering, 6*, 239–242.

Leach, N., Carlson, A., Khoshnevis, B., & Thangavelu, M. (2012). Robotic construction by contour crafting: The case of lunar construction. *International Journal of Architectural Computing, 10*, 423–438.

Leenen, L., & Meyer, T. (2021). Artificial intelligence and big data analytics in support of cyber defense. In M. Khosrow-Pour (ed.), *Research anthology on artificial intelligence applications in security* (pp. 1738–1753). IGI Global.

Li, J., & Pumera, M. (2021). 3D printing of functional microrobots. *Chemical Society Reviews, 50*(4), 2794–2838.

Mhlanga, D. (2022). The role of artificial intelligence and machine learning amid the COVID-19 pandemic: What lessons are we learning on 4IR and the sustainable development goals. *International Journal of Environmental Research and Public Health, 19*, 1879.

Mohajan, H. (2019). The second industrial revolution has brought modern social and economic developments, *Journal of Social Sciences and Humanities, 6*(1), 1–14.

Muhammad, A., Ali, M. A., Shanono, I. H., & Abdullah, N. R. H. (2021). A systematic and bibliometric analysis on 3D printing published in scientific citation index-expanded indexed journals between 1999 and 2019. *Materials Today: Proceedings, 44*, 1739–1743.

Ndlovu, T., Root, D., & Wembe, P. (2020). A review of the advantages and disadvantages of the use of automation and robotics in the construction industry. In C. Aigbavboa & W. Thwala (ed.), *The Construction Industry in the Fourth Industrial Revolution: Proceedings of 11th Construction Industry Development Board (CIDB) Postgraduate Research Conference 11*, pp. 197–204. Springer International Publishing.

Ngo, T. D., Kashani, A., Imbalzano, G., Nguyen, K. T. & Hui, D. (2018). Additive manufacturing (3D printing): A review of materials, methods, applications and challenges. *Composites Part B: Engineering, 143*, 172–196.

Ninpetch, P., Kowitwarangkul, P., Mahathanabodee S., Chalermkarnnon, P., & Ratanadecho, P. (2020). A review of computer simulations of metal 3D printing. *AIP Conference Proceedings, 2020*. AIP Publishing LLC, 050002.

Parmar, H., Khan, T., Tucci, F., Umer, R., & Carlone, P. (2021). Advanced robotics and additive manufacturing of composites: Towards a new era in Industry 4.0. *Materials and Manufacturing Processes, 37*(5), 483–517.

Popkin, B. M. (1999). Urbanization, lifestyle changes and the nutrition transition. *World Development, 27*, 1905–1916.

Prasad, L. K., & Smyth, H. (2016). 3D Printing technologies for drug delivery: A review. *Drug Development and Industrial Pharmacy, 42*, 1019–1031.

Rouhana, C. M., Aoun, M. S., Faek, F. S., Eljazzar, M. S., & Hamzeh, F. R. (2014). The reduction of construction duration by implementing contour crafting (3D printing). *Proceedings of the 22nd Annual Conference of the International Group for Lean Construction, 2014*.

Sathish, T., Vijayakumar, M., & Ayyangar, A. K. (2018). Design and fabrication of industrial components using 3D printing. *Materials Today: Proceedings, 5*, 14489–14498.

Schmidleithner, C., & Kalaskar, D. M. (2018). *Stereolithography*. IntechOpen.

Schuh, G., Anderl, R., Gausemeier, J., Ten Hompel, M., & Wahlster, W. (2017). Industrie 4.0 maturity index. In G. Schuh, R. Anderl, R. Dumitrescu, A. Krüger, M. T. Hompel (eds.), *Managing the digital transformation of companies.* Herbert Utz.

Shahidinejad, A., Ghobaei-Arani, M., & Masdari, M. (2021). Resource provisioning using workload clustering in cloud computing environment: A hybrid approach. *Cluster Computing, 24*, 319–342.

Shahrubudin, N., Lee, T. C., & Ramlan, R. (2019). An overview on 3D printing technology: Technological, materials, and applications. *Procedia Manufacturing, 35*, 1286–1296.

Sharma, S., Ahmed, S., Naseem, M., Alnumay, W. S., Singh, S., & Cho, G. H. (2021). A survey on applications of artificial intelligence for pre-parametric project cost and soil shear-strength estimation in construction and geotechnical engineering. *Sensors, 21*, 463.

Singh, S., Ramakrishna, S., & Berto, F. (2020). 3D printing of polymer composites: A short review. *Material Design & Processing Communications, 2*, e97.

Stansbury, J. W., & Idacavage, M. J. (2016). 3D printing with polymers: Challenges among expanding options and opportunities. *Dental Materials, 32*, 54–64.

Stearns, P. N. (2020). *The industrial revolution in world history.* Routledge.

Su, A., & Al'aref, S. J. (2018). History of 3D printing. In J. K. Min, B. Mosadegh, S. Dunham, & S. J. Al'Aref (eds.), *3D Printing Applications in Cardiovascular Medicine.* Elsevier.

Sutherland, E. (2020). The fourth industrial revolution–the case of South Africa. *Politikon, 47*, 233–252.

Tack, P., Victor, J., Gemmel, P., & Annemans, L. (2016). 3D-printing techniques in a medical setting: A systematic literature review. *Biomedical Engineering Online, 15*, 1–21.

Tay, Y. W., Panda, B., Paul, S. C., Tan, M. J., Qian, S. Z., Leong, K. F., & Chua, C. K. (2016). Processing and properties of construction materials for 3D printing. *Materials Science Forum, 2016.* Trans Tech Publications, 177–181.

Torabi, P., Petros, M., & Khoshnevis, B. (2014). Selective inhibition sintering: the process for consumer metal additive manufacturing. *3D Printing and Additive Manufacturing, 1*, 152–155.

Travitzky, N., Bonet, A., Dermeik, B., Fey, T., Filbert-Demut, I., Schlier, L., Schlordt, T., & Greil, P. (2014). Additive manufacturing of ceramic-based materials. *Advanced Engineering Materials, 16*, 729–754.

Tselegkaridis, S., & Sapounidis, T. (2021). Simulators in educational robotics: A review. *Education Sciences, 11*, 11.

Vicentini, F. (2021). Collaborative robotics: A survey. *Journal of Mechanical Design, 143*, 040802.

Vrontis, D., Christofi, M., Pereira, V., Tarba, S., Makrides, A., & Trichina, E. (2021). Artificial intelligence, robotics, advanced technologies and human resource management: a systematic review. *The International Journal of Human Resource Management, 33*(6), 1237–1266.

Wang, X., Jiang, M., Zhou, Z., Gou, J., & Hui, D. (2017). 3D printing of polymer matrix composites: A review and prospective. *Composites Part B: Engineering, 110*, 442–458.

Wei, Q., Li, H., Liu, G., He, Y., Wang, Y., Tan, Y. E., Wang, D., Peng, X., Yang, G., & Tsubaki, N. (2020). Metal 3D printing technology for functional integration of catalytic system. *Nature Communications, 11*, 1–8.

Wu, P., Wang, J., & Wang, X. (2016). A critical review of the use of 3-D printing in the construction industry. *Automation in Construction, 68*, 21–31.

Yahyaoui, A., Abdellatif, T., Yangui, S., & Attia, R. (2021). READ-IoT: Reliable event and anomaly detection framework for the internet of things. *IEEE Access, 9*, 24168–24186.

Yan, Q., Dong, H., Su, J., Han, J., Song, B., Wei, Q., & Shi, Y. (2018). A review of 3D printing technology for medical applications. *Engineering, 4*, 729–742.

Yuan, L., Ding, S., & Wen, C. (2019). Additive manufacturing technology for porous metal implant applications and triple minimal surface structures: A review. *Bioactive Materials, 4*, 56–70.

Yuan, X., Zhang, J., Zahiri, B., & Khoshnevis, B. (2016). Performance of sulfur concrete in planetary applications of contour crafting. *2016 International Solid Freeform Fabrication Symposium*. University of Texas at Austin.

Zhong, Q.-C. (2017). Synchronized and democratized smart grids to underpin the third industrial revolution. *IFAC-PapersOnLine, 50*, 3592–3597.

Zhou, L. Y., Fu, J., & He, Y. (2020). A review of 3D printing technologies for soft polymer materials. *Advanced Functional Materials, 30*, 2000187.

7 Materials and Manufacturers of 3D Printing Technology

7.1 MATERIALS FOR 3D PRINTING

The 3D printing industry has evolved over time, with new innovations being launched on a regular basis. The technology of 3D printing is a versatile one that has found applications in various sectors. There is a need to know the type of material to use, where to use them, and the type of 3D printer to use. Various types of materials called filament and resin are used for 3D printing.

New 3D printers are also being developed to print a variety of materials, including plastics, metals, composites, and many more. There are many materials to choose from when it comes to industrial 3D printing. These materials have distinct characteristics, strengths, and shortcomings. Furthermore, essential elements such as material type, texture, pricing, and others must be considered to avoid 3D printing errors. It can be tough to select the best material for a particular job. Attempts have been made to classify the materials used for 3D printing into metals and alloys, ceramics, polymers and composites, concrete, and hybrid, among others, as shown in Figure 7.1. The market is growing, and new materials are emerging.

7.2 POLYMER 3D PRINTING

A polymer is a natural or manmade substance made of long molecules called macromolecules that are multiples of smaller chemical units known as monomers. Polymers are found in many living organisms and are the building blocks of numerous minerals and manmade things. High strength or modulus to weight ratios (lightweight yet comparatively stiff and strong), toughness, resilience, corrosion resistance, lack of conductivity (heat and electrical), color, transparency, processing, and low cost are some of the important qualities of many engineering polymers (Brinson and Brinson, 2015; James, 2019). Extrusion, resin, and powder 3D printing technologies can be used to produce polymer materials, giving designers more options for material selection and supporting designs with a variety of topologies, responses, and layouts (Arefin et al., 2021). The 3D printing hardware market is dominated by polymer 3D printers. The polymer 3D printers are in the lead on all fronts, including shipment revenues, installed base, and the number of developments in the space. In 2020, it

DOI: 10.1201/9781003364481-7

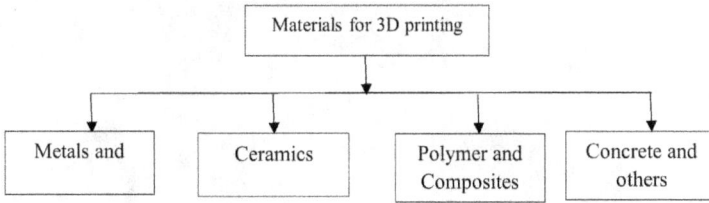

FIGURE 7.1 Classification of materials used for 3D printing.

is expected that polymer 3D printing will produce $11.7 billion in revenue, which includes hardware, materials, and 3D-printed parts sales.

Polymer 3D printing is a new technique that is seeing greater application in industry, particularly in the medical field, thanks to recent research (Egan, 2019; Ligon et al., 2017). Polymer printing has the advantage of allowing low-cost functional parts with a wide range of qualities and capabilities to be printed. Polymers with favorable mechanics and biocompatibility have been developed as a result of material research, with mechanical qualities tuned by changing printing process settings. Extrusion, resin, and powder 3D printing are suitable polymer printing methods that enable directed material deposition for the design of beneficial and customizable architectures. Design solutions such as hierarchical material distribution enable the balancing of competing features in tissue scaffolds, such as mechanical and biological requirements.

Polymers have merits beyond metal printing techniques, which result in metal implants that do not dissolve in the body and cause mechanical concerns including stress shielding (Seaman et al., 2017). Polymer-printed lattices offer efficient energy absorption in safety equipment with a quick fabrication approach that avoids the supply chain constraints of bulk manufacture (Erickson et al., 2020). Extrusion, resin, and powder 3D printing technologies can be used to produce polymer materials, giving designers more options for material selection and supporting designs with a variety of topologies, responses, and layouts (Mao et al., 2015). Because of the huge design freedom provided by 3D polymer printing, as well as its potential for improved medical applications, it is becoming increasingly popular. Polymer 3D printing research is critical for expanding engineering and medical frontiers.

Polymer material capabilities for 3D printing are influenced by their molecular structures, as well as how the material is processed during printing (Arefin et al., 2021). Choice of materials for design applications is frequently made based on measurable attributes, such as mechanical parameters, with ranges based on manufacturing and testing methods, which adds to the difficulty of predicting part performance during the system design.

Some examples of polymer 3D printed materials include acrylonitrile butadiene styrene (ABS), acrylic-based (Stratasys: MED 610), methacrylic acid, polycarbonate (PC), polyether ether ketone (PEEK), polyethylene terephthalate glycol (PETG), polylactic acid (PLA), and polyamide 12 (Nylon). Different 3D printing processes use different polymers as the 3D printing material. These include fused deposition modeling (FDM) (ABS, PETG, PC, PEEK, PLA) (Kannan and Ramamoorthy, 2020),

FIGURE 7.2   A picture of filaments used for 3D printing.

Multi Jet Fusion (nylon), polyjet (acrylic-based) (Egan et al., 2019), and stereo-lithography (epoxy-based and methacrylic acid) (Chantarapanich et al., 2013; Moniruzzaman et al., 2020).

7.2.1 Acrylonitrile Butadiene Styrene (ABS)

Figure 7.2 shows a picture of a filament used in 3D printing. In the field of 3D printing, ABS has a lengthy history. One of the earliest plastics to be utilized with industrial 3D printers was this substance. ABS is still a popular material due to its low cost and outstanding mechanical qualities years later. ABS is known for its toughness and impact resistance, allowing users to print long-lasting parts that can withstand heavy use. ABS also has a greater glass transition temperature, which implies it can resist far higher temperatures before deforming. ABS is a suitable choice for outdoor or high-temperature applications because of this. Because ABS has a slight odor, it should be printed in an open location with sufficient ventilation. ABS also tends to contract as it cools, so keeping an eye on the temperature of the build volume and the item within can be extremely beneficial. The advantage includes low-cost, high-impact, and wear-resistant material models with less oozing and stringing have a smoother finish. Heat resistance of ABS is excellent. Printing with ABS produces a lot of warping. It is necessary to have a heated bed or a heated chamber. While printing, it has a foul stench, and parts tend to shrink, resulting in dimensional inaccuracies.

7.2.2 Polylactic Acid (PLA)

It is manufactured from sugarcane or corn starch, both of which are renewable resources. It's also known as "green plastic." Because it is safe to use and print with, it is primarily utilized in primary and secondary schools.

PLA is a simple material to print with. It has a low printing temperature, doesn't require a heated bed (though one does assist), and is less prone to warping. Another advantage of PLA is that it does not emit an unpleasant odor when printing (unlike ABS). Furthermore, it is a 3D printing material that is acceptable for single-use food contact. PLA, on the other hand, is less durable than ABS or PETG and is more vulnerable to heat. As a result, you'll be better off with the latter for any form of engineering part. Avoid using it for constructing phone covers, high-wear toys, or tool handles that may be bent, twisted, or frequently dropped.

FDM desktop printing also makes use of it. PLA is one of the most extensively used materials in desktop 3D printing. Because it can be produced at a low temperature and does not require a heated bed, it is the default filament for most extrusion-based 3D printers. PLA has a temperature range of 190–220°C. No matter the geographic location, PLA will never melt in the sun. PLA, however, is less heat resistant than other filaments like ABS, PET, or PETG and is generally not suggested for applications that require prolonged exposure to the sun. The problem of the nonsticking of the layers is usually solved by heating up the hot end. Although, precaution should be taken so that the extruder does not become too hot, which could result to the PLA filament being too soft and frail.

PLA prints that are preserved and used indoors will endure almost indefinitely if they are not subjected to strong mechanical loads. When kept indoors, an object constructed with PLA should last at least 15 years, according to anecdotal evidence. Gifts and decorative items printed with PLA would not pose a difficulty under these settings. PLA is one of the least harmful fibers available, whereas nylon is one of the most toxic. By using an enclosure and an air filter, you can reduce the toxicity. PLA is degradable but not biodegradable. Enzymes that hydrolyze PLA are only found in relatively uncommon instances in the environment. PLA is Generally Recognized as Safe (GRAS) when used in contact with food, according to research. PLA does release a small quantity of lactic acid, according to the study, but it's a common food ingredient, and the amount isn't enough to harm humans. PLA cannot be dissolved by isopropyl alcohol. Polymaker, on the other hand, makes Ploysmooth, a PLA filament that can be dissolved in isopropyl alcohol. Table 7.1 gives the depiction of ABS and PLA filaments.

7.2.2.1 Demerit of PLA for 3D Printing

PLA has a low glass transition temperature (between 111°F and 145°F). As a result, it's not ideal for high-temperature applications. Even something as simple as a heated car in the summer can soften and deform parts.

7.2.3 POLYCARBONATE (PC)

It is the undisputed king of desktop 3D printing materials. The strength of polycarbonate is shocking. Polycarbonate's tensile strength of 9,800 psi, compared to nylon's 7,000 psi, makes it the best material for high-strength, functional components.

7.2.4 NYLON FOR 3D PRINTING

The most prevalent plastic substance is nylon (also known as polyamide), which is a synthetic thermoplastic linear polyamide (Chen, 2020). Because of its flexibility,

TABLE 7.1
A Depiction of ABS and PLA Filaments Features

ABS	PLA
• A thermoplastic filament made of acrylonitrile butadiene styrene is called ABS.	• PLA is an easier material to print with since it melts at a lower temperature, reduces warping.
• It is slightly more durable in general and is better suitable for instances where the printed object must withstand high temperatures.	• This improves bed adherence and enables for a finer printing detail.
• ABS also can be smoothed out with acetone.	• PLA is perfect for 3D printing that requires a high level of aesthetics.
• ABS is best used in applications that require a high level of strength, ductility, machinability, and thermal stability.	• It is easier to print with because of the lower printing temperature and hence more suited for pieces with fine details.
• ABS is more prone to warping than other materials.	
• More technical than PLA, PET-G	
• A beginner-friendly material.	
• It is a suitable material for functional prototypes since it can resist greater temperatures and has a chemical composition that's ideal for these applications.	

durability, minimal friction, and corrosion resistance, it is a well-known 3D printing filament. Nylon is also a widely utilized fabric in the production of clothing and accessories.

When crafting complicated and delicate shapes, nylon is a good choice. It's mostly utilized as filament in 3D printers that use FDM or FFF (fused filament fabrication). This is a low-cost substance that is known for being one of the toughest plastics. Although, it is also toxic.

7.2.4.1 Merits of Nylon for 3D Printing

The durability of nylon is well known. Its strength-to-flexibility ratio is good. Nylon contains a negligible amount of warpage. This type of material is simple to dye or color.

7.2.4.2 Demerits of Nylon for 3D Printing

Nylon should be kept dry because it is hygroscopic. It has a 12-month shelf life. Because this material shrinks when it cools, prints may be less exact. The suitability of a printer differs as well.

7.2.5 PETG 3D Printing

PETG stands for polyethylene terephthalate glycol, a long name for a glycol-enhanced version of regular PET plastic. The glycol in this material keeps the filament from becoming hazy and weak, as well as improving the aesthetics of the finished product. PETG, or glycolized polyester, is a thermoplastic that combines the ease of PLA 3D printing with the robustness of ABS in the additive manufacturing market. It is simple to print with because it does not require an enclosure or a heated bed, making it compatible with almost any FDM printer. It's an excellent filament for any print because it's easy to print and has a lot of strength and durability. PETG filaments are easy to print, have a smooth surface finish, and are resistant to water. When compared to the ever-popular PLA, PETG is a very popular 3D printing filament because of its great strength, relative flexibility, and temperature resistance. It is measured in the same ways as ABS, but it is a lot easier to work with and comes with the extra benefit of being food safe. PETG adheres extraordinarily well to PEI surfaces, making PEI an excellent choice for PETG printing. Because PETG can attach to PEI excessively in some situations, use a powder-coated PEI surface instead of a smooth one, or apply a separating agent (such as hair spray) on smooth PEI. Ender 3's bed temperature should be set between 50°C and 60°C due to the high melting temperature of PETG. Ender 3 standard develops the BuildTak surface area, which is the ideal surface area for printing PETG with Ender 3. A heated bed of 70°C to 80°C is needed for PETG printing. Heat the heated bed no higher than the PETG glass transition temperature (80°C). It is not necessary to print pieces in PETG in a heated chamber.

7.2.5.1   Merits of PETG 3D Printing

It is strong and affordable. PETG is perfect for glazing and high-strength display units because of its strength and impact resistance. It is recyclable. PETG is suitable for usage in food containers and beverage bottles. It's simple to shape. It's simple to color. Emissions from it are nontoxic and odorless. PETG's chemical resistance is the main cause for its widespread use in food, beverage, and consumer products packaging. PETG is both durable and flexible, making it preferable to materials like PLA in terms of durability. PETG can produce better results in direct sunlight than other filaments like PLA and ABS since it can survive UV rays considerably better. PETG can be utilized for a variety of purposes and can even be maintained in the car.

7.2.5.2   Demerits of PETG 3D Printing

It is inexpensive. Because PETG is such a robust and resistant material, achieving a clean finish during postprocessing might be more difficult. Heat treatment or manual sanding and polishing are two options because the product is naturally resistant to mechanical and chemical finishing processes. PETG has a softer exterior, which can have undesirable consequences like stickiness. When surplus material is melted and adheres to the layer, the stickiness can cause stringing on the product. This can be remedied by melting and removing the strings using a heat gun, or by slowing down the cooling process to prevent the strings from permanently adhering to the object.

PETG's stickiness can cause the material to fuse to the print bed. To avoid damage to the print bed, it is necessary to treat the printing surface with a releasing agent. The reactivity of PETG to temperature and humidity is another disadvantage.

7.3   METALS FOR 3D PRINTING

3D printing may be used to make objects out of a variety of various metals in powder form. Steel, stainless steel, aluminum, copper, cobalt chrome, titanium, tungsten, and nickel-based alloys, as well as precious metals like gold, platinum, palladium, and silver, are all accessible in powdered form for 3D printing. These distinct metals have different qualities, allowing them to be used in a variety of applications. Stainless steel, for example, has exceptional corrosion resistance and is hence good for printing pipes, valves, and steam turbine parts.

Any metal that is available as a powder can theoretically be utilized for 3D printing. Materials that burn rather than melt at high temperatures cannot be safely treated by sintering or melting, but they can be utilized for 3D printing when extruded through a nozzle. These techniques are unable to 3D print wood, fabric, or paper.

Sintering (forming inside a mold at high temperature and extremely high pressure) can also be used to make solid products from metal powders, and in the case of metals with exceptionally high melting temperatures, sintering is the only reliable way of producing items from these materials.

Metals may be 3D printed using a variety of processes. Direct metal laser sintering (DMLS), selective laser melting (SLM), and electron beam melting (EBM) are the most extensively utilized powder bed fusion processes for metal additive manufacturing. Selective laser sintering (SLS), laser metal deposition (LMD), binder jetting, and metal injection molding are some of the other techniques that can be used.

A laser is used to sinter metal powder layer by layer to build an object in the DMLS method. The method is used for prototyping and manufacturing final products, such as medical devices and instruments, without melting the metal.

In an inert gas environment, the SLM technique includes utilizing a laser to melt the material where it is needed within a layer of powder. This is done layer by layer to create objects with parameters that are comparable to those created by casting. Parts made of aluminum and titanium, such as those for the medical, automotive, and aerospace industries, are frequently manufactured using SLM.

The EBM technique is like SLM, except that instead of a laser, an electron beam is used to melt the material. EBM is widely utilized to create cobalt and titanium goods because it is thought to be faster and more exact than SLM. The aircraft sector uses EBM for a variety of things, including engine components.

In the aerospace, automotive, and medical industries, LMD is used to create items by depositing hot metal on a metallic substrate layer by layer. LMD is faster than other approaches because it permits diverse materials to be utilized to construct an object.

SLS sinters powdered materials with the use of a laser. It's been used to make things out of a variety of materials, including metal. SLS is now used for nylon and polyamide.

Binder jetting is less expensive than DMLS, SLM, or EBM since it employs a specific solvent to bond the powder material. This method's accuracy and strength are not perfect; thus postprocessing is frequently required. Hot isostatic pressing can be employed to improve the finished object's strength and rigidity, but it comes at a cost. Binder jetting is most used to create large-scale, complex prototypes.

Metal injection molding is a technique that combines injection molding with 3D printing to create small components in a variety of industries, including medical and defense. Metal powder is mixed with thermoplastic and wax binders in this technique. This mixture is heated until the binder melts and the powder is covered, after which it is granulated into pellets. Before the binder substance is removed, usually via solvent extraction, these pellets are heated and fed into a cavity to form the item. After that, the part is sintered, which evaporates any residual binder and compresses the piece into a thick solid. After that, the object can be finalized as needed.

Metal 3D printing technology has found applications across industry, from inserts with cooling channels to lightweight structures for the aerospace industry to complex parts for use in high-demand environments, thanks to its ability to combine the flexibility of 3D printing with the mechanical properties of metal. Fully functioning prototyping, creation of production tools, and tooling for molds or inserts, housings, ductwork, heat exchangers, and heatsinks are all common applications. Among the 49% of enterprises working in metal 3D printing, the number of companies selling 3D printing systems increased from 49 in 2014 to 97 in 2016 (Wohlers, 2017).

The various metals used in 3D printing, their unique properties, applications, and the type of 3D printing used are tabulated in Table 7.2.

Due to its excellent corrosion resistance, stainless steel is ideal for products that will come into contact with corrosive substances, water, or steam. Pump impellers and marine propellers, as well as fittings and more aesthetic things like vases, are all made of bronze. Gold is a metal that can be used to print jewelry. Nickel is a metal that can be used to make turbine engine parts and even coins. Aluminum is ideal for metal objects, particularly where lightness is required, such as in aircraft parts.

TABLE 7.2
Summary of 3D Metal Printing Process

Name of metal	Property	Application	Method of printing
Stainless steel (316L)	excellent corrosion resistance	Water, steam, etc.	DMLS, PBF, SLM
Bronze	Aesthetics	Pump impellers and marine propellers, fittings, vase	DMLS
Nickel and alloy	Durability	Turbine engine, coins	SLS, PBF
Aluminum and alloy	Lightness	Aircraft parts	SLM, PBF
Titanium (mainly Ti-6Al-4V)	Precision, high performance, lightweight and robustness	Medical implants, aerospace, etc.	SLM, EBM, PBF
Gold	Aesthetics	Jewelry	PBF
Cobalt	Lightweight	Defense and aviation	EBM
Steel (austenitic stainless steel, maraging, tool and precipitation hardenable stainless steel)	Corrosion resistance, ductility	Simple to complex parts	DMLS, Powder Bed Fusion

Titanium can make extremely robust and precise parts such as medical implants (for example, hip joints) and other solid fixtures and items.

7.4 MANUFACTURERS OF METAL 3D PRINTERS

Metal 3D printers have seen an increase in both supply and demand in recent years. Metal additive manufacturing machines are being introduced by manufacturers that are faster, easier to use, and more powerful, with a growing range of compatible metals. Many businesses are turning to 3D metal printing to generate low-cost metal components and prototypes, as well as to take advantage of the additional design freedom that additive manufacturing provides. They can be used in a range of industries, including aerospace, automotive, health, engineering, and others. Even though metal 3D printer prices have been steadily reducing, these machines are still relatively expensive, ranging from $80,000 to over $1 million. Table 7.3 shows the list of top metal 3D printers for 2022 adapted from (Cherdo, 2022).

Metal powder bed fusion 3D printing (SLS, SLM, DMP), directed energy deposition (DED), metal filament extrusion (FFF, FDM), and material jetting and binder jetting are the various kinds of 3D metal printing technology. It's not uncommon for similar technology to have distinct acronyms and names. Each company promotes its own set of methods. Some metal 3D printer businesses employ a combination of technologies. Figure 7.3 gives the distribution of metal 3D printers as adapted from Cherdo (2022).

The value in Figure 7.3 shows the 3D printer models per technology in existence. It can be seen that powder bed fusion dominates the 3D printer market.

TABLE 7.3
List of Top 3D Printer Manufacturers with 3D Printer Details

S/N	Product	Manufacturer	Country	Dimensions	Cost
1.	XM200C	Xact Metal	United States	127 × 127 × 127 mm	$ 65,000
2.	uPrinter	Additec	Germany	160 × 120 × 450 mm	$ 90,000
3.	Metal X (Gen 2)	MarkForged	United States	300 × 220 × 180 mm	$ 99,500
4.	CREATOR	Coherent	Germany	–	$ 100,000
5.	Studio 2	Desktop Metal	United States	300 × 200 × 200 mm	$ 110,000
6.	Pam Series MC	Pollen AM	France	300 × 300 mm	$ 140,000
7.	TruPrint 1000	TRUMPF	Germany	100 × 100 × 100 mm	$ 170,000
8.	Metal 3D printer	Rapidia	Canada	200 × 280 × 200 mm	$ 185,000
9.	L-Series	Formalloy	United States	1000 × 1000 × 1000 mm	$ 200,000
10.	DMP Flex 100	3D Systems	United States	100 × 100 × 80 mm	$ 245,000
11.	EOS M 100	EOS	Germany	100 × 100 × 95 mm	$ 350,000
12.	EVEMET 200	SISMA	Italy	200 × 200 mm	$ 350,000
13.	Carmel 700M	XJet	Israel	501 × 140 × 200 mm	$ 599,000

Source: Adopted from Cherdo, 2022.

Rapidia, based in Canada, offers a new and intriguing approach to 3D printing metal. They utilize a water-based metal paste that does not require chemical debinding. Because the water evaporates throughout the 3D printing process, the object just needs to go through the furnace to solidify and achieve its final shape. Stainless steels, Inconel, and a few ceramics have all been confirmed as accessible paste kinds. Copper, tungsten, chrome carbide, titanium, and a variety of other metals are currently being researched.

Metal 3D printing filament is typically made composed of metal particles mixed with a binding agent in extrusion-based metal 3D printers (FFF, FDM). The result of 3D printing is a raw item or part that requires various postprocessing stages to achieve its final form, such as debinding and sintering.

Powder bed fusion 3D printing is now the most widely used metal 3D printer. Simply defined, the 3D printer uses a powerful laser to produce objects out of a bed of powdered metal. Some of the top performers include DMP Flex 100, Creator, EOS M 100, and Arcam EBM Spectra L, among others. The DMP Flex 100 is a metal 3D printer that is quick, precise, and inexpensive. Part repeatability and surface finishes of roughly 20 m and 5 Ram, respectively, are excellent. DMP is the abbreviation for Direct Metal Printing. 3D Systems' 3DXpert All-in-One Software Solution for Metal Additive Manufacturing is included with the printer. Their metal 3D powders, LaserForm, are accredited. Before being acquired by Coherent in 2018, the CREATOR was developed by a German company called OR Laser. It is a reasonably priced metal 3D printer, costing roughly $100,000.

Material jetting 3D printers have a variety of inkjet printheads that jet material onto a surface (similar to 2D printing). After the material has hardened, a layer of "metal ink" is shot on top. Binder jetting is like powder jetting, except it involves jetting a binding agent on top of a coating of powder. This technology includes 3D printers such as Production System P-1 produced by Desktop Metal, DM P2500 by Digital Metal, and LIGHTSPEE3D by SPEE3D, among others.

DED is similar to filament extrusion. Like FFF/FDM, the metal material is driven via a specific nozzle, but at the application site, a strong laser beam hardens the substance. The printers are uPrinter by Additec and L-series by Formalloy are the known available at the moment. As a "low-cost" laser metal 3D printing device, through its 600W print head, it can 3D print both metal powder and metal wire (three 200W laser diodes). The printer has a small footprint and an inert chamber system, and it can handle wire widths of 0.6 mm to 1 mm. Metal DED 3D printers with up to 5 axes of motion are available from Formalloy. They are used for making and repairing or clad existing metal pieces. There are three alternative laser wavelengths and build volumes to choose from: 200 × 200 × 200 mm, 500 × 500 × 500 mm, and 1000 × 1000 × 1000 mm. Formalloy's metal 3D printers can be tailored to meet specific business needs. It costs around $200,000.

7.4.1 Merits of Metal 3D Printing Process

Items with complex shapes can be produced more quickly than with traditional manufacturing methods. For some items, it is less expensive than many traditional

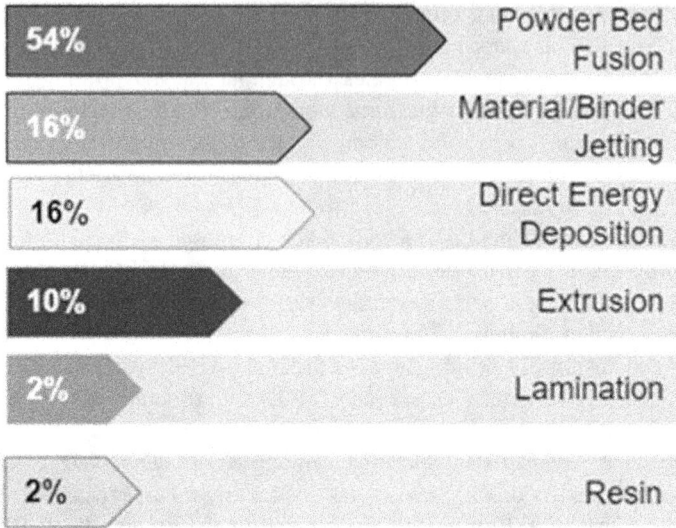

54%	Powder Bed Fusion
16%	Material/Binder Jetting
16%	Direct Energy Deposition
10%	Extrusion
2%	Lamination
2%	Resin

FIGURE 7.3   Distribution of metal 3D printers for 2019.

production methods. It is capable of being used to manufacture highly detailed and exact items. When compared to more traditional manufacturing methods, it can save time and money because details can be added throughout the assembly process. 3D metal printing allows for complex geometries to be formed in order to build lighter things without sacrificing strength, making it perfect for automotive, aeronautical, and space applications. There was very little waste of materials. A difficult assembly's multiple elements can be merged into a single component, decreasing part count and assembly costs.

7.4.2   DEMERITS OF 3D METAL PRINTING

Parts built for traditional manufacturing are slow to make, making high-volume production uncompetitive on price alone. Metals that have been powdered are more expensive than metals that have not been powdered (e.g., billet or bar). Metal 3D printers are not cheap. The typical cost of an SLS or SLM printer is $550,000. However, depending on the features, it can cost up to $2,000,000. 3D-printed objects may require surface finishing and postprocessing. Precision and tolerance are lower than with specialized CNC machining. To minimize internal tensions in a 3D-printed item or attain maximum metal strength, heat treatment may be required. The design of 3D metal parts can be difficult, necessitating the assistance of skilled CAD engineers. The build volume of the 3D printer limits the size of the parts.

Metal 3D printing material is available in a variety of formats to accommodate various metal 3D printing procedures. Powder, wire, and filament are the most frequent. Metal 3D printing resin and metal sheets for lamination-based 3D printers are also available.

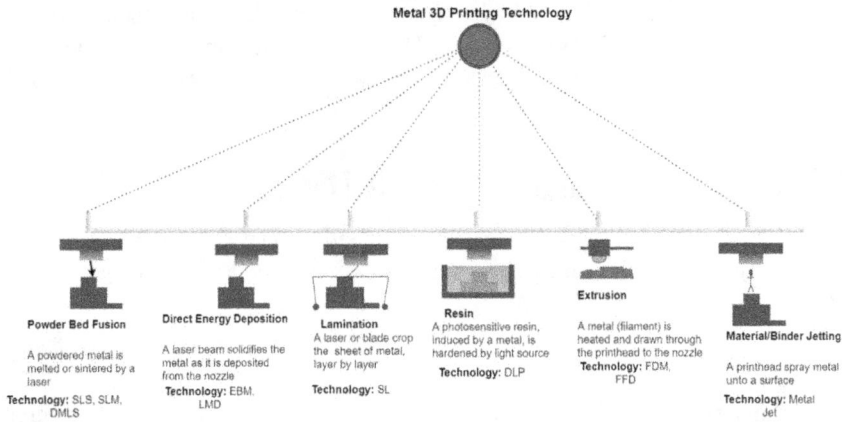

FIGURE 7.4 Summary of metal 3D printing technologies.

Metals and metal alloys can now be 3D-printed in increasing numbers. These are the most important: aluminum, nickel, titanium, Inconel, copper, bronze, cobalt, cobalt-chrome, precious metals (gold, silver, platinum), and steels (tooling, maraging, and stainless). Figure 7.4 shows the summary of metal 3D printing technologies adapted from Cherdo (2022).

7.4.3 STAINLESS STEEL FOR 3D PRINTING

Fusion or laser sintering is used to create stainless steel. This material can be produced using two different processes, viz. DMLS and SLM technologies. Stainless steel is ideal for miniatures, bolts, and key chains because it combines strength and detail.

7.4.3.1 Merits of Stainless Steel for 3D Printing

Heat treatment can be used to improve the strength and hardness of stainless steel. It works effectively in applications that require a lot of strength. It has a high level of corrosion resistance. It possesses a high degree of ductility.

7.4.3.2 Demerits of Stainless Steel for 3D Printing

The building time for 3D printing with these metals is significantly longer. Stainless steel printing is costly. The print size is restricted.

7.4.4 ALUMINUM AND ALLOYS

For a variety of reasons, only a few Al alloys are currently used in 3D printing. They are easy to process and have a cheap cost when compared to Ti alloys (Brice et al., 2015). As a result, business interest in their use in 3D printing has waned. Furthermore, some high-performance Al alloys are difficult to weld (because of the high volatility of some of their constituent components, such as Zn) (Bartkowiak et al., 2011), and Al has a high reflectivity for the laser wavelengths employed in 3D printing. Furthermore, due to the low viscosity of molten Al, a large melting pool is not possible; therefore, PBF is chosen over DED for manufacturing. On the plus side, Al's strong thermal conductivity decreases internal thermal stresses and speeds up 3D printing processes.

7.5 CERAMICS 3D PRINTING TECHNOLOGY

Ceramic resin is a photopolymer that contains silica. The photopolymer network burns out during firing, resulting in a genuine ceramic item. Fired parts have unique features.

7.5.1 RESINS

Ceramic materials offer unique qualities that are incredibly valuable in both high-tech manufacturing and art, even though there are hundreds of metals and plastics to 3D-print with. 3D printing adds to the appeal of this material because it not only allows the creation of parts with shapes that are not currently possible with any other method, but it is also far more cost-effective and faster than traditional methods of producing ceramic parts, especially for industrial applications.

One of the most used materials in 3D printing is resin. It's mostly employed in technologies like SLA, DLP, Multi Jet, and CLIP. There are many different types of resins that can be used in 3D printing, including castable resins, strong resins, and flexible resins, among others. UV-curing liquid plastics are known as resin or synthetic resin. Stereolithography 3D printers use these materials. The photosensitive resin can be turned into a solid model by curing it with a laser or light. The 3D printer exposes the liquid resin at the required spots, curing the resin.

3D printing ceramics has sparked a new generation of painters, sculptors, and architects who are drawn to the ornate and intricate shapes that would be impossible or too time-consuming to create using traditional clay methods. Designers are inventing new ways to work with ceramics, from pottery to architectural design applications. Ceramic artists like Jonathon Keep, Kate Blacklock, and Emre Can have been employing 3D printing technologies to create gorgeous and distinctive ceramic works for museum and private collections. Jonathon uses computer design tools to produce his art, which he then 3D-prints in clay on a custom-built delta-arm 3D printer. The ceramic work is fired and glazed in the conventional manner after printing.

A nozzle can extrude a wide range of clay materials, including ceramic and terracotta, to form final shapes. Everything from Kaolin and porcelain clay to stoneware and terracotta, and even concrete, can be extruded by 3D printers.

Extrusion, also known as FDM, is not the only way to 3D print ceramic art. Binder jetting, stereolithography, and DLP are used by designers to create intricate ceramic objects from a liquid ceramic slurry.

A variety of materials can be printed with 3D printers. All that is usually required is a toolhead change or material presetting, and you're good to go. Even though these machines are not specifically designed for ceramic 3D printing, they can "officially" generate ceramic parts right out of the box.

7.5.1.1 Characteristics

The material has the following features that distinguish it. It has a wide range of applications. It shrinks very little. Chemical resistance is high in resin materials. This is a hard and delicate material. Composite resins are less brittle than ceramics, but they wear out more quickly at the edges; thus they may not last as long as a bonded ceramic restoration. When restoring teeth with low biting pressures, composite resins are employed. They can also be used as intermediate restorations when planning whole-mouth restorative situations, while vat polymerization can generate extremely high-resolution prints.

In recent years, 3D printers have been used to extrude terracotta to construct one-of-a-kind roof tiles, artificial reefs, and even complete buildings. They have been used to print building facades, including a 3D-printed ceramic pavilion by Ceramic Morphologies that explores the design possibilities and expressive potential of ceramic 3D printing, and a blue ceramic archway in the historic Dutch city of Delft, which is world famous for its white and blue porcelain.

7.5.2 Manufacturers of Ceramic 3D Printers

A detailed description of the top-performing ceramic 3D printers including their manufacturers, technology used, pricing, and cost of origin is shown in Table 7.4.

The 3D PotterBot series is one of the first ceramic 3D printers created exclusively for the pottery industry. The 3D PotterBot series enormous capacity ram extruder produces large volumes of undiluted ceramic materials, allowing for a continuous flow and exact prints. By eliminating the traditional hose feeding method, latency and hysteresis induced by hoses are eliminated. The continual extrusion enables exact layer deposition throughout the print, and the extruder itself propels the paste without the necessity of additional power or air compressors. DeltaBots (based in Port Saint Lucie) created them specifically for larger ceramic pots. The Scara V4 can print to a height of 1 meter! Currently, there are only six ceramic 3D PotterBot printers available, ranging in price from $3250 to $27,950.

WASP, an Italian 3D printer company, is noted for its massive delta 3D printers. The Delta WASP 40100 Clay costs $10,000. This is a great buy considering the functionality (features, speed, and resolution). It attracted news when it was used to construct houses and shelters in underdeveloped nations utilizing their large delta systems. WASP first entered the ceramic printing market with the LDM clay extruder technology, which can be utilized with their printers as well as a few third-party 3D printers. This method, when correctly set up, provides some of the most exact paste extrusion ceramic 3D printing available. The WASP clay kit costs approximately $760. After some time, the company realized that some consumers would prefer an original ceramic 3D printer that worked right out of the box. The WASP 2040 Clay and WASP 40100 Clay are both clay 3D printing–specific extrusion-based delta machines.

StoneFlower, a German company, offers ceramic 3D printing for prototyping and small production in the arts, design, and architecture. The StoneFlower 3.0 Multimaterial printer can print objects up to 480 × 480 × 500 mm in a variety of

TABLE 7.4
Summary of Top-performing Ceramic 3D Printers

Product	Manufacturer	Country	Cost	Dimension	Technology
DeltaBots 3D PotterBot	DeltaBots	United States	$3250 to $27,950	490 × 490 × 710 mm	Extrusion
Delta WASP 2040 & 40100 Clay	WASP	Italy	$10,000	400 × 200 × 0.5 mm	Extrusion
StoneFlower 3.0	Maximo Soalheiro	Germany	$3850	500 × 500 × 800 mm	Extrusion
Cerambot Eazao & Delta	Eazao	Shandong, China		370 × 390 × 470 mm	Extrusion
VormVrij 3D Lutum (Lutum 4M and 5M)	VormVrij 3D	Netherlands	$3700 to $7400	430 × 450 × 500 mm	Extrusion
FORM 3	FORMLAB	Somerville, Massachusetts, USA	$3499	125 × 125 × 165 mm	SLA

paste-like or viscous liquids, including plaster, pulp, wax, and even soldering paste. It has a built size of $500 \times 500 \times 800$ mm. This printer, which comes with a 5-inch touchscreen control panel, can also be fully operated in real time through Wi-Fi.

A clay extruder add-on system is also available for use with third-party 3D printers. A ram extruder, a print head, and a control unit are included in each kit. The StoneFlower 3.0 out-of-the-box 3D printer costs $3850, while the regular kit costs roughly $680. It uses extrusion technology.

Even though it is still in the Kickstarter stage, it is now accepting orders to fund the development of its devices. The Cerambot Eazao appears to be a good candidate. Cerambot's second ceramic 3D printer has been crowdfunded. The Eazao has already surpassed its $10,000 fundraising goal. The original Cerambot is a delta printer; however, the business claims that the Eazao is more stable and faster than the prior model. The Cerambot Delta costs $499, but the Eazao starts at $799. Cerambot also provides a bigger Delta Pro variant.

The Formlabs Form 2 3D printer is unique in our list because it prints with the company's ceramic resin rather than an extruded material. This machine is a must-have for every engineer's office, dental lab, or jewelry designer's store, as it truly transforms thoughts into reality. It can even be used to make tiny amounts of products. It uses SLA technology (which is renowned to give more details than extruder technology) to make exact ceramic parts at a fraction of the cost of industrial equipment, thanks to the introduction of its ceramic resin.

7.5.2.1   Demerits of Resins for 3D printing

It's not cheap. This sort of filament also has an expiration date. Because of its great photoreactivity, it must be stored safely. It can produce early polymerization when exposed to heat. One major disadvantage is the resin expense. The cost of standard 3D printing resin for these machines varies a lot, but on average, it costs around $50 per kilogram.

7.5.3   TECHNICAL CERAMICS 3D PRINTING

Technical ceramics (also known as industrial or engineering ceramics) have little to do with clay, unlike design ceramics. They occur in a range of formulations, ranging from the most common, aluminum oxide (Al_2O_3), to zirconium oxide (ZrO_2), sometimes known as "ceramic steel." Tricalcium phosphate (TCP) is a functional substance used in implants and bone replacement, whereas silicon carbide (SiC) and boron carbide (B_4C) are utilized in oil and gas and automotive applications, respectively. Technical ceramics, unlike design ceramics, are rarely employed with extrusion technology. Binder jetting, stereolithography (SLA), and digital light processing (DLP), as well as ceramic powder-based selective laser sintering and selective laser melting, are becoming more popular. Table 7.5 shows the top-performing ceramics 3D printers with a detailed description, manufacturers, technology used, pricing, and cost of origin.

The 3DCeram Ceramaker is an industrial 3D printer manufactured by 3DCeram, a French company. The Ceramaker is primarily designed for jewelry and medical uses,

TABLE 7.5
Summary of Top Technical Ceramic 3D Printers

Product	Manufacturer	Country	Cost (USD)	Dimension	Technology
3DCeram Series	3DCeram	France	300,000	300 × 300 × 110 mm	DLP (SLA)
Lithoz CeraFab S Series (S25, S65, and S230)	Lithoz	Austria	585,000 (S230)	S25 is 64 × 40 × 320 mm. S65 is 102 × 64 × 320 mm. S230 is 192 × 120 × 320 mm	DLP
Tethon 3D Bison 1000	Tethon3D	USA	17,995	100 × 60 × 138 mm	Binder jetting
Admatic Admaflex 130 & 300	Admatec	Netherlands	250,000	200 × 200 × 300 mm (130), 1282 × 1900 × 1000 mm (300)	DLP
ExOne X1 160Pro	ExOne	Germany	150,000	800 × 500 × 400 mm	Binder jetting
Voxeljet VX200 and VX1000	Voxeljet	Germany	3250 to 27,950	1,000 × 600 × 500 mm (VX1000), 1,700 × 900 × 1,500 mm (VX200)	Binder jetting
XJet Carmel	XJet	Israel	599,000	3100 × 1850 × 2120 mm	NanoParticle jetting

but it can also print components for the electronics industry. Photocurable ceramic paste, such as alumina, zirconia, or hydroxyapatite, is 3D-printed using this industrial additive manufacturing technique. The Ceramaker uses stereolithography 3D printing technology to 3D-print this material. First, a thin coating of photocurable ceramic paste is applied to the construction platform by the 3D printer. The Ceramaker then uses a precision UV laser to selectively solidify the material. The process continues layer after layer, resulting in a green (unfinished) object. To reach its final form, the green portion must go through cleaning, debinding, and sintering operations. The connectivity includes Wi-Fi, USB, Ethernet, SD card, and Bluetooth.

Israel-based XJet uses a material jetting method called NanoParticle Jetting to take a novel approach to industrial ceramic 3D printing. Ceramic materials such as zirconia and alumina are available in sealed cartridges from XJet. The Carmel 1400C, unlike other ceramic printers, does not employ ceramic powder or a ceramic slurry vat. Instead, hundreds of tiny ink jets spray a ceramic liquid substance and a second soluble support liquid onto the build plate. The Carmel 1400C has a $1,400 \text{ cm}^2$ construction tray that allows many ceramic items to be produced at the same time. Parts are transferred into XJet's SMART (Support Material Automatic Removal Technology) station after printing, which automates the removal of soluble supports after printing. Finally, the pieces are sinterized overnight.

The Admatec 130 was the company's first machine, released in September 2016. It was created with the intention of 3D printing precise and finely detailed ceramic items utilizing the DLP 3D printing method. Later, the business produced a metal 3D printing add-on for the Admaflex 130. Admaflex 300 (released November 2019) improves on the prior system's formula by increasing the build volume. It is compatible with Admatec's metal add-on for metal DLP 3D printing and has a modular design. Users may create their own materials, set up a customized 3D printing process, and have complete control over the system thanks to the modular design. This will make it easier for Admatec to accept all future advances, such as multimaterial printing. Admatec's new machine features a proprietary automated feedstock system that allows for high-volume production with continuous throughput. It's made to work with materials that have a lot of viscosity, such as ceramic slurries. To enable 3D printing of massive components, the system also employs an autonomous vision-controlled dosage mechanism. The integrated in-process quality monitoring system, which was first launched by Admatec as an add-on for the Admaflex 130 earlier this year, is another major feature of the Admaflex 300 3D printer. It allows users to track the entire 3D printing process, including temperature, humidity, and foil usage, as well as view the print job in real time, layer by layer. A dual-camera system and integrated sensors are among the hardware and software components that enable this. Users can also halt a print operation if a problem is detected.

7.6   CONCRETE 3D PRINTING TECHNOLOGY

Cementitious 3D construction printing (3DCP), often known as 3D concrete printing, is a type of additive manufacturing that allows for the fabrication of wholly new geometries not previously feasible with standard concrete formwork. "Additive construction" is another title for this method (Kreiger et al., 2019). "3D Concrete" refers to concrete extrusion technologies, whereas autonomous robotic construction system (ARCS), Large Scale Additive Manufacturing "LSAM," and Freeform Construction "FC" relate to other sub-groups. Extrusion (concrete/cement, wax, foam, polymers), powder bonding (polymer bond, reactive bond, sintering), and additive welding are the most common 3D concrete printing procedures .

3D-printed concrete is a particular sort of concrete that can only be utilized in 3D printers for building. Because 3D printing does not require traditional technology such as formwork or vibration, 3D-printed concrete combines the benefits of spray concrete and self-compacting concrete (Zhang et al., 2019). ICON's patented concrete compound, Lavacrete, is used only in the machine. The Vulcan II, like most large-scale concrete 3D printers, is pricey, costing slightly under $250,000. A construction 3D printer might cost anywhere from $180,000 to over $1 million.

In recent years, the technology has grown in prominence, with numerous new companies springing up, some of which are supported by well-known figures from the construction industry and academics. This has resulted in several significant 3D printing milestones, including the first building, a printed bridge, part of a public building, a housing structure in Europe and the Commonwealth of Independent States (CIS), and a fully approved building in Europe (COBOD International).

A variety of technologies have been demonstrated, including building and construction component fabrication on-site and off-site, using industrial robots, gantry systems, and tethered autonomous vehicles. Fabrication of dwellings, building components (cladding and structural panels and columns), bridges and civil infrastructure (France-Presse, 2017), artificial reefs, follies, and sculptures have all been demonstrated utilizing construction 3D printing technologies.

7.6.1 HISTORY OF CONCRETE 3D PRINTING TECHNOLOGY

The history of concrete 3D printing spans several decades, starting from 1950 to the present, as shown in Figure 7.5. The 1950s heralded seeding technologies. This is followed by the early development era from 1995 to 2000. This gave way to the first generation from 2000 to 2010. The present generation started in 2010 and is called the second generation.

In the 1950s, robotic bricklaying was conceived and investigated, and in the 1960s, pumped concrete and isocyanate foams were used to create related technology for automated building (Papanek and Fuller, 1972). In the 1980s and 1990s, Shimizu and Hitachi pioneered the development of automated manufacturing of entire buildings utilizing slip-forming techniques and robotic assembly of components, like 3D printing, to solve the dangers of creating high-rise structures (Design, 2008). Because of the building "bubble," their inability to respond to novel architectures, and the challenges of feeding and preparing supplies to the site in congested regions, many of these early initiatives for on-site automation failed. The early development started around 1995. Joseph.B. Gardiner (2011) created a sand/cement forming technique that used steam to selectively connect the material in layers or solid portions, albeit this technology was never tested. The second technique, contour crafting by Behrohk Khoshnevis, was first patented in 1995 as a novel ceramic extrusion and shaping method as an alternative to emerging polymer and metal 3D printing techniques

(Khoshnevis, 2011; Khoshnevis and Hwang, 2006). Khoshnevis realized that this technique could go beyond these techniques, where "current methods are limited to fabrication of part dimensions that are generally less than one meter in each dimension." Around the year 2000, Khoshnevis's team at USC Vertibi began focusing on large-scale 3D printing of cementitious and ceramic pastes, which included and explored automated integration of modular reinforcement, built-in plumbing, and electrical utilities, all within one continuous build process. To date, this technique has only been tried on a lab scale, and it is rumored to have formed the basis for recent Chinese attempts.

This is followed by the first generation, which started in 2000 and ended in 2010. Rupert Soar received funds and established the freeform construction group at Loughborough University in the United Kingdom in 2003 to investigate the possibility of scaling up existing 3D printing technology for construction applications. Early research identified the difficulty of achieving any meaningful break-even for the technology at the construction scale but also suggested that there might be ways of application that dramatically improve the value proposition of integrated design (many functions, one component). The group received money in 2005 to create a large-scale construction 3D printing machine using "off the shelf" components (concrete pumping, spray concrete, gantry system) to see how complex such components may be and whether they could meet construction demands. Enrico Dini of Italy patented the D-Shape technology in 2005, which uses a massively scaled powder jetting/bonding approach over a $6 \times 6 \times 3$ m area (Dini, 2012; Russell et al., 2005). Although this process was developed with an epoxy resin bonding system, it was later modified to use inorganic bonding agents. Commercially, this technology has been employed for a variety of projects in construction and other industries. The printing of a bridge, the first of its kind in the world, in conjunction with IaaC and Acciona, is one of the most recent developments. In 2008, Richard Buswell and colleagues at Loughborough University in the United Kingdom started 3D Concrete Printing to expand the group's previous research and look to commercial applications, moving from a gantry-based technique to an industrial robot.

The current era is known as the second generation. In 2014, Buswell's team was successful in licensing that robotic technology to Skanska (Paolini et al., 2019). On January 18, 2015, the company unveiled two buildings with 3D-printed components: a mansion-style residence and a five-story tower (Rimmer, 2021; Yin et al., 2018). A close examination of the photographic documentation indicates that they were built using both traditional precast and 3D-printed components. They are thought to be the first fully functional buildings created using 3D printing technology. A new "office building" inaugurated in Dubai in May 2016, a 250-square-meter (2,700-square-foot) facility was billed as the world's first 3D-printed office

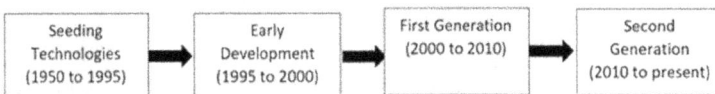

Seeding Technologies (1950 to 1995)	→	Early Development (1995 to 2000)	→	First Generation (2000 to 2010)	→	Second Generation (2010 to present)

FIGURE 7.5 Trends of concrete 3D printing technologies.

structure by Dubai's Museum of the Future (Holt et al., 2019). In 2017, the United Arab Emirates unveiled an ambitious initiative to construct a 3D-printed skyscraper (Pekuss and García de Soto, 2020; Wang et al., 2020). Cazza Building would assist in the construction of the structure (Bogue, 2017).

James B Gardiner and Steven Janssen of Laing O'Rourke invented FreeFAB Wax construction company (Gardiner and Janssen, 2014). Since March 2013, the unique technique has been in development (Gardiner and Janssen, 2014). The method employs construction-scale 3D printing to manufacture a "quick and dirty" 3D-printed mold for precast concrete, glass fiber-reinforced concrete (GRC), and other sprayable/castable materials by producing high quantities of engineered wax (up to 400 L/h) (Jipa and Dillenburger, 2021). To generate a high-quality mold, the casting surface is then five-axis machined, removing roughly 5 mm of wax (with approximately 20-micron surface roughness). When compared to traditional mold technologies, the mold is crushed or melted after curing, with the wax purified and reused, resulting in less waste (Teymouri, 2017). Fast mold fabrication, increased production efficiencies, decreased labor, and virtual waste elimination through material reuse are all advantages of custom-made molds.

In September 2015, the US Army Corps of Engineers' Engineer Research Development Center in Champaign, Illinois, supervised by the Construction Engineering Research Laboratory (ERDC-CERL), began research into deployable construction 3D printer technology. As a result of the success of this effort, ERDC-CERL has developed an additive construction program. Automated Construction for Expeditionary Structures (ACES) was a pilot project that focused on concrete 3D printing and encompassed a wide range of research topics (Diggs et al., 2021). Printing systems, printable concrete materials, structural design and testing, and construction procedures were among the topics covered. The US Army Maneuver Support, Sustainment, and Protection Experiments produced three demonstrations: an Entry Control Point, the first Reinforced Additively Constructed Concrete Barracks, and the printing of civil and military infrastructure (Jersey barriers, T-walls, culverts, bunkers, and fighting positions) at the ACES project (MSSPIX) (Jagoda et al., 2020). In 2017, ERDC-CERL began working with the US Marine Corps, which resulted in the first military display of concrete 3D printing, a substantially strengthened reinforced 3D printed concrete Barracks Hut built in 40 hours (Guimarães et al., 2021), the first 3D-printed bridge in America (Ali et al., 2021) and the first demonstration of printing with a 3-inch nozzle (Khandelwal et al., 2021). ERDC and the Marines used the project to ascertain the structural performance of reinforced 3D printed concrete wall assemblies and bridge beams, as well as print system resilience and maintenance cycles, extended printing operations, and the well-publicized 24-hour building claim.

MX3D Metal, founded by Loris Jaarman and his colleagues, has developed two 6-axis robotic 3D printing technologies (Grozdanic, 2014). The first employs an extruded thermoplastic to fabricate freeform nonplanar beads. The second is an additive welding-based technology (essentially spot welding on previous spot welds). Various groups have created additive welding technologies in the past, but the MX3D metal system is the most complete to date. For 6 years, MX3D worked on the manufacturing and installation of the metal bridge in Amsterdam (Evjemo et al., 2017).

The pedestrian and cycling bridge were constructed and opened in July 2021. The bridge has a 12 m (39 ft.) span and a stainless-steel final mass of 4,500 kg (9,900 lb) (Gardner et al., 2020).

BetAbram is a 3D printer for concrete extrusion that uses a simple gantry system and was created in Slovenia (Molitch-Hou, 2020). Since 2013, this system has been commercially available, with three types (P3, P2, and P1) offered to consumers (Zukas and Zukas, 2015). The P1 can print objects up to $16 \times 9 \times 2.5$ m in size (Nieto and Molina, 2020). Rudenko developed a total custom concrete 3D printer that uses concrete for the deposition process installed on a gantry (Kreiger et al., 2015). The device produces similar results to Shanghai-based Winsun and other concrete 3D printing technologies; however, it uses a lightweight truss-type gantry. In the Philippines, the technique was utilized to build a backyard-scale version of a castle and a hotel room (Florea et al., 2019) with the castle having a dimension of 10 m high and 30 m wide and Lewis Grand Hotel of dimension 130 m^2 (Pastia, 2020).

SPECAVIA, situated in Yaroslavl, launched the world's first serial manufacturing of construction printers (Poluektova, 2020). The firm announced the commencement of sales for the first model of a construction 3D printer in May 2015. Since the beginning of 2018, the "AMT-SPECAVIA" group of firms has produced seven different types of portal construction printers, ranging from tiny format (for printing modest architectural forms) to big scale (for printing buildings up to three floors). Today, Russian-made building 3D printers bearing the "AMT" trademark are in use in several countries, including Europe, where the first construction printer was delivered in August 2017 to 3DPrinthuset (Denmark) (Poluektova, 2019). This printer was used in Copenhagen to manufacture the EU's first 3D-printed building (office-hotel 50 m^2) (Manju et al., 2019; Pessoa et al., 2021; Sharanova and Dmitrieva, 2019). This office is known as Europe's first 3D-printed office space (Adeeb Fahmy Hanna, 2019; Pessoa et al., 2021).

XtreeE has created a multicomponent printing system that is attached to a 6-axis robotic arm (Kuzmenko et al., 2019). XtreeE is a 3D printer company that focuses on architectural design and highly customizable parts (Youssef et al., 2021). For large-scale additive manufacturing, XtreeE provides an integrated digital design-build system. Saint-Gobain, Vinci, and LafargeHolcim collaborated on the project, which began in July 2015 and boasts strong names in the construction sector as collaborators and investors (Barjot, 2013; Issa, 2021). With its sister firm COBOD International, a successful Danish 3D printing startup, which created its own gantry-based printer in October 2017, 3DPrinthuset, has also moved into construction (Bos et al., 2022; Weger et al., 2022). The company's spin-off immediately gained traction when it constructed the first 3D-printed house in Europe, thanks to the partnership of prominent companies in the Scandinavian region, such as NCC and Force Technology. The Building on Demand (BOD) project, as the structure is known, is a modest office hotel in Copenhagen's Nordhavn district, with entirely printed walls and part of the foundation, and traditional construction for the rest of the structure (Manju et al., 2019). The COBOD technology is capable of being applied in the energy sector for the actualization of the green economy (Bogue, 2021).

SQ4D with its first-of-its-kind infinite footprint design S-Squared ARCS VVS NEPTUNE with its Gantry system 9.1×4.4 from the United States was named the best 3D homebuilder of 2019. S-Squared 3D Printers is a Long Island, New York–based 3D printer manufacturer and retailer. The company, which began operations in 2014, manufactures 3D printers for hobbyists, libraries, and STEM programs. In 2017, the business established S-Squared 4D Commercial, a new branch that uses its ARCS 3D printing gear to construct houses and commercial buildings (De Schutter et al., 2018). Robert Smith and Mario Szczepanski cofounded this bootstrapped business, which now employs 13 people.

The ARCS is a 20-by-40-foot environmentally friendly concrete printer that can construct a 1,490-square-foot home in 36 hours. The system can construct houses, businesses, roads, and bridges. ARCS can handle projects ranging in size from 500 square feet to over 1 million square feet. Mario Cucinella Architects and WASP, a 3D printing company, showed the first 3D printing of a house constructed of a clay mixture called Tecla in 2021 (Romani et al., 2021; Schweiker et al., 2021).

7.6.2   CLASSIFICATION OF CONCRETE 3D PRINTING

Contour crafting (CC), concrete printing, and D-shape are three of the most used concrete 3D printing technologies (Buswell et al., 2018; Duballet et al., 2017; Perrot and Amziane, 2019). Some innovative approaches, such as rock printing, have recently been proposed (Ali et al., 2022). The first multistory residence was printed by the Eindhoven University of Technology in the Netherlands in 2015 (Gijsbers and Lichtenberg, 2014), and it met all applicable national regulations. In addition, the first office building printed in Dubai, which was completed in 2016, has been put to use. Table 7.6 gives the parameters of the various concrete 3D printing process as adapted from Zhang et al. (2019).

The cost structure of traditional construction has changed because of 3D printing. Labor, materials, and equipment are the key components of traditional construction costs. Labor costs contribute to more than half of the cost of traditional construction, with material and equipment costs each accounting for roughly 20%. The material cost is increased to approximately half of the entire cost, the labor cost is decreased to about 35%, and the equipment cost is lowered to about 20% for 3D-printed concrete. The varied requirements of formwork are mostly to blame for the differences in cost ratios. The use of formwork necessitates a significant investment, resulting in

TABLE 7.6
A Comparison of Parameters for Concrete 3D Printing

S/N	Parameter	Concrete printing	Contour crafting	D-shape
1.	Printing process	Extruding	Extruding	Spreading
2.	Nature of material	Mixed 3D printing concrete	Cementitious or mortar	Powder or chemical agent
3.	Speed of printing	High	Low	Medium
4.	Dimension	Large scale	Large scale	Medium size
5.	Resolution	4 to 6 mm	Smooth	Approx. 13 mm

expensive labor, material, and equipment expenses. 3D printing, on the other hand, does not require formwork, which lowers the cost significantly.

Because there is no overengineering or material waste, the cost of 3D printed concrete may be cheaper than conventional concrete; but it may be more because it employs a lot of expensive, fine additives like nanoclay and unique chemical admixtures. Because there is so much equipment, such as the production of printing heads, the cost of equipment is primarily generated in the early days of application. However, once the 3D printing technology is widely adopted, the cost of equipment will be lower than in traditional construction. Overall, when compared to traditional building, the cost of 3D-printed concrete is quite low, which is directly tied to the development of 3D-printed concrete.

7.6.3 FUTURE AND MARKET OUTLOOK

Weinstein and Nawara (2015) used four important characteristics to study and analyze the appeal of 3D printed objects in developing and developed countries: affluence, size, likelihood to consume, and concrete consumption per capita. They claimed that low-income countries would be more likely to develop 3D-printed concrete and 3D-printed buildings, with China being one of the best candidates. According to the survey, 3D-printed concrete is expected to see considerable growth in the following years since it decreases construction waste by 30–60%, labor expenses by 50–80%, and production time by 50–70%. By 2021, the global market for 3D-printed concrete will have grown at a compound yearly growth rate of 15.02%, reaching a value of $56.4 million (Teymouri, 2017; Zhang et al., 2019). The use of 3D-printed walls in the building industry is expected to grow in the near future.

It is being used by the automobile industry to quickly prototype new car models. In industries like aerospace, 3D printing is utilized to create spare and replacement parts. Molds in dentistry, prosthesis, and 3D-printed models for intricate surgeries are all examples of 3D printing applications in healthcare.

7.7 CONCLUSION

The technology of 3D printing is a versatile one that has found applications in various sectors. There is a need to know the type of material to use, where to use them, and the type of 3D printer to use. This chapter was able to discuss the various types of materials (filament or resin) used for 3D printing. It was found that different materials, including plastic, metals and alloys, ceramics, polymers and composites, concrete, and hybrid, can all be used for printing. It means a lot of things can be 3D-printed depending on the level of expertise and finishing available. Biscuit and other food have also been printed using SLA (resin). The chapter also discusses the types of 3D printers and the major manufacturers. This will help in guiding a novice and an expert in choosing the type of 3D printer to use depending on the application, speed, finishing, and other factors needed. The chapter is rich and useful for beginners and experts alike.

REFERENCES

Adeeb Fahmy Hanna, H. (2019). 3D printing technology in construction industry on-site: Analysis of the implications. *JES. Journal of Engineering Sciences, 47*(3), 426–450.

Ali, H. T., Akrami, R., Fotouhi, S., Bodaghi, M., Saeedifar, M., Yusuf, M., & Fotouhi, M. (2021). Fiber reinforced polymer composites in bridge industry. *Structures, 30,* 774–785.

Ali, M. H., Issayev, G., Shehab, E., & Sarfraz, S. (2022). A critical review of 3D printing and digital manufacturing in construction engineering. *Rapid Prototyping Journal, 28*(7), 1312–1324.

Arefin, A. M., Khatri, N. R., Kulkarni, N., & Egan, P. F. (2021). Polymer 3D printing review: Materials, process, and design strategies for medical applications. *Polymers, 13*(9), 1499.

Barjot, D. (2013). "Why was the world construction industry dominated by European leaders?" The development of the largest European firms from the late 19th to the early 21st centuries. *Construction History, 28*(3), 89–114.

Bartkowiak, K., Ullrich, S., Frick, T., & Schmidt, M. (2011). New developments of laser processing aluminium alloys via additive manufacturing technique. *Physics Procedia, 12,* 393–401.

Bogue, R. (2017). What are the prospects for robots in the construction industry? *Industrial Robot: An International Journal, 45*(1), 1–6.

Bogue, R. (2021). The role of robots in the green economy. *Industrial Robot: the International Journal of Robotics Research and Application, 48*(5), 637–642.

Bos, F. P., Menna, C., Pradena, M., Kreiger, E., da Silva, W. L., Rehman, A. U., ... Ferrara, L. (2022). The realities of additively manufactured concrete structures in practice. *Cement and Concrete Research, 156,* 106746.

Brice, C., Shenoy, R., Kral, M., & Buchannan, K. (2015). Precipitation behavior of aluminum alloy 2139 fabricated using additive manufacturing. *Materials Science and Engineering: A, 648,* 9–14.

Brinson, H. F., & Brinson, L. C. (2015). Characteristics, applications and properties of polymers. In: *Polymer Engineering Science and Viscoelasticity* (pp. 57–100). Springer.

Buswell, R. A., De Silva, W. L., Jones, S. Z., & Dirrenberger, J. (2018). 3D printing using concrete extrusion: A roadmap for research. *Cement and Concrete Research, 112,* 37–49.

Chantarapanich, N., Puttawibul, P., Sitthiseripratip, K., Sucharitpwatskul, S., & Chantaweroad, S. (2013). Study of the mechanical properties of photo-cured epoxy resin fabricated by stereolithography process. *Songklanakarin J. Sci. Technol, 35*(1), 91–98.

Chen, A. (2020). Top 10 Materials Used for Industrial 3D Printing. Retrieved from www.cmac.com.au/blog/top-10-materials-used-industrial-3d-printing

Cherdo, L. (2022). *Metal 3D printers in 2022: a comprehensive guide.* Retrieved from www.aniwaa.com/buyers-guide/3d-printers/best-metal-3d-printer/

De Schutter, G., Lesage, K., Mechtcherine, V., Nerella, V. N., Habert, G., & Agusti-Juan, I. (2018). Vision of 3D printing with concrete—Technical, economic and environmental potentials. *Cement and Concrete Research, 112,* 25–36.

Design, A. (2008). *Versatility and Vicissitude: Performance in Morpho-ecological Design,* guest edited by Michael Hensel and Achim Menges. Wiley Academy.

Diggs, B. N., Liesen, R. J., Hamoush, S., Megri, A. C., & Case, M. P. (2021). *Automated Construction of Expeditionary Structures (ACES): Energy Modeling.* Retrieved from https://erdc-library.erdc.dren.mil/jspui/bitstream/11681/39759/1/ERDC-CERL%20TR-21-6.pdf

Dini, E. (2012). Method for automatically producing a conglomerate structure and apparatus therefor: Google Patents.

Duballet, R., Baverel, O., & Dirrenberger, J. (2017). Classification of building systems for concrete 3D printing. *Automation in Construction, 83*, 247–258.

Egan, P. F. (2019). Integrated design approaches for 3D printed tissue scaffolds: Review and outlook. *Materials, 12*(15), 2355.

Egan, P. F., Bauer, I., Shea, K., & Ferguson, S. J. (2019). Mechanics of three-dimensional printed lattices for biomedical devices. *Journal of Mechanical Design, 141*(3), 031703.

Erickson, M. M., Richardson, E. S., Hernandez, N. M., Bobbert II, D. W., Gall, K., & Fearis, P. (2020). Helmet modification to PPE with 3D printing during the COVID-19 pandemic at Duke University Medical Center: A novel technique. *The Journal of Arthroplasty, 35*(7), S23–S27.

Evjemo, L. D., Moe, S., Gravdahl, J. T., Roulet-Dubonnet, O., & Gellein, L. T. (2017). *Additive manufacturing by robot manipulator: An overview of the state-of-the-art and proof-of-concept results.* Paper presented at the 2017 22nd IEEE International Conference on Emerging Technologies and Factory Automation (ETFA).

Florea, V., Păuleţ-Crăiniceanu, F., Luca, S.-G., & Pastia, C. (2019). *3D printing of buildings. Limits, design, advantages and disadvantages. Could this technique contribute to sustainability of future buildings?* Paper presented at the International Conference on Critical Thinking in Sustainable Rehabilitation and Risk Management of the Built Environment.

France-Presse, A. (2017). World's first 3D-printed bridge opens to cyclists in Netherlands-Crossing printed from 800 layers of concrete could take weight of 40 trucks, designers say. *The Guardian, 18.*

Gardiner, J. B. (2011). *Exploring the Emerging Design Territory of Construction 3D Printing.* (PhD), RMIT University.

Gardiner, J. B., & Janssen, S. R. (2014). FreeFab *Robotic fabrication in architecture, art and design 2014* (pp. 131–146). Springer.

Gardner, L., Kyvelou, P., Herbert, G., & Buchanan, C. (2020). Testing and initial verification of the world's first metal 3D printed bridge. *Journal of Constructional Steel Research, 172*, 106233.

Gijsbers, R., & Lichtenberg, J. (2014). Demand driven selection of adaptable building technologies for flexibility-in-use. *Smart and Sustainable Built Environment, 3*(3), 237–260.

Grozdanic, L. (2014). MX3D-metal 3D printer creates complex metal objects in thin air. *Inhabitat, February*, 21.

Guimarães, A. S., Delgado, J. M., & Lucas, S. S. (2021). Advanced manufacturing in civil engineering. *Energies, 14*(15), 4474.

Holt, C., Edwards, L., Keyte, L., Moghaddam, F., & Townsend, B. (2019). Construction 3D printing. In J. G. Sanjayan, A. Nazari, & B. Nematollahi (eds.), *3D Concrete Printing Technology* (pp. 349–370). Elsevier.

Issa, M. (2021). *Additive manufacturing in the engineering and construction industry: Development and potentiality.* Universitat Politècnica de Catalunya.

Jagoda, J., Diggs-McGee, B., Kreiger, M., & Schuldt, S. (2020). The viability and simplicity of 3D-printed construction: A military case study. *Infrastructures, 5*(4), 35.

James, M. B. (2019). Polymers in civil engineering: Review of alternative materials for superior performance. *Journal of Applied Science and Computations, 6*(5), 1770–1773.

Jipa, A., & Dillenburger, B. (2021). 3D printed formwork for concrete: State-of-the-art, opportunities, challenges, and applications. *3D Printing and Additive Manufacturing, 9*(2), 85–107.

Kannan, S., & Ramamoorthy, M. (2020). Mechanical characterization and experimental modal analysis of 3D printed ABS, PC and PC-ABS materials. *Materials Research Express, 7*(1), 015341.

Khandelwal, V., Bhatia, V., Dogra, V., Sharma, S., Chhabra, V., Singh, R., ... Bera, T. (2021). *3-D printer robot for civil construction: A bond graph approach.* Paper presented at the 2021 8th International Conference on Signal Processing and Integrated Networks (SPIN).

Khoshnevis, B. (2011). Gantry robotics system and related material transport for contour crafting. Google Patents.

Khoshnevis, B., & Hwang, D. (2006). Contour crafting. In A. Kamrani & E. A. Nasr (eds.), *Rapid prototyping* (pp. 221–251). Springer.

Kreiger, E. L., Kreiger, M. A., & Case, M. P. (2019). Development of the construction processes for reinforced additively constructed concrete. *Additive Manufacturing, 28,* 39–49.

Kreiger, M. A., MacAllister, B. A., Wilhoit, J. M., & Case, M. P. (2015). *The current state of 3D printing for use in construction.* Paper presented at the Proceedings of the 2015 Conference on Autonomous and Robotic Construction of Infrastructure. Ames, Iowa.

Kuzmenko, K., Gaudillière, N., Feraille, A., Dirrenberger, J., & Baverel, O. (2019). *Assessing the environmental viability of 3D concrete printing technology.* Paper presented at the Design Modelling Symposium Berlin.

Ligon, S. C., Liska, R., Stampfl, J., Gurr, M., & Mülhaupt, R. (2017). Polymers for 3D printing and customized additive manufacturing. *Chemical Reviews, 117*(15), 10212–10290.

Manju, R., Deepika, R., Gokulakrishnan, T., Srinithi, K., & Mohamed, M. (2019). A research on 3D printing concrete. *International Journal of Recent Technology and Engineering, 8*(2).

Mao, Y., Yu, K., Isakov, M. S., Wu, J., Dunn, M. L., & Jerry Qi, H. (2015). Sequential self-folding structures by 3D printed digital shape memory polymers. *Scientific Reports, 5*(1), 1–12.

Molitch-Hou, M. (2020). BetAbram set to 3D print Two-Story house this summer. Retrieved from https://3dprintingindustry.com/news/betabram-set-to-3d-print-two-story-house-this-summer-50826/

Moniruzzaman, M., O'Neal, C., Bhuiyan, A., & Egan, P. F. (2020). Design and mechanical testing of 3D printed hierarchical lattices using biocompatible stereolithography. *Designs, 4*(3), 22.

Nieto, D. M., & Molina, S. I. (2020). Large-format fused deposition additive manufacturing: A review. *Rapid Prototyping Journal, 26*(5), 793–799.

Paolini, A., Kollmannsberger, S., & Rank, E. (2019). Additive manufacturing in construction: A review on processes, applications, and digital planning methods. *Additive Manufacturing, 30,* 100894.

Papanek, V., & Fuller, R. B. (1972). *Design for the Real World.* Bantam Books.

Pastia, C. (2020). *3D printing of buildings. Limits, design, advantages and disadvantages. Could this technique contribute to sustainability of future buildings?* Paper presented at the Critical Thinking in the Sustainable Rehabilitation and Risk Management of the Built Environment: CRIT-RE-BUILT. Proceedings of the International Conference, Iaşi, Romania, November 7–9, 2019.

Pekuss, R., & García de Soto, B. (2020). *Preliminary productivity analysis of conventional, precast and 3D printing production techniques for concrete columns with simple geometry.* Paper presented at the RILEM International Conference on Concrete and Digital Fabrication.

Perrot, A., & Amziane, S. (2019). 3D printing in concrete: General considerations and technologies. In A. Perrot (ed.), *3D Printing of Concrete: State of the Art and Challenges of the Digital Construction Revolution* (pp. 1–40). Wiley.

Pessoa, S., Guimarães, A. S., Lucas, S. S., & Simões, N. (2021). 3D printing in the construction industry-A systematic review of the thermal performance in buildings. *Renewable and Sustainable Energy Reviews, 141,* 110794.

Poluektova, V. (2019). Electrokinetic properties and aggregative stability of polymer–Mineral dispersions for 3D printing in building. *Russian Journal of Physical Chemistry A, 93*(9), 1783–1788.

Poluektova, V. (2020). Designing the composition of a cement-based 3D construction printing material. *Inorganic Materials: Applied Research, 11*(5), 1013–1019.

Rimmer, M. (2021). Automating fab cities: 3D printing and urban renewal. In B. T. Wang & C. M. Wang (eds.), *Automating Cities* (pp. 255–272). Springer.

Romani, A., Rognoli, V., & Levi, M. (2021). Design, materials, and extrusion-based additive manufacturing in circular economy contexts: From waste to new products. *Sustainability, 13*(13), 7269.

Russell, D., Hernandez, A., Kinsley, J., & Berlin, A. (2005). Methods and apparatus for 3D printing. Google Patents.

Schweiker, M., Endres, E., Gosslar, J., Hack, N., Hildebrand, L., Creutz, M., …. Mehnert, J. (2021). Ten questions concerning the potential of digital production and new technologies for contemporary earthen constructions. *Building and Environment, 206*, 108240.

Seaman, S., Kerezoudis, P., Bydon, M., Torner, J. C., & Hitchon, P. W. (2017). Titanium vs. polyetheretherketone (PEEK) interbody fusion: Meta-analysis and review of the literature. *Journal of Clinical Neuroscience, 44*, 23–29.

Sharanova, A., & Dmitrieva, M. (2019). *Selection of compositions for additive technologies in construction.* Paper presented at the E3S Web of Conferences.

Teymouri, A. (2017). Potentialities and restrictions of construction 3D printing. Bachelor Thesis of Karelia University of Applied Sciences, Finland.

Wang, B., Wang, C., & Rimmer, M. (2020). Automating fab cities: 3D printing and urban renewal. In B. T. Wang & C. M. Wang (eds.), *Automating Cities: Design, Construction, Operation and Future Impact* (pp. 255–272). Springer.

Weger, D., Gehlen, C., Korte, W., Meyer-Brötz, F., Scheydt, J., & Stengel, T. (2022). Building rethought–3D concrete printing in building practice. *Construction Robotics, 5*(3–4), 203–210.

Weinstein, D., & Nawara, P. (2015). Determining the applicability of 3D concrete construction (Contour Crafting) of low income houses in select countries. Cornell Real Estate Review, *13*(1), 94–111. Retrieved from http://scholarship.sha.cornell.edu/crer/vol13/iss1/11

Wohlers, T. (2017). 3D Printing and Additive Manufacturing State of the Industry, Annual Worldwide Progress Report Wohlers Report: Chapters Titles: The Middle East, and Other Countries, Wohlers Associates. *Inc., Fort Collins, Colorado, USA*, 344.

Yin, H., Qu, M., Zhang, H., & Lim, Y. (2018). 3D printing and buildings: a technology review and future outlook. *Technology|Architecture + Design, 2*(1), 94–111.

Youssef, N., Rabenantoandro, A. Z., Lafhaj, Z., Dakhli, Z., Hage Chehade, F., & Ducoulombier, L. (2021). A novel approach of geopolymer formulation based on clay for additive manufacturing. *Construction Robotics, 5*(2), 175–190.

Zhang, J., Wang, J., Dong, S., Yu, X., & Han, B. (2019). A review of the current progress and application of 3D printed concrete. *Composites Part A: Applied Science and Manufacturing, 125*, 105533.

Zukas, V., & Zukas, J. A. (2015). *An introduction to 3D* printing. First Edition Design Pub.

8 Emerging Applications of 3D Printing Technology

8.1 INTRODUCTION

Additive manufacturing is gaining ground in various applications. It has been deployed in aviation and other key areas of humanity. However, some technologies are just gaining relevance and application. Technical advancements that indicate progressive innovations within a field for competitive advantage are known as emerging technologies. Emerging technologies are those whose development, practical applications, or both are yet substantially unreached, to the point that they are symbolically emerging into prominence from obscurity. These technologies are mostly new, but they do include some older ones.

8.1.1 Factors That Make a Technology Emerging

The following are the requirements for inclusion as an emerging technology. First, exist in some way; entirely hypothetical technologies should be included in a list of hypothetical technologies rather than being deemed emergent. Actively researched and prototyped technologies, on the other hand, are permitted. They have a Wikipedia page dedicated to them. Currently, it is not extensively utilized. Emerging technologies cannot be called mainstream or commercialized.

8.1.2 Classification of Emerging Technologies

The following are some of the emerging technologies as classified based on the sector.

8.1.2.1 Agriculture

Agriculture is rapidly evolving into a high-tech industry that is attracting new professionals, businesses, and investors. Technology is continually evolving, boosting not only farmers' production capacity but also robotics and automation technology as we know it. The quest for higher yield and growing population are major drivers for new technology in agriculture. According to the United Nations, the global population will increase from 7.3 billion today to 9.7 billion by 2050. The world will require far more food, and farmers will be under extreme pressure to meet demand. The following are the major emerging technologies in the agricultural sector. They are

DOI: 10.1201/9781003364481-8

agricultural robotics or drone, cultured meat, vertical farming, and closed ecological systems. An example of closed ecological systems includes Greenhouse, Biosphere 2, Eden Project, Bioshelter, Seawater greenhouse, and Perpetual harvest greenhouse system.

8.1.2.2   Vertical Farming

Vertical farming is the process of cultivating crops in layers that are piled vertically. It frequently integrates soilless farming techniques such as hydroponics, aquaponics, and aeroponics, as well as controlled-environment agriculture, which tries to optimize plant growth. The vertical farming market was worth US$3.1 billion in 2021, and it is predicted to grow at a rate of 25.0% between 2021 and 2026, reaching US$9.7 billion (Moghimi and Asiabanpour, 2021). In 2019, the global vertical farming market by crop type was valued at US$1.57 billion, with a forecasted value of US$11.55 billion by 2025 (Ares et al., 2021). Vertical farming is expected to develop at a rate of 22.1% in Asia Pacific, with a market size of US$2101.0 million (Teo and Go, 2021).

In 1915, American geologist Gilbert Ellis Bailey created the term "vertical farming" (Bailey, 1915). Dickson Despommier, a professor at Columbia University in New York, popularized the current concept of vertical farming in 1999, working on it with his students. Vertical farming is good for corn, okra, Brussels sprouts, and sunflowers (Barker, 2010). They grow vertically spontaneously and do not require any support. In fact, these tall plants can operate as a support system for vines that aren't too heavy. Greens such as lettuce, kale, and basil can also be grown vertically.

8.1.2.3   Agricultural Robots

It is also called AgBot. They are highly specialized pieces of technology that may aid farmers with a variety of tasks. They can assess, consider, and carry out a wide range of tasks, and they can be designed to grow and evolve to meet the demands of varied tasks. Farmers can focus more on boosting overall crop yields by using agricultural robots to automate slow, repetitive, and boring jobs. Agricultural robots help farmers increase output yields in a variety of ways. Drones, autonomous tractors, and robotic arms are all examples of how technology is being used in unique and innovative ways. Farmers use agricultural robots to automate slow, repetitive, and boring jobs. Eli Whitney first introduced robots to agriculture on March 14, 1974. Her machine can separate the cotton seed from the cotton fiber, and it has produced up to 50 pounds of cotton in a single day.

8.1.2.4   Agricultural Robots Applications

Robotic weeding AgBots monitor crop fields, identify and eradicate weeds. Weeds will be detected by low-speed robots. Thereafter, algorithms utilized to categorize them using computer vision. Boxing, weed control, spraying and thinning, phenotyping, pruning, harvesting and picking, autonomous mowing, planting, sorting and packing, data collection, and utility platforms are the major applications of robots in agriculture. Harvesting and picking is one of the most prominent robotic applications in agriculture, because of the precision and speed with which robots can increase yields and reduce waste from crops left in the field. However, automating these applications can be tricky. A robotic system meant to harvest sweet peppers, for example, faces numerous challenges. In difficult conditions, such as dust, fluctuating light intensity, temperature changes, and wind activity, vision systems must assess the position and ripeness of the pepper. Picking a pepper, however, requires more than powerful vision technologies. To precisely grasp and position a pepper, a robotic arm must navigate situations with just as many difficulties. This is not the same as picking and positioning metal parts on a production line. The agricultural robotic arm should be adaptable in a changing environment and precise enough to avoid damaging the peppers while picking them. There is a slew of other creative ways the agriculture business is using robotic automation to boost output yields.

AgBot may drive through a paddock, recognizing weeds and spraying herbicides directly where they are needed, down to an individual leaf, using robotic vision. The robot also learns what is not considered a weed and avoids spraying on a healthy crop. AgBot2 has an accuracy of 93.9% (Jose et al., 2021). Optical sensors are used in agriculture to analyze the amount of reflected light on the growing regions of the crop in real-time to understand the qualities of the soil and crop. The Rosphere, Merlin Robot Milker, Harvest Automation, lettuce bot, Orange Harvester, and weeder are some models and prototypes of robots. The milk bot is an example of a large-scale application of robots in agriculture. The AgBot2 was a scientific experimental robot that was created between 2013 and 2017 to show off essential skills like maneuvering and vision-based weed recognition. It was developed in Australia by QUT with funding from the Queensland Government. It has the potential to save Australia's farm sector US$1.3 billion per year by halving the cost of weeding crops by 90%.

Axbom and Ralsgård (2019) developed an AgBot to address issue of herbicide-resistant weeds in large farms. This was achieved by combining the existing FarmBot (Murcia et al., 2021, Pandya et al., 2019) stationary agricultural robot with a light-weight, cost-effective, visually beautiful, and robust frame on wheels that can operate independently on farms and is powered by solar energy. It assessed many concepts before using CAD to create a final product. Parts that can be ordered off the shelf as well as parts that can be 3D printed were used to construct the finished vehicle. It weighs 154 kg, costs roughly US$5300 in parts, and is designed to function independently for at least a day. It's simple to put together and features a modular design that makes future development easier. Also, an AgBot called MONTe (Halle and Hickle, 2011) has been deployed in New Zealand in pipfruit farms (Victor Rueda-Ayala et al., 2019).

Additive manufacturing has been applied in manufacturing of different categories of robots including agricultural robots. Key components of such robots are printed using 3D printer for agility, cost reduction, and effectiveness. A 3D-printed mobile robot is shown in Figure 8.1.

8.1.2.5 Aerospace

Plasma propulsion, pulse detonation engine. The plasma propulsion is still at the research and development stage with potential application in spacecraft propulsion. The pulse detonation engine is still in R&D stage although a prototype was flown in 2008 and the area of application is in propulsion.

8.1.2.6   Medical

Artificial uterus, body implants, prosthesis, cryonics, human DNA vaccination and mRNA vaccinations, enzybiotic, genetic engineering of organisms and viruses, hibernation or suspended animation.

8.1.2.7   Neuroscience

Brain-computer interface, head transplant, neuroprosthetics, electroencephalography, brain-reading neuroinformatics are emerging technology in neuroscience.

8.1.2.8   Space

Artificial gravity, asteroid mining, Starshot, reusable launch system, Stassis chamber, inflatable space habitat.

8.2   APPLICATION OF 3D PRINTING IN ELECTRICITY

What would you do if you could print your electricity from the comfort of your home at reduced cost? This is the promise that 3D printing offers in application in several renewable energy. 3D printing, if properly harnessed with solar energy, wind and hydro turbines can unlock the issues of initial cost of manufacturing associated with most renewable energy sources such as wind and hydro turbines. Additive manufacturing using 3D printing is an emerging technology that can change many human lives, including electricity generation using wind and hydro turbines. It offers the promise of users being able to print their electricity from the comfort of their home. It could potentially end load-shedding, electricity shortage, and other issues of electricity experienced globally, especially in the global south. This will greatly improve the quality of life, reduce greenhouse gas emissions, and meet other sustainable development goals (SDG).

Figure 8.1 shows solar panel and Pelton hydro turbine used for electricity generation in Africa. The growing global population has resulted in affordable, clean, and stable electricity demand. There are still some countries, especially developing countries, where the citizens lack access to an uninterrupted power supply. In some cases, there is no electricity at all. Although, such regions have the requirement for turbine usage. They can comfortably generate electricity using a wind turbine, tidal turbine, hydro turbine, etc. However, the cost and other reasons such as inadequate generation using the turbines have hitherto hindered the full implementation and domestication of wind turbines in such areas. In some regions with wind turbines installed, the annual energy production from the wind turbine is not yet sufficient, with a gain of around 1 to 2%. Numerous research works are focusing on reducing the cost and, most especially, increasing the efficiency of wind turbines and other forms of turbines used for electricity generation.

There was a 6% increase in demand for electricity globally in 2021. This may be attributed to the attendant effect of climate change, colder and warmer summers, and economic recovery. The 2022 mid-year report by IEA suggests a 2.4% increase in global electricity demand, which is expected to grow on a similar path in 2023 (www.iea.org/news/global-electricity-demand-growth-is-slowing-weighed-down-by-economic-weakness-and-high-prices). The increase in energy prices may not

TABLE 8.1
List of Emerging Technologies and Their Applications

Emerging technology	Status	Potential application	Related
Artificial uterus	Hypothetical, research	Space travel, extracorporeal pregnancy, Reprogenetics, same-sex procreation	
Body implants, prosthesis	Trials, from animal (e.g., brain implants) to human clinical (e.g., insulin pump implant), to commercial production (e.g., pacemaker, joint replacement, cochlear implant)	Brain implant, retinal implant	Prosthetics, prosthetics in fiction, cyborg
Cryonics	Hypothetical, research, commercialization (e.g., Alcor, Cryonics Institute)	Life extension	
De-extinction	Research, development, trials	Animal husbandry, pets, zoos	Revival of the woolly mammoth
Electronic medical records	Deployment	Paper medical records	
Human DNA vaccination and mRNA vaccinations	Implementation in 2021 to combat the COVID-19 pandemic	Disease vaccinations, cancer therapy	COVID-19 pandemic
Enzybiotics	Successful first trials		
Genetic engineering of organisms and viruses	R&D, commercialization	Creating and modifying species (mainly improving their physical and mental capabilities), bio-machines, eliminating genetic disorders (gene therapy), new materials production, healthier and cheaper food, creating drugs and vaccines, research in natural sciences, bioremediation, detecting arsenic, CO_2-reducing superplant	Biopunk, genetically modified food, superhuman, human enhancement, transhumanism, gene doping, designer baby, genetic pollution
Hibernation or suspended animation	Research, development, animal trials	Organ transplantation, space travel, prolonged surgery, emergency care	
Life extension, strategies for engineered negligible senescence	Research, experiments, animal testing	Increased life spans	Immortality, biological immortality

Technology	Status	Applications	Examples
Nanomedicine	Research, experiments, limited use		
Nanosensors	Research and development		
Omni processor	Research and development; some prototypes		
Robotic surgery	Research, diffusion		
Senolytic	Under investigation		
Stem cell treatments	Research, experiments, phase I human trial spinal cord injury treatment (GERON), cultured cornea transplants	Treatment for a wide range of diseases and injuries	Stem cell, stem cell treatments, skin cell gun
Synthetic biology, synthetic genomics	Research, development, first synthetic bacteria created May 2010	Creating infinitely scalable production processes based on programmable species of bacteria and other life forms	BioBrick, iGEM, synthetic genomics
Tissue engineering	Research, diffusion	Organ printing, tooth regeneration	
Tricorder	Research and development	Diagnosing medical conditions	Medical tricorder
Virotherapy	Research, human trials	Gene therapy, cancer therapy	Virotherapy, oncolytic virus
Vitrification or cryoprotectant	Hypothetical, some experiments	Organ transplantation, cryonics	

FIGURE 8.1   Solar energy and hydro turbine evolving sources of electricity generation.

be unconnected to the Russia-Ukraine war. This increment has prompted European countries to advance the clean energy transition and reduce fuel importation. This has given rise to solar, small modular nuclear, and wind energy. Wind and solar constituted about 10% of the total electricity generated by 50 countries. Solar and wind electricity generation increased by 23% and 14% in 2021, making a combined 10% of global electricity generation. It is the decade of wind and solar energy to cut global emissions, cut electricity cost and instability, and increase performance.

There is a need for innovation and forthrightness to achieve this lofty dream of massive usage of wind energy. The manufacturing process and testing of wind turbines are crucial in increasing efficiency, reducing cost, and reducing the manufacturing time. Additive manufacturing using 3D printing ticks these boxes. 3D printing is creating 3D objects by "printing" the computer-aided design equivalent files of that object. In the case of the wind turbine blade, the blade is either modeled from the first principle or CAD firmware is used to create the 3D files. The process requires an expert or little knowledge of CAD. After that, the CAD file is converted to the file (.stl) that the 3D printer understands. The 3D printer outputs the final 3D object. Technology has been applied in various spheres of humanity. It has been used to print human organs. It has been used to build houses within a week and is even researched to build houses and structures in space. Biome Renewables, Canada, proved that 3D printing could be combined with a turbine (wind turbine) to increase the efficiency of the wind turbine and increase the annual energy production (www.biome-renewables.com/powercone) as shown in Figure 8.2.

Wind turbines usually have issues with efficiency and cost. This prompted a company in Canada (Biome Renewables) to develop a retrofit for existing wind turbines. The retrofit is made using 3D printing technology. Fiberglass was used as

FIGURE 8.2   PowerCone retrofit for wind turbine made using 3D printing. (Courtesy www. biome-renewables.com/powercone)

the material for device manufacturing. This can be reciprocated and improved upon to solve the issues of unstable, fossil-powered, and expensive electricity generation worldwide.

The growing interest in wind and solar may not be unconnected to the following analysis. The cost of purchasing a 5–6 kW diesel-powered generating set may be less than US$1400 and a hydro turbine cost about US$1000–1500 per kW. However, the cost of fueling the diesel-powered generating set, the unreliability of the supply of diesel, and the negative impact on the environment cause a shift toward the hydro turbine. The hydro turbine's cost can be reduced further by 3D printing of the major components, if not all the components.

There are pockets of rivers across the world capable of generating electricity using hydro turbines. Small kinetic hydro turbines can generate enough electricity for rural power communities. This kinetic turbine only requires a small amount of flowing water to produce about 5 kW of electricity. An improvement is the small-scale modular hydrokinetic that involves adding up to five kinetic turbines to produce 25 kW. Hydro turbines are capable of serving large, medium, and small electricity users. However, the initial cost is usually not "affordable" to the average user. Additive manufacturing is capable of improving the efficiency of hydro turbine and reducing the initial cost.

The cost of 3D printers is plummeting and it is becoming user-friendly. If this is implemented in the hydro turbine industry, it means users can almost print critical components of hydro turbines at an affordable rate. The critical component of the kinetic turbine and other hydro turbines can be 3D printed to reduce manufacturing time and cost and increase efficiency. The major components are shown in Figure 8.2. They include a propeller made from a composite that can be easily manufactured using a 3D propeller. Also, the diffuser, which is plastic and tubular, can be 3D printed. The propeller socket is capable of being manufactured cost-effectively using 3D printing.

Similarly, Biome Renewables, Canada, has implemented 3D printing of a wind turbine retrofit called PowerCone to increase wind turbine efficiency. A glass fiber 3D printed retrofit gave 10–13% increased AEP, less noise that helped the turbine reach the rated power sooner, reduced loads with smooth incoming gusts, and aligned the airflow with the turbine blade. This gave reduced turbulence, vibrations, and the associated loads.

One would ask what is the environmental implication and benefit of this. Using renewable energy sources like wind and solar reduces global emissions causing untold hardship and endangering the planet's existence. A 5 kW produced by kinetic hydro turbine results in 30,000 kWh per year, subsequently preventing about 12,000 liters per year from being burnt to power diesel generators. That is about 34,000 kg of CO_2.

The material selection and methodology of implementation are subject to the location. The process will require knowing the region's locations and weather conditions. This will guide the material selection with special attention given to corrosion. An acidic or salty region will require proper material tuning to infuse the appropriate inhibitor into the final component. Detailed analysis and computer-aided testing were done before 3D printing and after 3D printing.

The application of 3D printing in solving the world's electricity needs, especially in developing countries with pressing electricity challenges, may be the solution that has eluded humanity for a long time. 3D can reduce the manufacturing time, the cost of manufacturing the components, and the final turbine's resultant efficiency. This is in addition to positive environmental benefits. Newer models of 3D printers now allow different materials for the filament ranging from metal to composite, ceramics, and the like. Such models are now being made with users being able to print already-made free versions of several equipment. This means users can easily print hydro turbine, solar, winds, and other renewable energy components and equipment.

8.3 APPLICATION OF 3D PRINTING IN EMERGING APPLICATIONS

The five of the most popular 3D printing applications include education, manufacturing, medicine, construction, and jewelry and art. In Education, more schools are adopting 3D printing into their curricula every day. Manufacturing and prototyping 3D printing started off to speed up prototyping.

8.3.1 APPLICATION OF 3D PRINTING IN BIOMEDICAL APPLICATION

The use of biological sciences, particularly biochemistry, molecular biology, and genetics, to the understanding, treatment, and prevention of disease is referred to as biomedical (Abubakre et al., 2023). This includes bioprinting, pharmaceutical formulation, and medical devices.

8.3.1.1 Medical Application

Medical applications require a wide range of material qualities, which can be achieved with 3D printing. Medical applications frequently necessitate the use of discrete capabilities, such as energy-absorbing materials in impact resistance, colorful parts with appropriate textures for simulating surgical anatomies, or specific material qualities to replicate biological tissues.

Surgical applications of 3D printing-centric therapies date back to the mid-1990s when anatomical modeling was used to plan bone reconstructive surgery (Hoang et al., 2016). Surgeons were more prepared, and patients received better treatment after training on a tactile model before surgery. Patient-matched implants were a natural outgrowth of this research, resulting in genuinely individualized implants that fit a single person. Virtual surgery planning and guidance employing 3D-printed, individualized instruments have been used with remarkable success in a variety of surgical procedures, including total joint replacement and craniomaxillofacial reconstruction (Dimitroulis et al., 2018; Zheng et al., 2019). The use of models for planning heart and solid organ surgery has expanded because of more research (Ganguli et al., 2018; McDaniel, 2017). Hospital-based 3D printing is gaining popularity, and several institutions are considering incorporating it within individual radiology departments (Bastawrous et al., 2021; Mitsouras et al., 2020). The technology is being utilized to build one-of-a-kind, patient-matched equipment for rare diseases. The bioresorbable tracheal splint created at the University of Michigan to treat babies with tracheobronchomalacia is an example of this (Les et al., 2019). Several medical device companies have begun to use 3D printing to create patient-specific surgical guides (polymers). Due to the capacity to efficiently construct porous surface features that allow osseointegration, the application of additive manufacturing for serialized manufacture of orthopedic implants (metals) is also expanding. Broken bone-printed casts are tailored and open, allowing the wearer to scratch itches, cleanse, and breathe the affected area. They're also recyclable.

Microstructures with a 3D internal geometry have been created using fused filament fabrication (FFF) (Han et al., 2019, Malinauskas et al., 2014). There is no requirement for sacrificial structures or additional support materials. Structures made of polylactic acid (PLA) can have totally controlled porosity ranging from 20% to 60%. These scaffolds could be used as biomedical templates for cell culture or as biodegradable tissue engineering implants. Patient-specific implants and medical devices have been manufactured using 3D printing. A titanium pelvis was implanted into a British patient, a titanium lower jaw was transplanted into a Dutch patient (Goriainov et al., 2018), and a plastic tracheal splint was put into an American infant (Fessenden, 2013). The hearing aid and dentistry industries are projected to be the biggest beneficiaries of custom 3D printing technology in the future (Ventola, 2014).

Surgeons in Swansea utilized 3D-printed components to reconstruct the face of a motorcyclist who was gravely damaged in a road accident in March 2014 (Soliman et al., 2015). Methods to bioprint replacements for lost tissue due to arthritis and cancer are also being researched. Organ replicas may now be created using 3D printing technology (Kusaka et al., 2015; Radenkovic et al., 2016). The printer creates layers of rubber or plastic using images from patients' MRI or CT scan images as a template.

Personal protective equipment, often known as PPE, is worn by medical and laboratory personnel to safeguard themselves from infection when treating patients using 3D printing technology. Face masks, face shields, connections, gowns, and goggles are examples of PPE. Face masks, face shields, and connectors are the most common 3D-printed PPE items (Salmi et al., 2020, Spake et al., 2021). Additive Manufacturing is now being used in the pharmaceutical sciences as well. Different 3D printing processes (e.g., FDM, SLS, inkjet printing, etc.) are used for diverse medication delivery applications based on their unique benefits and downsides.

8.3.1.2 Bioprinting

Three-dimensional bioprinting is the use of 3D printing–like procedures to produce biomedical parts using cells, growth factors, and/or biomaterials, with the goal of mimicking natural tissue features. 3D bioprinting is a new approach for fabricating tissues and organs by accurately controlling the periodic arrangement of diverse biological materials, such as biochemicals and biocells (Ahmad et al., 2019). Bioprinters produce structures that are fundamentally different. Bioprinters, unlike 3D printers, are designed to print liquid and gel-based materials, as well as noncontact droplet printing (Akiah et al., 2020; Gu et al., 2020; Xie et al., 2021).

3D bioprinting creates a microenvironment in which cells can grow and differentiate in tissue architectures by combining cells, growth factors, and biomaterials. Biomaterials are produced layer by layer in 3D bioprinting to create structures that resemble a target organ or tissue (Iordache, 2019). In the United States, the first patent for this invention was proposed in 2003 and granted in 2006 (Rodríguez-Salvador et al., 2017; Rodriguez-Salvador and Ruiz-Cantu, 2019; Thomas et al., 2016). Selecting biomaterial, generating a bioprinting model using CAMs software, printing, and analyzing the printed constructions are all processes in the bioprinting process.

Bioprinting is a technology for fabricating biological tissues and organs that uses 3D automated printers to deposit cells embedded in a biomaterial utilizing a computer-aided layer-by-layer methodology.

8.3.2 The Working Process of Bioprinter

The CT or MRI scans of the intended organ are the first step in the bioprinting process (Zhang et al., 2017). The photograph is then fed into a computer, which uses a software program to create a 3D layout of the organ (Gao et al., 2018). To create a layer-by-layer model of the organ, the 3D data is coupled with histology information based on microscopic inspection. After that, the data is fed into the printer. Other information, such as the type of material to be utilized, is also fed into the printer. The printer then reads the pattern and deposits the biomaterial layer by layer onto the receiver. It is accomplished by moving the print head in all directions to generate the desired depth and thickness. When a layer reaches the platform, it either cools or undergoes a chemical reaction to solidify (Aljohani et al., 2018). A new layer is placed on top of the solidified layer to build a sturdy structure. The organ is then removed from the printer and placed in an incubator to settle and stabilize the structures.

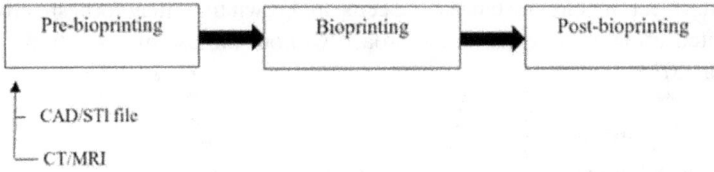

FIGURE 8.3 Schematic of the bioprinting process.

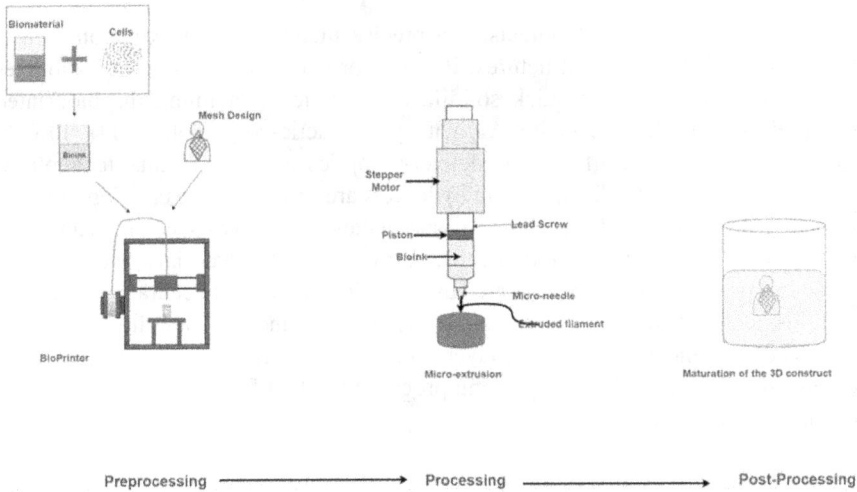

FIGURE 8.4 Steps for bioprinting process.

The fabrication process is grouped into three, namely prebioprinting, bioprinting, and postbioprinting as shown in Figure 8.3.

Biological inks (also known as bioinks) such as cytocompatible hydrogels are used in specialized bioprinters. Hydrogels are a type of insoluble hydrophilic polymeric network that can swell and absorb a large amount of water without dissolving.

8.3.3 THE STEP FOR 3D BIOPRINTING PROCESS

The step for bioprinting includes preprocessing followed by processing and postprocessing as illustrated in Figure 8.4, adapted from Sigaux et al. (2019).

8.3.3.1 Prebioprinting

The model to be printed is created in this stage. Models can be constructed by importing computed tomography (CT) and magnetic resonance pictures that have been turned into STL (stereolithography) files, or by importing 2D or even 3D structures generated by CAM software. In addition, the cells that will be printed are cultured and encased in hydrogels, with the quantity, type, and conditions of cultivation taken into mind. In certain circumstances, the cells are grown until they form spheroids and then printed utilizing an extrusion approach without the use of a scaffold (Murphy and Atala, 2014).

8.3.3.2 Bioprinting

Bioprinting technologies can be divided into four basic modalities depending on their working mechanisms: (1) extrusion-based bioprinting; (2) inkjet-based bioprinting; (3) laser-based bioprinting; and (4) stereolithography as shown in Figure 8.5 (Cui et al., 2012, Klebe, 1988, Shor et al., 2009). The different types of bioprinting are shown in Figure 8.5.

Bioink is deposited as filaments in a precise manner during extrusion, generating targeted 3D bespoke structures. Physical or chemical crosslinking stabilizes the structure, allowing for quick solidification while maintaining the bioprinted structure's geometrical accuracy. Alginate poly (lactic-co-glycolic acid) (PLGA) scaffolds are employed for drug delivery applications using this technology. Collagen, hyaluronic acid, and gelatin hydrogels are utilized in the creation of blood arteries, heart valves, and myocardial tissue in cardiovascular tissue engineering. It creates a bioink solution by mixing cells, nutrients, and biomaterials, which is then placed in a printer cartridge and printed using various ways. For accurate and defined scaffolds, technical factors are critical. These variables include the following: speed rate, needle diameter, number of layers, biomaterial viscosity, and temperature. A detailed difference of the bioprinting process modified from Loai et al. (2019) is shown in Figure 8.6.

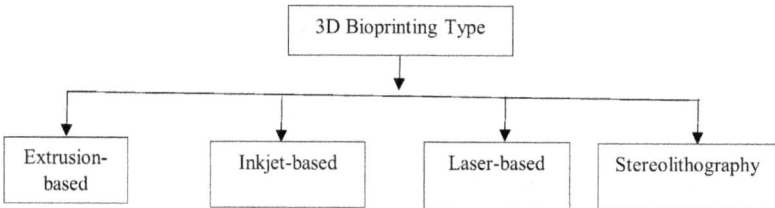

FIGURE 8.5   Different types of 3D bioprinting.

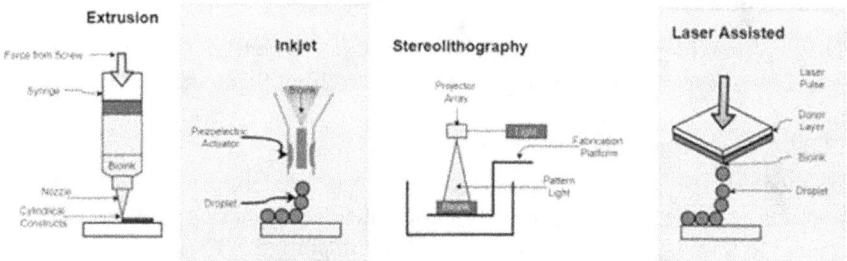

FIGURE 8.6   Schematic representation of different 3D bioprinting.

8.3.3.3 Postbioprinting

The constructs should be kept in good condition during this step so that cells can grow and proliferate through the scaffold and form the tissue-specific structure. The creations are placed in bioreactors that provide nutrition, oxygen, dynamic flow, and microgravity in an attempt to simulate the creatures' natural surroundings (Hinton et al., 2015). Bioprinted scaffolds can be classed as (1) hydrogel scaffolds, (2) fibrous scaffolds, or (3) porous scaffolds in general, but a more helpful classification is depending on the application (e.g., soft tissue, hard tissue, connective tissue) or biomaterial composition (e.g., organic, inorganic, polymeric, natural). In order to construct the ideal scaffolds, 3D bioprinting requires a comprehensive strategy that incorporates knowledge and methodology from a variety of domains, including engineering, tissue engineering, stem cell biology, and biomaterials research. Biocompatible, nonimmunogenic, nontoxic, antithrombotic, with vasoactive qualities, compliant, and allowing for postimplantation modification of the host tissue are all desirable characteristics of an excellent scaffold. Geometry, surface characteristics, pore size, adhesion, degradation, and biocompatibility are all physicochemical parameters that must be considered when making such a scaffold (Datta et al., 2017).

In the fabrication of 3D functional human structures that replicate native tissues/organs, 3D bioprinting has gained traction. Biocompatible and biodegradable scaffolds based on carbohydrates, proteins, and nucleic acids, as well as nanocomposites, are being developed to promote cell adhesion and proliferation in tissues produced by 3D printing (Poluri, 2019). Polysaccharides are being investigated to produce bioinks in regenerative medicine because they are biocompatible, inexpensive, and renewable. The effects of extracellular matrix proteins and other structural proteins on tissue mechanics and cell behavior are well understood. They have the capacity to give cellular imitation while enabling scaffold creation with good mechanical strength and porosity. For 3D bioprinting, DNA-based hydrogels and scaffold materials with self-assembling and hybridization capabilities are desirable (Allen et al., 2015, Budharaju et al., 2022).

8.3.4 Basic Principle and Approach for Bioprinting

An agreed standard approach for fabricating living organs using 3D printing has been developed with special consideration of the mechanical and biological properties. This approach starts with biomimicry, followed by autonomous self-assembly and thereafter combination of both biomimicry and autonomous self-assembly into mini-tissue (Ashraf et al., 2018, Murphy and Atala, 2014).

Biomimicry is the first approach to bioprinting. The major purpose of this method is to produce artificial structures that are equivalent to the natural structures seen in human tissues and organs. The structure, architecture, and microenvironment of organs and tissues must all be duplicated in biomimicry (Reddi, 1998). The use of biomimicry in bioprinting entails the creation of identical cellular and extracellular organ parts (Miri et al., 2019; Rubtsova and Onat, 2017). The tissues must be recreated in a tiny size for this method to work. As a result, it's critical to comprehend the microenvironment, the nature of biological forces operating within it, the precise structure of functional and supportive cell types, solubility variables, and the extracellular matrix's composition.

Autonomous self-assembly is the second bioprinting method (Vyas and Udyawar, 2019). The physical process of embryonic organ development is used as a model to duplicate the tissues of interest in this method (Pati et al., 2016). To supply the needed biological activities and microarchitecture, cells construct their own extracellular matrix building block, correct cell signaling, and independent arrangement and patterning during their early development. Autonomous self-assembly necessitates detailed knowledge of the embryo's tissues and organs' developing procedures. A "scaffold-free" model that uses self-assembling spheroids that are subjected to fusion and cell arrangement to simulate emerging tissues has been developed. The cell is the fundamental driver of histogenesis, controlling the building blocks, structural, and functional features of these tissues. Autonomous self-assembly is dependent on the cell as the fundamental driver of histogenesis (Iram et al., 2019). It necessitates a better understanding of how embryonic tissues develop as well as the microenvironment in which they develop in order to construct bioprinted tissues.

The third way to bioprinting is termed mini tissues, and it combines both biomimicry and self-assembly approaches. Organs and tissues are made up of a small number of functional components. The mini-tissue technique manufactures and arranges these little bits into bigger frameworks. Small functional components of tissues and organs, known as mini tissues, are created using this bioprinting method. Mini tissues, such as the kidney neuron, are the smallest structural and functional unit of the organs. These mini tissues can then be created using either autonomous self-assembly or biomimicry techniques. The bioprinting process starts with the assembly of mini tissues into macro-tissues using biologically inspired organization, followed by the replication of tissue units that can self-assemble into functional structures.

8.3.5 Benefits of 3D Printing Technology in Biomedical

The benefit of 3D printing cannot be overemphasized. This includes production of custom-made parts, highly cost-efficient, improved productivity, large options of materials. The freedom to create custom-made medical devices and equipment is the main advantage that 3D printers bring to medical applications (Banks, 2013). The use of 3D printing to personalize prosthetics and implants, for example, can be extremely beneficial to both patients and physicians. Furthermore, 3D printing can be used to create custom jigs and fixtures for use in operating rooms (Mertz, 2013). Custom-made implants, fixtures, and surgical instruments can help reduce surgery time, improve patient recovery time, and increase the success of the surgery or implant. 3D printing technologies are also expected to allow medicine dosage forms, release profiles, and dispensing to be personalized for each patient in the future (Ursan et al., 2013). Another significant advantage of 3D printing is the capacity to make goods at a low cost. For large-scale production, traditional manufacturing processes remain less expensive; nevertheless, the cost of 3D printing is becoming increasingly competitive for short production runs. This is especially true for small conventional implants or prosthesis used for spinal, dental, or craniofacial diseases.

Custom-printing a 3D object is inexpensive, with the first piece costing the same as the last. This is especially beneficial for businesses with modest production numbers, as well as those who create highly complicated parts or products that require frequent adjustments. By reducing the usage of unneeded materials, 3D printing can help save manufacturing costs (Gebler et al., 2014; Hornick and Roland, 2013). A pharmaceutical tablet weighing 10 mg, for example, might be custom-fabricated as a 1-mg tablet on demand. Some pharmaceuticals may also be printed in dose forms that make delivery to patients easier and more cost-effective. It saves cost and reduces time (Mansour et al., 2020). Also, in 3D printing medical application, "fast" denotes that a product may be created in a matter of hours. As a result, 3D printing is far faster than traditional techniques of producing prosthetics and implants, which require milling, forging, and a lengthy delivery period. Aside from speed, other features of 3D printing technologies are improving as well, such as resolution, accuracy, dependability, and repeatability. A growing number of materials are becoming available for use in 3D printing, and their prices are falling. This enables more people, especially those in medical disciplines, to develop and build unique goods for personal or commercial use with little more than a 3D printer and their imaginations. The nature of 3D printing data sets also provides a once-in-a-lifetime chance for scholars to collaborate (Gross et al., 2014). Rather than attempting to replicate settings specified in scientific articles, researchers can use open-source databases to get downloadable .stl files. They can utilize a 3D printer to manufacture an exact reproduction of a medical model or equipment in this way, allowing for precise design sharing (Gross et al., 2014).

8.3.6 Materials Printed with 3D Bioprinting Technology

Bioprinting tissues and organs, customized implants and prostheses, anatomical models for surgical preparation, custom 3D-printed dosage forms and drug delivery devices, and complex drug-release profiles have been 3D printed as biomedical applications. One of the most studied applications of 3D printing is the production of pharmaceuticals with complex drug-release characteristics. Compressed dosage formulations are frequently manufactured from a blend of active and inactive substances that is homogenous, and as a result, they are often constrained to a single drug-release profile. 3D printers, on the other hand, may print binder onto a matrix powder. Layers of 200 micrometers thick are used to create a barrier. between the active ingredients to make it easier to administer a controlled substance release. In addition, 3D-printed dosage forms can be made in a variety of ways. porous, complicated shapes that are loaded with many medicines all over, surrounded by modulating barrier layers release.

The body's basic structural units are tissues and organs. Organs are collections of tissues, whereas tissues are collections of cells that serve the same purpose (Datta, 2018). Organ systems are made up of groupings of organs that have the same function and purpose. Tissues are spread evenly throughout the body and have comparable functions. Organs in plants and animals are made up of tissues that are structured and execute specialized functions. Epithelial tissue, connective tissue, muscle tissue, and nerve tissue are the four main tissue types found in humans and other big multicellular organisms (Chiquet-Ehrismann and Tucker, 2004; Kamrani et al., 2019). 3D printing has been used to print organs and tissues.

One of the most well-known uses of 3D bioprinting is tissue engineering. It allows for the creation of complex tissues and organs that can be used to replace tissues that have failed or been lost. A variety of tissue engineering techniques are used to fabricate skin tissue. Tissue engineering can be used to create autologous split-thickness skin grafts, allografts, acellular dermal substitutes, and cellularized graft-like commercial items, among other things. An eight-channel valve-based bioprinter can be used to bioprint skin tissue, resulting in a 13-layer tissue made of collagen hydrogel. To build structures with tightly packed cells in epidermal layers, keratinocytes are bioprinted on top of alternating layers of human foreskin fibroblasts and acellular collagen layers. After around 10 days in the stratified epidermis, the tissue constructs are engrafted with the host. The stratum corneum, as well as certain blood vessels, show early signs of differentiation and development because of this. The biomaterial employed in the procedure varies; however, keratinocytes and fibroblasts are the most common cells used. Furthermore, skin with infections or disorders can be used as biomaterials for bioprinting to investigate the disease's pathogenesis.

Because the nature of such hard tissues is uncomplicated and primarily made up of inorganic materials, bone and cartilage creation is the most advanced use of bioprinting. Even though various procedures like as gas foaming, salt leaching, and freeze-drying have been used to create hard tissues, 3D bioprinting creates the most precise architectures. Polymethacrylate scaffolds are made from bone marrow-derived human mesenchymal stem cells using a thermal inkjet bioprinter. To manage the spatial location of cells, the cells are coprinted with bioactive glass nanoparticles. A printed bioink is made by combining nano fibrillated cellulose and alginate with human chondrocytes as living soft tissue in cartilage tissue engineering.

Many metabolic, endocrine, and exocrine activities are controlled by the liver. Liver failure is responsible for almost 2 million fatalities per year. Modern 3D bioprinting techniques combined with autologous induced pluripotent stem cell (iPS)-derived grafts could be a viable tissue engineering treatment option for individuals with end-stage liver disease. However, bioprinting protocols that accurately replicate the epithelial parenchyma of the liver are still being developed. Liver cells have a high potential to regenerate which makes it difficult to bioprint the liver tissue. However, the number of healthy donors is limited, and the regeneration time for such a liver is lengthy. Cells such as primary hepatocytes and stem-cell-derived hepatocytes are used as bioink in this process. 3D printing technology can create a liver that is the exact size and shape that the patient requires (da Silva Morais et al., 2020). Canaliculi are formed by bioprinting and are joined together by the collagen matrix to build bigger structures. Researchers have attempted and made progress in 3D printing of liver cells (Lewis and Shah, 2016, Zhong et al., 2016). X. Ma et al. (2018) used DLP for 3D bioprinting of decellularized extracellular matrix with regionally varied mechanical properties and biomimetic microarchitecture. Goulart et al. (2019) studied 3D bioprinting of liver spheroids derived from human induced pluripotent stem cells used to sustain liver function and viability in vitro. The process of bioprinting liver cells was outlined in vitro liver tissue construction (Ma et al., 2020).

Series of research has been done on blood vessels fabricated by 3D bioprinting (El-Ghazali et al., 2021, Fazal et al., 2021, Li et al., 2018, Papaioannou et al., 2019). In the elderly, cardiovascular diseases (CVDs) are the main cause of death (Mc Namara et al., 2019). The replacement of blocked or narrowed arteries, which is currently the best vascular transplant connected with autograft transplantation, is a frequent medical surgery for the treatment of CVDs (Fortier et al., 2014). In general, the mammary gland's saphenous veins and radial arteries are thought to be the selected channels for vascular replacement. Artificial blood vessels (ABVs) are not employed in many cardiac patients for a variety of reasons, including the patient's age, the tiny size of the veins, past impressions, and excessively large veins (Matarneh et al., 2018). The fabrication of tissues and organs relies on vascularization to give oxygen and media to the bioprinted constructions, hence bioprinting of vascular networks is critical. Extrusion- and laser-assisted bioprinting techniques are among the bioprinting techniques utilized to create bioprinted vascular networks. Hydrogen gels, such as sodium alginates and chitosan, are bioprinted directly in tubular form with encapsulated cells during bioprinting. The tubular structures formed as a result of this process have improved metabolic transport and cellular viability. Noor et al. (2019) describe a straightforward method for 3D-printing thick, vascularized, and perfusable cardiac patches that perfectly match the patient's immunological, biological, biochemical, and morphological characteristics. Patients are given a sample of omental tissue for this purpose. The extracellular matrix is processed into a customized hydrogel while the cells are reprogrammed to become pluripotent stem cells and differentiated into cardiomyocytes and endothelial cells. Bioinks for parenchymal heart tissue and blood vessels are made by combining the two cell types with hydrogels separately. It is proved that functioning vascularized patches may be printed according to the anatomy of the patient. The patches' structure and function are investigated in vitro, and the morphology of cardiac cells is evaluated following transplantation, revealing elongated cardiomyocytes with extensive actinin striation. Cellularized human hearts with natural architecture are printed as a proof of concept. Their findings show the method's potential for creating tailored tissues and organs, as well as drug screening in a patient-specific anatomical structure and biochemical milieu. Also, Ulag et al. (2019) did a 3D printing of artificial blood vessel constructs using PCL/chitosan/hydrogel biocomposites.

Pinnock et al. (2016) developed a customizable engineered blood vessels using 3D printed inserts which mimicked a crucial structural component of blood arteries – the smooth muscle layer, or tunica media – a novel method for constructing customized, tissue designed, self-organizing vascular constructions. The study uses a one-of-a-kind technology that combines 3D printed plate inserts to control construct size and shape, as well as cell sheets supported by a temporary fibrin hydrogel to induce cellular self-organization into a tubular form that resembles a natural artery. 3D printed inserts are glued to tissue culture plates, fibrin hydrogel is deposited around the inserts, and human aortic smooth muscle cells are seeded above the fibrin hydrogel to produce the vascular construct. The gel collects toward the center post insert, facilitated by the smooth muscle cells' natural contractile characteristics, forming a tissue ring of smooth muscle cells. Esmaeili et al. (2019) examined the application of artificial vasculature' biomechanical and chemical properties in coronary artery bypassing as a soft tissue engineering procedure in atherosclerosis. To make the ABVs, the researchers used thermoplastic polyurethane (TPU) made from nanocrystalline hydroxyapatite (HA) nanopowder and the extrusion process. The ideal specimen was investigated using X-ray diffraction (XRD) and scanning electron microscopy (SEM). The capacity to determine the elastic modulus, wettability, and porosity of the veins via fused deposition modeling and 3D printing was a key aspect of the ABVs. Esmaeili et al. (2019) developed an artificial blood vessel fabricated by 3D printing for pharmaceutical application. Their study developed a novel artificial biocomposite blood vessel using polymer-reinforced and bioceramic nanoparticles. The lateral view of SEM image constructed by 3D printer had a pore size between 50 and 100 microns.

Drug discovery is costly and takes time hence requiring a significant financial and human resource investment (Shaker et al., 2021, Tamimi and Ellis, 2009). As a result, developing a technique to improve the ability to anticipate the efficacy and toxicity of newly produced medications earlier in the drug discovery process saves time and money. Bioprinting can create 3D tissue models that look like genuine tissue and can be used in high-throughput tests (Peng et al., 2016). The most popular tissues used to construct tissue models for drugs are liver and tumor tissues. Furthermore, tissue models of such cells can be constructed and tested based on the target cells of produced medications. Epithelial cells constitute the lining through which the medication diffuses into the bloodstream, therefore tissue constructions of these cells are first made. The path of medications and their action on target cells can be predicted based on investigations on such constructions. Bioprinting can also be utilized to develop prescription medications as an alternative method. By using a set of biological inks to generate suitable amounts of medication print, the pharmaceuticals can even be personalized for each patient. Instead of taking many tablets throughout the day, 3D-printed composite pills comprising multiple medications with different release rates can be employed. Series of studies have been performed on drug discovery (Awad et al., 2018, Lee and Dai, 2015, Svanström et al., 2021, Yao et al., 2019).

The process of finding potential detrimental effects of chemicals on people or the environment is known as toxicology screening or testing (Allen and McWhinney, 2019, Ankley and Edwards, 2018). Pharmaceutical substances, cosmetic ingredients, home chemicals, and industrial chemicals are examples of chemicals. Some studies investigating the toxicity of certain compounds may necessitate a higher number of human participants with different metabolisms, which may appear unethical. Animals can be used in some research. However, they may not be able to accurately or reliably anticipate human responses. Instead, 3D bioprinting may be used to create structures that imitate the form and function of human tissues in a highly automated and advanced manner. Real-time monitoring and high-throughput screening of diverse compounds are made easier with the usage of such structures. For a long time, cosmetic chemicals have been tested in human-relevant skin tissue models. On models that replicate human tissue architecture, these experiments look at skin absorption, skin irritation, skin corrosion, and skin sensitization. Pirosa et al. (2018) review gave an overview of existing models and methodologies for generating in vitro stem cell-based vascularized bone, with a focus on the major vasculature engineering problems. These difficulties originate from the selection of biomaterials, scaffold-building processes, and cells as well as the types of culturing conditions needed, particularly the use of dynamic culture systems in bioreactors. Using their freeform reversible embedding of suspended hydrogels technology, Adam Feinberg and his colleagues created a full-size 3D bioprinted human heart model (Mirdamadi et al., 2020).

In cancer research, 3D bioprinting is replacing 2D tumor models. For a long time, 2D tumor models have been utilized in cancer research, but they do not accurately depict the physiologically relevant environment because they lack cell-cell interactions. 3D bioprinting, on the other hand, enables for accurate simulation of the disease microenvironment to investigate cancer development and spread. With a regulated cell density and cell-cell distance and a sparsely controlled microenvironment, many cell types can be bioprinted at the same time to construct multicellular structures in a repeatable manner.

In human clinical trials, cancer-fighting compounds that appear promising in the lab frequently fail. One reason is that standard tumor models involve cancer cells growing in a flat plastic dish, which pales in comparison to the complex architectural development and ongoing feedback from other tissues that occur in a living body. Scientists at OHSU's Knight Cancer Institute are creating 3D models of human tumors in the laboratory, shown in Figure 8.7 as adapted from OHSU.

Layers of living cells are deposited by the printer, resulting in tissues made up of a variety of cell types. Cancer cells develop and exchange signals with different cell types within these "bioprinted" tissues. They mature, secrete extracellular matrix, and self-organize to form features similar to those found in real tumors, such as blood vessel networks.

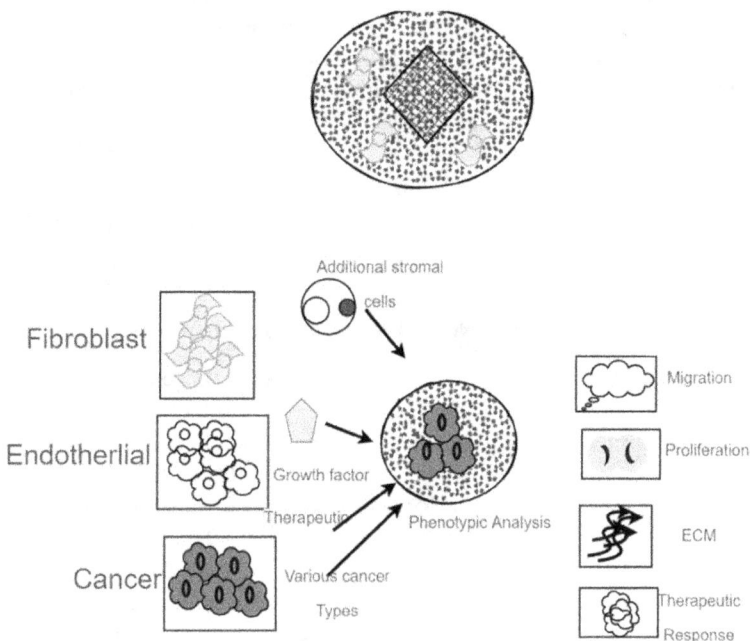

FIGURE 8.7   3D printing of live cells for cancer study.

To explore cell aggregation, HeLa cells can be bioprinted in a gelatin-alginate composite hydrogel. These tissues can be utilized to research cancer progression and changes in tissue structure and function throughout time. Tissue models can also be used to test the efficacy of different treatment strategies against different carcinogens. 3D printing accelerates cancer research development (Haleem et al., 2020).

Serrano et al.'s (2018) review focused on the major uses of 3D printing in cancer, including the production of in vitro cancer models as well as individualized cancer treatments for patients, with an emphasis on hydrogels and therapeutic implants. Tumor heterogeneity and interindividual variability are key issues in cancer treatment since each patient reacts differently to existing therapeutic regimens. 3D printing is a technique that can help cancer patients by allowing for individualization of treatment through the creation of in vitro models with microenvironments that more closely match real-life cancer circumstances, making complex medicines more accessible. A 3D printed, 3D CAD model, and AR kidney cancer models with the kidney-clear, tumor-white (3D print and computer), tumor-purple (AR), artery-red, vein-blue, collecting system-yellow. A 3D printed, 3D computer model, and AR prostate cancer models (sagittal view) with the prostate-clear, tumor-blue, rectal wall-white, bladder neck and urethra-yellow, and neurovascular bundles-pink were diagrammatically represented by Wake et al. (2019).

8.3.7 MARKET OUTLOOK

There are now companies that employ 3D printing for commercial medical uses. Helisys, Ultimateker, and Organovo, a startup that fabricates living human tissue using 3D printing, are among them (Klein et al., 2013). However, 3D printing's significance in medicine is currently limited. Only US$11 million (1.6%) of the US$700 million 3D printing business is invested in medical applications (Laverty et al., 2016, Schubert et al., 2014). However, 3D printing is forecast to expand into a US$8.9 billion sector in the next ten years, with US$1.9 billion (21%) expected to be spent on medical applications (Kumar Gupta et al., 2022, Li et al., 2017).

8.3.8 BARRIERS TO 3D PRINTING IN BIOMEDICAL APPLICATIONS

The issue of safety, regulatory concerns, unrealistic expectations, patents, and copyright concerns continues to act as hindrances to the technology. Bioinks with good biocompatibility and mechanical strength are the key hurdles in bioprinting. The present bioprinter technology has a lesser resolution and speed than previous generations, posing a hurdle for future improvement. Similarly, bioprinters should be able to work with a variety of biomaterials. Because the current speed of the bioprinting technique is modest, it should be enhanced to mass-produce biomaterials at a commercially acceptable level. Tissue construct vascularization is a significant hurdle in 3D bioprinting since tissues demand constant oxygen and nutrients.

8.3.9 FUTURE OF BIOPRINTING

The evolution of 3D printing in medical started in 1984 with conventional 3D printing process. This has evolved from 2010 with bioprinting to 2013 that heralded 4D printing. The year 2014 saw the advent of 4D bioprinting and the latest is 5D printing which began in 2016. A modified schematic of the evolution of 3D printing in biomedical adapted from Ghilan et al. (2020) is shown in Figure 8.8.

Even though 3D bioprinting has had different degrees of success, according to B. Gao et al. (2016), it has a key downside. The biggest disadvantage is that 3D printing only considers the printed structure's initial condition and assumes it is static and inanimate. In 2013, a novel idea known as 4D printing was presented to solve this challenge, allowing bioengineered objects to be preprogrammed to evolve in a certain way after printing (Wu et al., 2018). Unlike prior technology, 4D printing is based on the integration of smart biomaterials and leverages the ability of shape and functionality modification over time when exposed to intrinsic/external stimuli, allowing a more exact reproduction of the dynamics of original tissues. Self-healing polymers, thermally activated polymers, smart/nanocomposites, piezoelectric materials, shape memory alloys, and shape memory polymers are among the potential responsive materials (Khan et al., 2018).

In addition, 4D bioprinting, also known as laser-aided bioprinting, has lately become popular. 4D bioprinting is a specialized extension of 3D bioprinting that uses stimuli-responsive biomaterials and cells to reproduce the biochemical and biophysical content, as well as the hierarchical shape, of diverse tissues (Yang et al., 2019). Hydrogels can be employed in 4D printing because of their propensity to swell when exposed to an appropriate solvent. Although, the following limitations need to be considered; it has a flaw, can lose bulk during the hydration/dehydration cycle, and the actuation shape may not be stable due to water's volatility. These characteristics

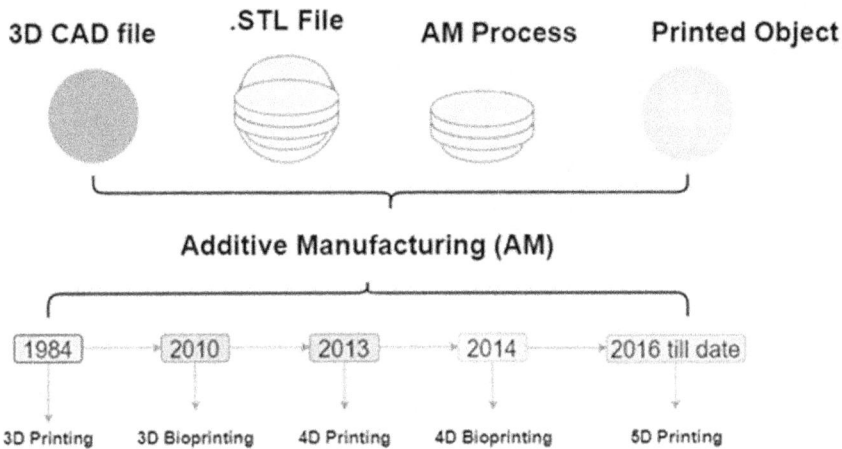

FIGURE 8.8   Schematic of the evolution of 3D printing in biomedical.

may jeopardize the printed construct's integrity. Novel ideas such as double networks, bilayered structures, and hydrogel composites have been created to overcome this issue (Nkomo, 2018).

Since 4D bioprinting is a new technology, there are several important challenges that must be addressed first, (1) the 4D constructs require cell-bioink synergy; (2) the cells must survive both the printing process and the stimuli required for shape change; and (3) the 4D fabricated construct's response capability must not be harmed by the inclusion of cells (Li et al., 2016). In Figure 8.9 adapted from Ghilan et al. (2020), a broad comparison of the printing technologies is provided.

William Yerazunis of Mitsubishi Electric Research Laboratories (MERL) introduced five-dimensional (5D) printing in 2016 (Ghazal et al., 2022). Due to a moveable plateau, five-axis 3D printing is an extension of 3D printing in which the print head may move around from five different angles (Isa and Lazoglu, 2019; Shen et al., 2018). This enables the creation of curved layers that are more durable than typical 3D fat layers (Anas et al., 2022). This also means that stronger curved-shaped items or implants can be created, with potential uses in orthopedics and dentistry.

There are also some ethical concerns about the technology of 3D bioprinting. The high cost of 3D bioprinting threatens the successful deployment and implementation of the technology for all sundry, especially the poor. Bioprinting is a new technology, hence it needs to be thoroughly researched to verify that it is safe for humans. Personalized 3D printing technology may result in a slew of regulatory issues in terms of overseeing printed products.

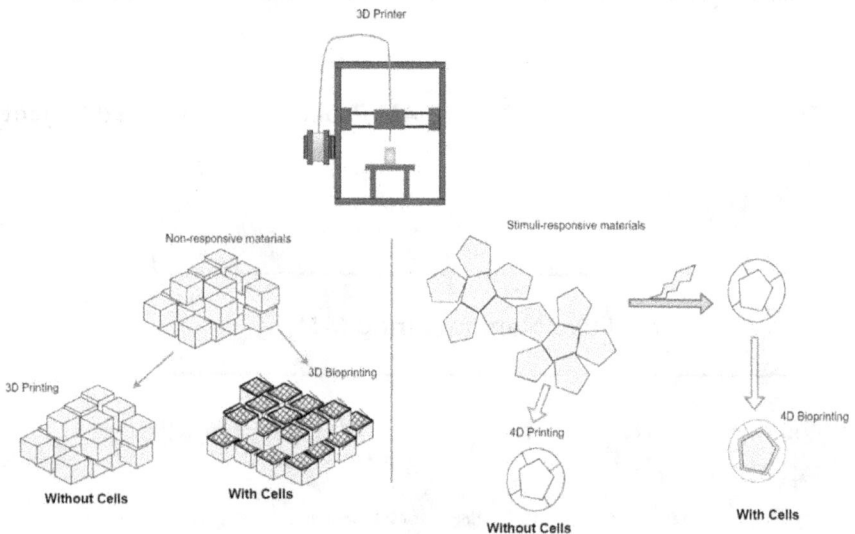

FIGURE 8.9 Depiction of key differences between 4D and 3D (bio)printing.

8.4 APPLICATION OF 3D PRINTING FOR KIDS TOYS

This is an amazing area that a lot of people have used to keep kids entertained while making fortune. The toy industry is a billion-dollar industry generating revenue in high volumes. 3D printing has been used to print different types of toys for kids.

Figure 8.10 shows some toys that could be modified and 3D printed with ease. Extrusion 3D printing technology is the major beneficiairy of this sector. Different filaments with beautiful designs and postprocessing can be used to make it more aesthetic and appealing. Some will require additional assembling and others may not need assembling. The toy gun shown in Figure 8.11 requires each of the components to be 3D printed and assembled.

8.4.1 APPLICATION OF 3D PRINTING IN AVIATION AND AEROSPACE/SPACE IN ERA OF 4IR

The aviation and aerospace/space sector is one that continues to benefit from the discovery of 3D printing. The fuselage and other parts of an aircraft are some of the components that are being 3D printed using different filaments and resin 3D printing technologies. There is also the use of cement 3D printing technologies in space. NASA and other space agencies are exploring ways of using this technology to build structures in space. The bulky nature of the cement 3D printing technology is being researched to miniature it to ease transportation to space.

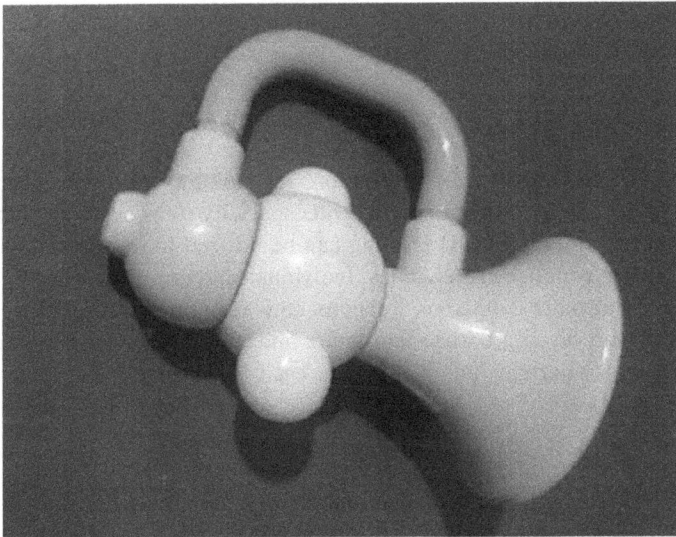

FIGURE 8.10   Picture of kids toy that can be 3D printed.

8.4.2 Application of 3D Printing in Construction

Although the first steps were taken nearly three decades ago, 3D building printing has had a long struggle to gain traction. Contour Crafting and D-Shape were the first technologies to receive public notice, with a few irregular pieces from 2008 to 2012 and a 2012 TV report. D-Shape was also featured in "The Man Who Prints Houses," an independent documentary about its designer Enrico Dini. The introduction of the first 3D printed structure, utilizing prefabricated 3D printed components created by Winsun, which claimed to be able to print 10 houses in a day using its technology, was a significant breakthrough. Despite the fact that the allegations have yet to be verified, the tale has gained widespread attention and sparked increased interest in the topic. Many new businesses sprouted up in a short of months. Many new ventures were publicized as a result, including the first pedestrian 3D-printed bridge and the first cycling 3D-printed bridge, as well as an early structural element built with 3D printing in 2016 (Roberts, 2019, Zastrow, 2020).

COBOD International, formerly 3DPrinthuset (its sister firm), has recently gotten a lot of press for its first permanent 3D-printed building, which is the first of its kind in Europe. The project set a significant precedent as the first 3D-printed structure to have a building permit and documentation in place, as well as complete approval from the city authorities, a critical milestone for wider adoption in the construction field (Elasad and Amirov, 2020). The story received widespread national and international publicity, with segments airing on television in countries such as Denmark, Russia, Poland, and Lithuania, among others.

A 3D-printed home can be completed in just 12 hours. It reduces the structure's carbon footprint by 70%. A six-room RDP house at the University of Johannesburg to address social housing issues built in a day is now one of the educational awareness for 3D printing in Africa. A school and house already printed in Malawi with Kenya and Zimbabwe next in line. Although, start-up costs are high, modest houses can be printed for under US$8000.

8.5 WARFARE IN ERA OF 4IR

The era of 4IR ushers a lot of innovation, technology, and sophistication. Any munitions maker can use 3D printing to create specialty rounds as part of their R&D program. On the other hand, 3D printing can be utilized to manufacture less lethal rounds for usage by police and civilians. Individuals hitherto were using 3D printing to print guns and other dangerous weapons until restrictions were imposed. A lot can be made with the appropriate legislature and government regulation using 3D printing. Figure 8.11 is a toy gun that may be easily 3D printed.

8.5.1 Education

Students learn faster with aid of visuals and objects. 3D printing is a key technology that can crash the cost associated with instructional materials, scientific and other hardware needed for better assimilation by students across the globe, especially those

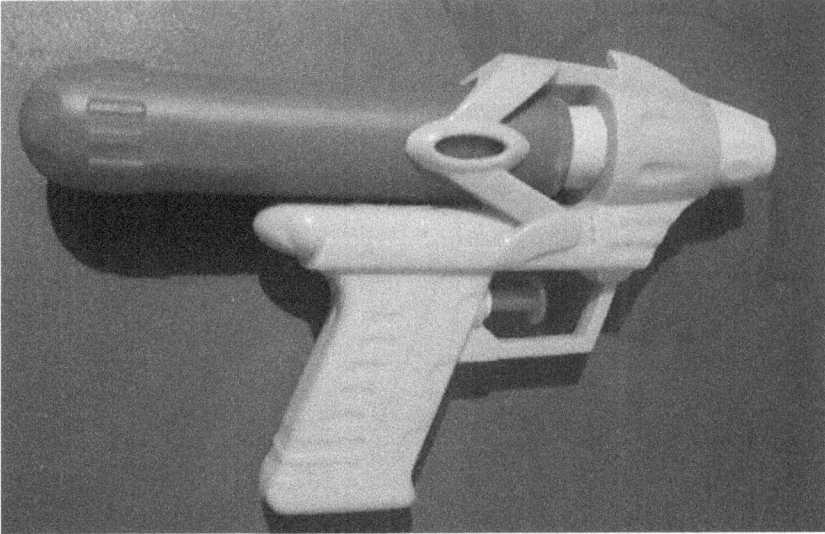

FIGURE 8.11 A picture of toy gun.

in the global south of Africa and beyond. A couple of work has been done in mini-aturing chemistry and other scientific equipment using 3D printing.

8.6   CONCLUSION

This chapter looked at some of critical emerging technologies helping to shape the world and space. It also discusses the application of 3D printing in some of the emerging sectors. Electricity is a crucial part of humanity. 3D printing technology will play a vital role in most of the renewable energy such as wind and hydro turbines manufacturing. It could be the game changer for every individual to print their own electricity using small hydro turbine. It also looks at application of 3D printing in improving and reducing cost of aircraft and other aerospace. It also discusses using cement 3D printing in outer space for construction of structures.

REFERENCES

Abubakre, O. K., Medupin, R. O., Akintunde, I. B., Jimoh, O. T., Abdulkareem, A. S., Muriana, R. A., James, J. A., Ukoba, K. O., Jen, T. C., & Yoro, K. O. (2023). Carbon nanotube-reinforced polymer nanocomposites for sustainable biomedical applications: A review. Journal of Science: Advanced Materials and Devices, 8(2), 1–17.

Ahmad, N., Gopinath, P., & Dutta, R. (2019). *3D printing technology in nanomedicine.* Elsevier.

Akiah, M.-A., Ahmad, M. A., Mustafa, Z., Alkahari, M. R., Khuen, C. C., Kasim, M. S., & Pembuatan, F. K. (2020). Preliminary development of modular-based hydrogel extruder for 3D-bioprinting. *Proceedings of Mechanical Engineering Research Day, 2020,* 69–71.

Aljohani, W., Ullah, M. W., Zhang, X., & Yang, G. (2018). Bioprinting and its applications in tissue engineering and regenerative medicine. *International Journal of Biological Macromolecules, 107*, 261–275.

Allen, P. B., Khaing, Z., Schmidt, C. E., & Ellington, A. D. (2015). 3D printing with nucleic acid adhesives. *ACS Biomaterials Science & Engineering, 1*(1), 19–26.

Allen, D. R., & McWhinney, B. C. (2019). Quadrupole time-of-flight mass spectrometry: A paradigm shift in toxicology screening applications. *The Clinical Biochemist Reviews, 40*(3), 135.

Anas, S., Khan, M.Y., Rafey, M., & Faheem, K. (2022). Concept of 5D printing technology and its applicability in the healthcare industry. Materials Today: Proceedings, *56*, 1726–1732.

Ankley, G. T., & Edwards, S. W. (2018). The adverse outcome pathway: A multifaceted framework supporting 21st century toxicology. *Current Opinion in Toxicology, 9*, 1–7.

Ares, G., Ha, B., & Jaeger, S. R. (2021). Consumer attitudes to vertical farming (indoor plant factory with artificial lighting) in China, Singapore, UK, and USA: A multi-method study. *Food Research International, 150*, 110811.

Ashraf, H., Meer, B., Naz, R., Saeed, A., Sajid, U., Nisar, K., ... Anwar, P. (2018). 3D-bioprinting: A stepping stone towards enhanced medical approaches. *Advancements in Life Sciences, 5*(4), 143–153.

Awad, A., Trenfield, S. J., Goyanes, A., Gaisford, S., & Basit, A. W. (2018). Reshaping drug development using 3D printing. *Drug Discovery Today, 23*(8), 1547–1555.

Axbom, S., & Ralsgård, L. (2019). *Design of an autonomous weeding vehicle.* Master thesis submitted to Division of Product Development, Department of Design Sciences Faculty of Engineering LTH, Lund University, 1–147.

Bailey, G. E. (1915). *Vertical farming.* EI Dupont de Nemours Powder Company.

Banks, J. (2013). Adding value in additive manufacturing: Researchers in the United Kingdom and Europe look to 3D printing for customization. *IEEE Pulse, 4*(6), 22–26.

Barker, C. C. (2010). *American urban agriculture.* [Senior Thesis]. Maryville College.

Bastawrous, S., Wu, L., Strzelecki, B., Levin, D. B., Li, J.-S., Coburn, J., & Ripley, B. (2021). Establishing quality and safety in hospital-based 3D printing programs: Patient-first approach. *Radiographics, 41*(4), 1208–1229.

Budharaju, H., Zennifer, A., Sethuraman, S., Paul, A., & Sundaramurthi, D. (2022). Designer DNA biomolecules as a defined biomaterial for 3D bioprinting applications. *Materials Horizons, 9*(4), 1141–1166.

Chiquet-Ehrismann, R., & Tucker, R. P. (2004). Connective tissues: Signalling by tenascins. *The International Journal of Biochemistry & Cell Biology, 36*(6), 1085–1089.

Cui, X., Boland, T., D'Lima, D., & Lotz, M. (2012). Thermal inkjet printing in tissue engineering and regenerative medicine. *Recent Patents on Drug Delivery & Formulation, 6*(2), 149–155.

da Silva Morais, A., Vieira, S., Zhao, X., Mao, Z., Gao, C., Oliveira, J. M., & Reis, R. L. (2020). Advanced biomaterials and processing methods for liver regeneration: State-of-the-art and future trends. *Advanced Healthcare Materials, 9*(5), 1901435.

Datta, A. K. (2018). *Principles of general anatomy.* Current Books International.

Datta, P., Ayan, B., & Ozbolat, I. T. (2017). Bioprinting for vascular and vascularized tissue biofabrication. *Acta Biomaterialia, 51*, 1–20.

Dimitroulis, G., Austin, S., Lee, P. V. S., & Ackland, D. (2018). A new three-dimensional, print-on-demand temporomandibular prosthetic total joint replacement system: Preliminary outcomes. *Journal of Cranio-Maxillofacial Surgery, 46*(8), 1192–1198.

El-Ghazali, S., Khatri, M., Hussain, N., Khatri, Z., Yamamoto, T., Kim, S. H., ... Kim, I. S. (2021). Characterization and biocompatibility evaluation of artificial blood vessels prepared from pristine poly (Ethylene-glycol-co-1, 4-cyclohexane dimethylene-co-isosorbide terephthalate), poly (1, 4 cyclohexane di-methylene-co-isosorbide terephthalate) nanofibers and their blended composition. *Materials Today Communications, 26,* 102113.

Elasad, M., & Amirov, D. (2020). 3D printed construction and implementation in Cyprus: Discussion and overview. *International Journal of Advanced Engineering, Sciences and Applications, 1*(2), 26–32.

Esmaeili, S., Shahali, M., Kordjamshidi, A., Torkpoor, Z., Namdari, F., Samandari, S. S., … Khandan, A. (2019). An artificial blood vessel fabricated by 3D printing for pharmaceutical application. *Nanomedicine Journal, 6*(3), 183–194.

Fazal, F., Sanchez, F. J. D., Waqas, M., Koutsos, V., Callanan, A., & Radacsi, N. (2021). A modified 3D printer as a hybrid bioprinting-electrospinning system for use in vascular tissue engineering applications. *Medical Engineering & Physics, 94,* 52–60.

Fessenden, M. (2013). 3-D printed windpipe gives infant breath of life. *Nature, 28,* 1–2.

Fortier, A., Gullapalli, V., & Mirshams, R. A. (2014). Review of biomechanical studies of arteries and their effect on stent performance. *IJC Heart & Vessels, 4,* 12–18.

Ganguli, A., Pagan-Diaz, G. J., Grant, L., Cvetkovic, C., Bramlet, M., Vozenilek, J., … Bashir, R. (2018). 3D printing for preoperative planning and surgical training: A review. *Biomedical Microdevices, 20*(3), 1–24.

Gao, G., Huang, Y., Schilling, A. F., Hubbell, K., & Cui, X. (2018). Organ bioprinting: Are we there yet? *Advanced Healthcare Materials, 7*(1), 1701018.

Gao, B., Yang, Q., Zhao, X., Jin, G., Ma, Y., & Xu, F. (2016). 4D bioprinting for biomedical applications. *Trends in Biotechnology, 34*(9), 746–756.

Gebler, M., Uiterkamp, A. J. S., & Visser, C. (2014). A global sustainability perspective on 3D printing technologies. *Energy Policy, 74,* 158–167.

Ghazal, A. F., Zhang, M., Mujumdar, A. S., & Ghamry, M. (2022). Progress in 4D/5D/6D printing of foods: Applications and R&D opportunities. *Critical Reviews in Food Science and Nutrition,* 1–24. https://doi.org/10.1080/10408398.2022.2045896

Ghilan, A., Chiriac, A. P., Nita, L. E., Rusu, A. G., Neamtu, I., & Chiriac, V. M. (2020). Trends in 3D printing processes for biomedical field: Opportunities and challenges. *Journal of Polymers and the Environment, 28*(5), 1345–1367.

Goriainov, V., McEwan, J. K., Oreffo, R. O., & Dunlop, D. G. (2018). Application of 3D-printed patient-specific skeletal implants augmented with autologous skeletal stem cells. *Regenerative Medicine, 13*(3), 283–294.

Goulart, E., de Caires-Junior, L. C., Telles-Silva, K. A., Araujo, B. H. S., Rocco, S. A., Sforca, M., … Assoni, A. F. (2019). 3D bioprinting of liver spheroids derived from human induced pluripotent stem cells sustain liver function and viability in vitro. *Biofabrication, 12*(1), 015010.

Gross, B. C., Erkal, J. L., Lockwood, S. Y., Chen, C., & Spence, D. M. (2014). *Evaluation of 3D printing and its potential impact on biotechnology and the chemical sciences.* ACS Publications.

Gu, Z., Fu, J., Lin, H., & He, Y. (2020). Development of 3D bioprinting: From printing methods to biomedical applications. *Asian Journal of Pharmaceutical Sciences, 15*(5), 529–557.

Haleem, A., Javaid, M., & Vaishya, R. (2020). 3D printing applications for the treatment of cancer. *Clinical Epidemiology and Global Health, 8*(4), 1072–1076.

Halle, S., & Hickle, J. (2011). *The design and implementation of a semi-autonomous surf-zone robot using advanced sensors and a common robot operating system.* Master thesis to Naval Postgraduate School, Monterey, California, USA, 1–131. Retrieved from https://citeseerx.ist.psu.edu/document?repid=rep1&type=pdf&doi=007dccbea39be6a46f2a7 da0fd42a62f1be72d7e

Han, X., Sharma, N., Xu, Z., Scheideler, L., Geis-Gerstorfer, J., Rupp, F., … Spintzyk, S. (2019). An in vitro study of osteoblast response on fused-filament fabrication 3D printed PEEK for dental and cranio-maxillofacial implants. *Journal of Clinical Medicine, 8*(6), 771.

Hinton, T. J., Jallerat, Q., Palchesko, R. N., Park, J. H., Grodzicki, M. S., Shue, H.-J., … Feinberg, A. W. (2015). Three-dimensional printing of complex biological structures by freeform reversible embedding of suspended hydrogels. *Science Advances, 1*(9), e1500758.

Hoang, D., Perrault, D., Stevanovic, M., & Ghiassi, A. (2016). Surgical applications of three-dimensional printing: A review of the current literature & how to get started. *Annals of Translational Medicine, 4*(23), 456. https://doi.org/10.21037/atm.2016.12.18

Hornick, J., & Roland, D. (2013). 3D printing and intellectual property: Initial thoughts. *The Licensing Journal, 33*(7), 12.

Iordache, F. (2019). Bioprinted scaffolds. In A.-M. Holban, & A. M. Grumezescu (Eds.), *Materials for Biomedical Engineering* (pp. 35–60). Elsevier.

Iram, D., a Riaz, R., & Iqbal, R. K. (2019). 3D Bioprinting: An attractive alternative to traditional organ transplantation. *Biomedical Science and Engineering, 5*(1), 007–018.

Isa, M. A., & Lazoglu, I. (2019). Five-axis additive manufacturing of freeform models through buildup of transition layers. *Journal of Manufacturing Systems, 50*, 69–80.

Jose, A., Nandagopalan, S., & Akana, C. M. V. S. (2021). Artificial intelligence techniques for agriculture revolution: A survey. *Annals of the Romanian Society for Cell Biology, 25*(4), 2580–2597.

Kamrani, P., Marston, G., & Jan, A. (2019). *Anatomy, connective tissue. Study guide.* StatPearls Publishing.

Khan, F. A., Celik, H. K., Okan, O., & Rennie, A. E. (2018). A short review on 4d printing. *International Journal of 3D Printing Technologies and Digital Industry, 2*(2), 59–67.

Klebe, R. J. (1988). Cytoscribing: a method for micropositioning cells and the construction of two-and three-dimensional synthetic tissues. *Experimental Cell Research, 179*(2), 362–373.

Klein, G. T., Lu, Y., & Wang, M. Y. (2013). 3D printing and neurosurgery—ready for prime time? *World Neurosurgery, 80*(3–4), 233–235.

Kumar Gupta, D., Ali, M. H., Ali, A., Jain, P., Anwer, M. K., Iqbal, Z., & Mirza, M. A. (2022). 3D printing technology in healthcare: Applications, regulatory understanding, IP repository and clinical trial status. *Journal of Drug Targeting, 30*(2), 131–150.

Kusaka, M., Sugimoto, M., Fukami, N., Sasaki, H., Takenaka, M., Anraku, T., … Hoshinaga, K. (2015). Initial experience with a tailor-made simulation and navigation program using a 3-D printer model of kidney transplantation surgery. Paper presented at the *Transplantation Proceedings, 47*(3), 596–599.

Laverty, D. P., Thomas, M. B., Clark, P., & Addy, L. D. (2016). The use of 3D metal printing (direct metal laser sintering) in removable prosthodontics. *Dental Update, 43*(9), 826–835.

Lee, V. K., & Dai, G. (2015). Three-dimensional bioprinting and tissue fabrication: Prospects for drug discovery and regenerative medicine. *Advanced Health Care Technologies, 1*, 23.

Les, A. S., Ohye, R. G., Filbrun, A. G., Ghadimi Mahani, M., Flanagan, C. L., Daniels, R. C., … Green, G. E. (2019). 3D-printed, externally-implanted, bioresorbable airway splints for severe tracheobronchomalacia. *The Laryngoscope, 129*(8), 1763–1771.

Lewis, P. L., & Shah, R. N. (2016). 3D printing for liver tissue engineering: Current approaches and future challenges. *Current Transplantation Reports, 3*(1), 100–108.

Li, V. C.-F., Dunn, C. K., Zhang, Z., Deng, Y., & Qi, H. J. (2017). Direct ink write (DIW) 3D printed cellulose nanocrystal aerogel structures. *Scientific Reports, 7*(1), 1–8.

Li, X., Liu, L., Zhang, X., & Xu, T. (2018). Research and development of 3D printed vascula-ture constructs. *Biofabrication, 10*(3), 032002.

Li, Y.-C., Zhang, Y. S., Akpek, A., Shin, S. R., & Khademhosseini, A. (2016). 4D bioprinting: The next-generation technology for biofabrication enabled by stimuli-responsive materials. *Biofabrication, 9*(1), 012001.

Loai, S., Kingston, B. R., Wang, Z., Philpott, D. N., Tao, M., & Cheng, H.-L. M. (2019). Clinical perspectives on 3D bioprinting paradigms for regenerative medicine. *Regenerative Medicine Frontiers, 1*(1), 1–40.

Ma, L., Wu, Y., Li, Y., Aazmi, A., Zhou, H., Zhang, B., & Yang, H. (2020). Current advances on 3D-bioprinted liver tissue models. *Advanced Healthcare Materials, 9*(24), 2001517.

Ma, X., Yu, C., Wang, P., Xu, W., Wan, X., Lai, C. S. E., … Chen, S. (2018). Rapid 3D bioprinting of decellularized extracellular matrix with regionally varied mechanical properties and biomimetic microarchitecture. *Biomaterials, 185*, 310–321.

Malinauskas, M., Lukoševičius, L., Mackevičiūtė, D., Balčiūnas, E., Rekštytė, S., & Paipulas, D. (2014). Multiscale 3D manufacturing: Combining thermal extrusion printing with additive and subtractive direct laser writing. Paper presented at the *Laser Sources and Applications II, 9135*, 124–135.

Mansour, A., Alabdouli, H., Alqaydi, H., Al, K., Ahmed, W., & Al, J. (2020). Evaluating the 3D printing capabilities. Paper presented at the *Proceedings of the International Conference on Industrial Engineering and Operations Management, 10th Annual International IEOM Conference*, Dubai, UAE.

Matarneh, R., Sotnik, S., & Lyashenko, V. (2018). Polymers in cardiovascular surgery. *Asian Journal of Pharmaceutical and Clinical Research, 11*(5), 1–63.

Mc Namara, K., Alzubaidi, H., & Jackson, J. K. (2019). Cardiovascular disease as a leading cause of death: How are pharmacists getting involved? *Integrated Pharmacy Research & Practice, 8*, 1.

McDaniel, L. (2017). 3D printing in medicine: Challenges beyond technology. Paper presented at the *Frontiers in Biomedical Devices*, American Society of Mechanical Engineers, 40672, V001T11A018.

Mertz, L. (2013). Dream it, design it, print it in 3-D: What can 3-D printing do for you? *IEEE Pulse, 4*(6), 15–21.

Mirdamadi, E., Tashman, J. W., Shiwarski, D. J., Palchesko, R. N., & Feinberg, A. W. (2020). FRESH 3D bioprinting a full-size model of the human heart. *ACS Biomaterials Science & Engineering, 6*(11), 6453–6459.

Miri, A. K., Mostafavi, E., Khorsandi, D., Hu, S.-K., Malpica, M., & Khademhosseini, A. (2019). Bioprinters for organs-on-chips. *Biofabrication, 11*(4), 042002.

Mitsouras, D., Liacouras, P. C., Wake, N., & Rybicki, F. J. (2020). RadioGraphics update: med-ical 3D printing for the radiologist. *Radiographics, 40*(4), E21–E23.

Moghimi, F., & Asiabanpour, B. (2021). *Economics of vertical farming: Quantitative decision model and a case study for different markets in the USA.* Research Square preprint, 1–34.

Murcia, V. A., Palacios, J. F., & Barbieri, G. (2021). FarmBot simulator: Towards a virtual environment for scaled precision agriculture. Paper presented at the *International Workshop on Service Orientation in Holonic and Multi-Agent Manufacturing* (pp. 234–246). Springer International Publishing.

Murphy, S. V., & Atala, A. (2014). 3D bioprinting of tissues and organs. *Nature Biotechnology, 32*(8), 773–785.

Nkomo, N. (2018). A review of 4D printing technology and future trends. Paper presented at the *11th South African Conference on Computational and Applied Mechanics*, 1–10. Vanderbijlpark, South Africa, September 17–19, 2018.

Noor, N., Shapira, A., Edri, R., Gal, I., Wertheim, L., & Dvir, T. (2019). 3D printing of personalized thick and perfusable cardiac patches and hearts. *Advanced Science, 6*(11), 1900344.

Pandya, J. R., Nagchaudhuri, A., Nindo, C., & Mitra, M. (2019). FarmBot-A platform for backyard precision farming: Installation and initial experimental layout. Paper presented at the *2019 ASABE Annual International Meeting*, 1–5. Boston, Massachusetts July 7–10, 2019.

Papaioannou, T. G., Manolesou, D., Dimakakos, E., Tsoucalas, G., Vavuranakis, M., & Tousoulis, D. (2019). 3D bioprinting methods and techniques: Applications on artificial blood vessel fabrication. *Acta Cardiologica Sinica, 35*(3), 284.

Pati, F., Gantelius, J., & Svahn, H. A. (2016). 3D bioprinting of tissue/organ models. *Angewandte Chemie International Edition, 55*(15), 4650–4665.

Peng, W., Unutmaz, D., & Ozbolat, I. T. (2016). Bioprinting towards physiologically relevant tissue models for pharmaceutics. *Trends in Biotechnology, 34*(9), 722–732.

Pinnock, C. B., Meier, E. M., Joshi, N. N., Wu, B., & Lam, M. T. (2016). Customizable engineered blood vessels using 3D printed inserts. *Methods, 99*, 20–27.

Pirosa, A., Gottardi, R., Alexander, P. G., & Tuan, R. S. (2018). Engineering in-vitro stem cell-based vascularized bone models for drug screening and predictive toxicology. *Stem Cell Research & Therapy, 9*(1), 1–23.

Poluri, K. M. (2019). Fabrication of biopolymer-based organs and tissues using 3D bioprinting. *3D Printing Technology in Nanomedicine, 43*, 43–62.

Radenkovic, D., Solouk, A., & Seifalian, A. (2016). Personalized development of human organs using 3D printing technology. *Medical Hypotheses, 87*, 30–33.

Reddi, A. H. (1998). Role of morphogenetic proteins in skeletal tissue engineering and regeneration. *Nature Biotechnology, 16*(3), 247–252.

Roberts, I. (2019). Trump, Twitter, and the First Amendment. *Alternative Law Journal, 44*(3), 207–213.

Rodríguez-Salvador, M., Rio-Belver, R. M., & Garechana-Anacabe, G. (2017). Scientometric and patentometric analyses to determine the knowledge landscape in innovative technologies: The case of 3D bioprinting. *PLoS One, 12*(6), e0180375.

Rodriguez-Salvador, M., & Ruiz-Cantu, L. (2019). Revealing emerging science and technology research for dentistry applications of 3D bioprinting. *International Journal of Bioprinting, 5*(1), 170.

Rubtsova, Y., & Onat, A. P. (2017). The 3D bioprinting as novel realm in it medicine. *ББК, 32 1 74*, 15.

Rueda-Ayala, V., Rasmussen, J., & Gerhards, R. (2019). Developing Automated and Autonomous Weed Control Methods On Pipfruit Orchards In New Zealand. The influence of post-emergence weed harrowing on selectivity, crop recovery and crop yield in different growth stages of winter wheat. *Weed Research, 51*(5), 478–488.

Salmi, M., Akmal, J. S., Pei, E., Wolff, J., Jaribion, A., & Khajavi, S. H. (2020). 3D printing in COVID-19: Productivity estimation of the most promising open source solutions in emergency situations. *Applied Sciences, 10*(11), 4004.

Schubert, C., Van Langeveld, M. C., & Donoso, L. A. (2014). Innovations in 3D printing: A 3D overview from optics to organs. *British Journal of Ophthalmology, 98*(2), 159–161.

Serrano, D. R., Terres, M. C., & Lalatsa, A. (2018). Applications of 3D printing in cancer. *Journal of 3D Printing in Medicine, 2*(3), 115–127.

Shaker, B., Ahmad, S., Lee, J., Jung, C., & Na, D. (2021). In silico methods and tools for drug discovery. *Computers in Biology and Medicine, 137*, 104851.

Shen, H., Diao, H., Yue, S., & Fu, J. (2018). Fused deposition modeling five-axis additive manufacturing: Machine design, fundamental printing methods and critical process characteristics. *Rapid Prototyping Journal, 24*(3), 548–561.

Shor, L., Güçeri, S., Chang, R., Gordon, J., Kang, Q., Hartsock, L., ... Sun, W. (2009). Precision extruding deposition (PED) fabrication of polycaprolactone (PCL) scaffolds for bone tissue engineering. *Biofabrication, 1*(1), 015003.

Sigaux, N., Pourchet, L., Breton, P., Brosset, S., Louvrier, A., & Marquette, C. (2019). 3D Bioprinting: Principles, fantasies and prospects. *Journal of Stomatology, Oral and Maxillofacial Surgery, 120*(2), 128–132.

Soliman, Y., Feibus, A. H., & Baum, N. (2015). 3D printing and its urologic applications. *Reviews in Urology, 17*(1), 20.

Spake, C. S., Carruthers, T. N., Crozier, J. W., Kalliainen, L. K., Bhatt, R. A., Schmidt, S. T., & Woo, A. S. (2021). 3D printed N-95 masks during the COVID-19 pandemic: Lessons learned. *Annals of Biomedical Engineering, 49*(12), 3666–3675.

Svanström, A., Rosendahl, J., Salerno, S., Leiva, M. C., Gregersson, P., Berglin, M., ... Chinga-Carrasco, G. (2021). Optimized alginate-based 3D printed scaffolds as a model of patient derived breast cancer microenvironments in drug discovery. *Biomedical Materials, 16*(4), 045046.

Tamimi, N. A., & Ellis, P. (2009). Drug development: From concept to marketing! *Nephron Clinical Practice, 113*(3), c125-c131.

Teo, Y. L., & Go, Y. I. (2021). Techno-economic-environmental analysis of solar/hybrid/storage for vertical farming system: A case study, Malaysia. *Renewable Energy Focus, 37*, 50–67.

Thomas, D. B., Hiscox, J. D., Dixon, B. J., & Potgieter, J. (2016). 3D scanning and printing skeletal tissues for anatomy education. *Journal of Anatomy, 229*(3), 473–481.

Ulag, S., Kalkandelen, C., Oktar, F. N., Uzun, M., Sahin, Y. M., Karademir, B., ... Gunduz, O. (2019). 3D printing artificial blood vessel constructs using PCL/chitosan/hydrogel biocomposites. *ChemistrySelect, 4*(8), 2387–2391.

Ursan, I. D., Chiu, L., & Pierce, A. (2013). Three-dimensional drug printing: A structured review. *Journal of the American Pharmacists Association, 53*(2), 136–144.

Ventola, C. L. (2014). Medical applications for 3D printing: Current and projected uses. *Pharmacy and Therapeutics, 39*(10), 704.

Vyas, D., & Udyawar, D. (2019). A review on current state of art of bioprinting. In L. Kumar, P. Pandey, & D. Wimpenny (Eds.), *3D printing and additive manufacturing technologies* (pp. 195–201). Springer.

Wake, N., Rosenkrantz, A. B., Huang, R., Park, K. U., Wysock, J. S., Taneja, S. S., ... Chandarana, H. (2019). Patient-specific 3D printed and augmented reality kidney and prostate cancer models: Impact on patient education. *3D Printing in Medicine, 5*(1), 1–8.

Wu, J.-J., Huang, L.-M., Zhao, Q., & Xie, T. (2018). 4D printing: History and recent progress. *Chinese Journal of Polymer Science, 36*(5), 563–575.

Xie, Z.-T., Kang, D.-H., & Matsusaki, M. (2021). Resolution of 3D bioprinting inside bulk gel and granular gel baths. *Soft Matter, 17*(39), 8769–8785.

Yang, G. H., Yeo, M., Koo, Y. W., & Kim, G. H. (2019). 4D bioprinting: Technological advances in biofabrication. *Macromolecular Bioscience, 19*(5), 1800441.

Yao, W., Li, D., Zhao, Y., Zhan, Z., Jin, G., Liang, H., & Yang, R. (2019). 3D printed multifunctional hydrogel microneedles based on high-precision digital light processing. *Micromachines, 11*(1), 17.

Zastrow, M. (2020). 3D printing gets bigger, faster and stronger. *Nature, 578*(7793), 20–24.

Zhang, Y. S., Yue, K., Aleman, J., Mollazadeh-Moghaddam, K., Bakht, S. M., Yang, J., … Shin, S. R. (2017). 3D bioprinting for tissue and organ fabrication. *Annals of Biomedical Engineering, 45*(1), 148–163.

Zheng, J., Liu, X., Chen, X., Jiang, W., Abdelrehem, A., Zhang, S., … Yang, C. (2019). Customized skull base–temporomandibular joint combined prosthesis with 3D-printing fabrication for craniomaxillofacial reconstruction: A preliminary study. *International Journal of Oral and Maxillofacial Surgery, 48*(11), 1440–1447.

Zhong, C., Xie, H.-Y., Zhou, L., Xu, X., & Zheng, S.-S. (2016). Human hepatocytes loaded in 3D bioprinting generate mini-liver. *Hepatobiliary & Pancreatic Diseases International, 15*(5), 512–518.

9 Demystifying 4IR and Emerging Modern Technology (Machine Learning and Energy)

9.1 INTRODUCTION

There has been a boom in interest in applying machine learning, and more broadly, artificial intelligence, to scientific applications in recent years. The advancement of human computational capabilities, combined with the creation of a broad variety of algorithms and machine learning frameworks, has sparked interest in performing complicated analysis of massive datasets in ways that were previously impossible. This prompted studies into techniques to speed up the process by combining ex situ and in situ experimental data, both of which are easily available in many laboratories.

Unsupervised, supervised, and reinforcement learning is used in different industries to fast-track and avoid repetitive process. It helps in optimizing processes for profit-taking and efficient service and product delivery. The energy sector finds applications in every sphere of human endeavors. From residential to industrial and commercial usage, energy will benefit from machine learning optimization, efficiency improvement, and cost reduction. It is safe to say that machine learning algorithms that optimize, reduce cost, and improve efficiency will find application in energy.

9.2 HISTORY OF MACHINE LEARNING: HISTORY, ALGORITHM, AND APPLICATION

Machine learning (ML) is a sort of artificial intelligence (AI) that allows software applications to improve their prediction accuracy without being expressly designed to do so. To forecast new output values, machine learning algorithms use historical data as input. Machine learning is a branch of artificial intelligence that is defined as a machine's ability to mimic intelligent human behavior. Artificial intelligence systems are utilized to complete complex jobs in a similar manner to how humans solve problems. Three kinds of artificial intelligence exist known as narrow, general, and superintelligence. Artificial narrow intelligence (ANI) has a limited range of abilities; artificial general intelligence (AGI) has human-like capabilities; artificial super intelligence (ASI) has human-like powers. Artificial narrow intelligence (ANI), often known as narrow AI or weak AI, is a type of artificial intelligence.

Machine learning algorithms are divided into four categories: supervised, semisupervised, unsupervised, and reinforcement. Python, Java, C++, JavaScript, and

DOI: 10.1201/9781003364481-9

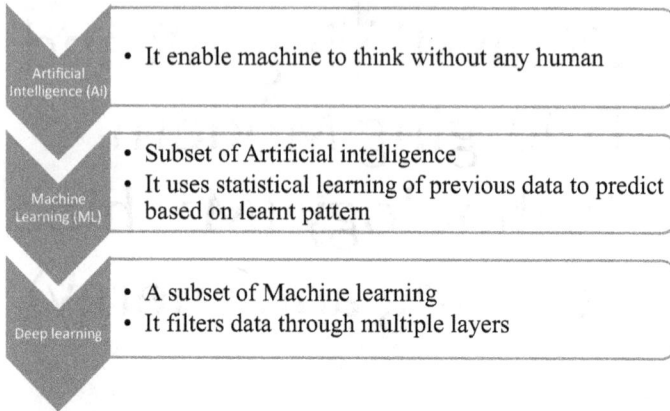

- Artificial Intelligence (AI)
  - It enable machine to think without any human
- Machine Learning (ML)
  - Subset of Artificial intelligence
  - It uses statistical learning of previous data to predict based on learnt pattern
- Deep learning
  - A subset of Machine learning
  - It filters data through multiple layers

FIGURE 9.1 Schematic difference between artificial intelligence, machine learning, and deep learning.

R are among the top five programming languages and libraries for machine learning. Artificial intelligence began as a strictly mathematical and scientific endeavor, but as it evolved into a computing process, it began to solve human problems. Artificial intelligence includes machine learning as a subset. ML is the study of computer algorithms that improve themselves over time. ML is the study and development of algorithms that can learn from data and generate predictions based on that data. Machine learning can adjust actions and responses based on more data, making it more efficient, adaptive, and scalable. Deep learning is a method for putting machine learning algorithms into practice. To achieve extremely promising decision-making, it uses artificial neural networks for training data. The neural network can handle tasks like humans by performing microcomputations with several layers of computing.

Figure 9.1 shows the difference between artificial intelligence, machine learning, and deep learning. Python is the language of choice for machine learning engineers, with more than 60% of them adopting and prioritizing it for development since it is simple to learn. A little coding knowledge is required for effective deployment of machine learning.

9.2.1 HISTORY OF MACHINE LEARNING

The word machine learning was first used by an IBMer known as Arthur Samuel around 1959 (Samuel, 1959). The American was responsible for starting several computer games and creating artificial intelligence. This led to discovery of self-taught computers that can receive training and translate the same into result. A key text that expanded the concept of machine learning at the early stage is that of Nilsson that shed light on pattern recognition in the 1960s. Pattern recognition took the center of attention in machine learning from 1973 (Duda and Hart, 1973) until around 1981 when it was replaced by teaching methodologies in neural network to identify 40 characters from the terminal of a computer. The 40 characters comprise 4 special

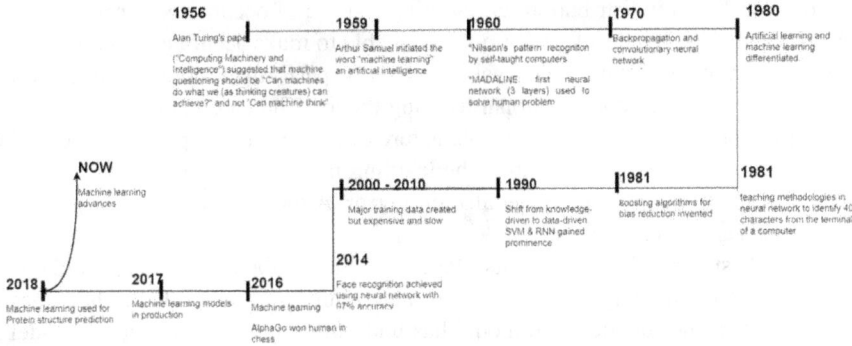

FIGURE 9.2   Abridge history of machine learning.

symbols, 10 digits, and 26 letters (Bozinovski, 2020). Tom M. Mitchell gave a formal definition of machine learning algorithms on how computer programs are taught to produce what they learned from experience. Tom's definition was operational description instead of providing it in a cognitive way (Mitchell, 1982, 1997). A similar approach was adopted by Alan Turing in his paper entitled "Computing Machinery and Intelligence." The paper opined the type of questioning to be adopted with machine learning that should be "Can machines do what we (as thinking creatures) can achieve?" and not "Can machine think" (Turing, 1956). Figure 9.2 summarizes the most common occurrence in the history of machine learning.

The objective of machine learning includes classification of data and generation of predictions of future outcomes. Also, it includes training of models (for supervised machine learning).

9.2.2   BENEFITS OF MACHINE LEARNING

Machine learning is a technology that simplifies, improves efficiency, reduces resource wastage, and reduces cost. It helps in learning and predicting using existing data. It is used for work automation, increased e-commerce sales, and powerful predictive ability. Additionally, ML is used in the medical field for medical discovery, performing simple and complex surgery, and improved diagnosis of diseases. In social media, it helps in detecting fake contents and facial recognition, among others. Economically, machine learning grows revenue, aids in secure business transactions, reduces resource wastage, and increases productivity, among others. The merit and benefit of machine learning is multifaceted and multidisciplined. Every day, it is being deployed in various existing and emerging fields based on research and understanding.

9.2.3   THEORY OF MACHINE LEARNING

The ability to adapt what was learned for specific application is the main objective of a learner. In this context, adapting represent the ability of a machine to replicate accurately what it was taught from previous data set. The machine is trained using unknown probability distribution representing a space of occurrence. Thereafter, the machine being trained develops a model to be able to make accurate predictions.

The aspect of computer science that examines the performance of machine learning algorithm is known as computational learning theory. The fact that data set used for training is limited and the unpredictable nature of the future, the performance of the algorithms is highly dependent upon the learning model. The bias-variance decomposition helps to quantify the generalization error. Although, probabilistic performance bounds are commonplace.

For the best generalization results, the complexity of the hypothesis should coincide with the complexity of the function underpinning the data. If the hypothesis is simpler than the function, the model has underfitted the data. When the model's complexity is increased in response, the training error decreases. The model will be vulnerable to clustering if the hypothesis is too complex, which will lead to weak generalization. Temporal complexity alongside learning feasibility and performance bound is usually examined by learning theorist in making inference. The ability of a task to be completed in polynomial time makes it a viable computation.

9.3 CLASSIFICATION OF MACHINE LEARNING APPROACH

Machine learning is classified into three distinct categories using the signal and feedback approach. This is shown in Figure 9.3 to be reinforcement, supervised, and unsupervised.

9.3.1 SUPERVISED LEARNING

In a supervised learning model, the computer learns from a labeled dataset in order to develop predicted responses to new input. It develops predictive models using both input and output data obtained from previous datasets. For example, to anticipate

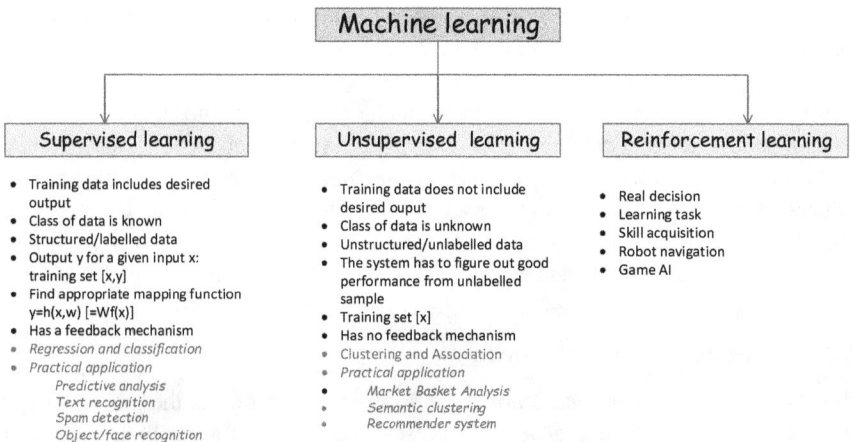

FIGURE 9.3 Classification of machine learning.

property prices, we need information on houses such as square footage, number of rooms, if the house has a garden, and other characteristics. The values of these houses, or class labels, must then be determined. We can now train a supervised machine learning model to forecast the price of a new house based on the model's previous experiences using data from thousands of properties, their attributes, and prices. A "teacher" presents the computer with sample inputs and desired outputs, with the purpose of learning a general rule that maps inputs to outputs.

Supervised learning is further divided into classification and regression. A computer program is trained on a training dataset and then categorizes the data into different class labels based on the training for classification supervised learning. This algorithm predicts discrete values like male|female, true|false, spam|not spam, and so on. For example, email spam detection, speech recognition, and cancer cell identification. This type is further grouped into decision trees Naive Bayes classifier, logistic regression, support vector machine, K-nearest neighbors, and random forest classification.

The aim of the regression method, on the other hand, is to discover the mapping function that will transfer input variables (x) to a continuous output variable (y). Continuous values such as price, salary, age, and marks are predicted using regression techniques. Regression is used for weather prediction and fake news detection, among others. The types include simple linear, polynomial, multiple linear, random forest, decision tree, and ensemble method.

9.3.2 Unsupervised Learning

This type of learning group interprets dataset based on the input only. It is when the learning algorithm is not given labels and is left to uncover structure in the data on its own. Unsupervised learning can be a goal in and of itself (finding hidden patterns in data) or a means to an end (finding hidden patterns in data) (feature learning). Good applications are detection of anomalies, including fraud detection. Another example is opening emergency rooms in the most accident-prone locations. K-means clustering will aggregate these locations of maximum-prone areas into clusters and assign each cluster a cluster center (hospital) (i.e., accident-prone areas). The types include neural network, clustering, association, anomaly detection, latent variables models, and autoencoder.

9.3.3 Reinforcement Learning

It is the third type of learning in which a computer program interacts with a dynamic environment in order to achieve a specific goal (such as driving a vehicle or playing a game against an opponent). The software receives input in the form of incentives as it navigates its issue space, which it strives to maximize.

9.4 PROCESS OF MACHINE LEARNING

The steps from start to finish of machine are grouped into six as discussed here. The first step is to identify and collate dataset for machine learning. The next step is preprocessing the data and exploratory data analysis. In this stage, the data is checked for missing, duplicate, and invalid values with the aid of various analytical techniques. The dataset is preprocessed extraction, analysis, and visualization. The third stage is to train the model for the task. This is done to enable the model to perform the pattern recognition, features, and rules needed to perform the task. The fourth stage is to test and evaluate the model. The testing is implemented by using a set of data to train the model. This is followed by deploying the model into the intended application. Techniques such as Kubernetes, Docker, MLFlow, AWS SageMaker, and Azure Machine Learning service can be used for deploying machine learning models. The last stage is monitoring and evaluating the performance of the model. This is done to check for latency, errors, crashes, and areas needing improvement and ascertain if the model is performing optimally. It is a very crucial stage needed to prevent degradation of the model over time.

9.5   MODELS OF MACHINE LEARNING

Machine learning entails building a model that has been trained on some training data and can subsequently process more data to produce predictions. For machine learning systems, various types of models have been utilized and investigated. The models include artificial neural networks, decision trees, support vector machines, regression analysis, genetic algorithms, Bayesian networks, training models, and federated learning.

9.6   APPLICATIONS OF MACHINE LEARNING

Automatic language translation in Google Translate, faster route selection in Google Maps, driverless/self-driving cars, smartphones with face recognition, speech recognition, ads recommendation system, Netflix recommendation system, and auto friend tagging suggestion in Facebook are just a few of the applications of machine learning. Other fields of study of machine learning include stock market trading, fraud detection, weather forecasting, and medical diagnosis.

9.7   LIMITATIONS OF MACHINE LEARNING

Although machine learning has demonstrated its ability to disrupt many industries, it frequently falls short of expectations.

The challenge associated with machine learning is the huge number of variable predictors. The governing equation is usually difficult to obtain as the data are usually complex. There is no exact solution for multiple problems, rather an iterative approach that is time consuming caused by experimenting with multiple algorithms. Machine learning requires a significant amount of expertise to obtain a near-perfect solution.

There are many causes for this, including a dearth of (suitable) data, problems with data access, bias in the data, privacy concerns, improper tools and employees, a lack of resources, and problems with evaluation. In 2018, a pedestrian was killed as a result of an Uber self-driving car's failure to recognize the individual (Kohli and Chadha, 2019). Even after spending billions of dollars and years of work, IBM Watson's attempts to use machine learning in healthcare fell short.

The growth of the biomedical literature has raised the strain on reviewers and necessitated the use of machine learning to update the evidence with respect to these issues. Students may feel let down if they "learned the incorrect lesson." For instance, an image classifier that has only been trained on images of brown horses and black cats may come to the conclusion that all brown patches are most likely those of horses (Mostajabi et al., 2015). Contrary to people, existing image classifiers frequently learn associations between pixels that humans are unaware of but that nonetheless correlate with images of types of real objects. Instead of making decisions based on the spatial relationship between picture components, these associations are often learned by existing image classifiers. If these patterns are changed on a legal image, the system can mistakenly label the image as "adversarial." The presence of nonpattern disturbances or nonlinear systems may expose adversary weaknesses. Some systems are so delicate that even a single hostile pixel change might lead to misclassification.

9.8   ETHICS OF MACHINE LEARNING

Machine ethics (also known as machine morality, computational morality, or computational ethics) is a branch of artificial intelligence ethics concerned with enhancing or ensuring the moral behavior of man-made machines that employ artificial intelligence, also known as artificial intelligent agents (Moor, 2006). Machine ethics is distinct from other topics of engineering and technology ethics. Computer ethics, which focuses on human usage of computers, must not be mistaken with machine ethics. It should also be separated from technology philosophy, which is concerned with the larger social consequences of technology (Boyles, 2018). Because of computing and artificial intelligence (AI) constraints, the ethics of machines had mostly been the theme of science fiction literature prior to the 21st century. Even though the meaning of "machine ethics" has changed over time, Mitchell Waldrop created the term in the 1987 *AI Magazine* article "A Question of Responsibility". Whether or not their programmers intend it, intelligent robots will embody values, assumptions, and purposes (Waldrop, 1987).

The AAAI Workshop on Agent Organizations on Theory and Practice (Wooldridge and Jennings, 1995) in 2004 featured a paper titled "Towards Machine Ethics" (Anderson et al., 2004). The paper laid out the theoretical basis for machine ethics. For the first time, researchers gathered at the AAAI Fall 2005 Symposium on Machine Ethics to discuss the application of an ethical dimension in autonomous systems (Cassimatis et al., 2005). The collected edition "Machine Ethics" (Anderson and Anderson, 2020), which derives from the AAAI Fall 2005 Symposium on Machine Ethics, contains a variety of perspectives on this new area. Machine Ethics: Creating an Ethical Intelligent Agent (Anderson and Anderson, 2007a) was published in *AI Magazine* in 2007, and it explored the importance of machine ethics, the necessity for machines that explicitly embody ethical ideas, and the obstacles faced by people working on machine ethics.

Moral Machines, Teaching Robots Right from Wrong was published by Oxford University Press in 2009, and it was billed as "the first book to tackle the problem of developing artificial moral agents, looking deeply into the nature of human decision making and ethics." It referenced over 450 sources, with over 100 of them addressing key machine ethical issues.

Michael and Susan Leigh Anderson, who also edited a special issue of IEEE Intelligent Systems on the topic in 2006, produced a volume of writings regarding machine ethics for Cambridge University Press in 2011 (Anderson and Anderson, 2007b). The issues of incorporating ethical standards into machines are featured in this collection (Siler, 2015).

9.9   HARDWARE OF MACHINE LEARNING

Since the 2010s, advances in computer technology and machine learning algorithms have led to more efficient techniques for training deep neural networks. It is a specific subfield of machine learning that includes multiple layers of nonlinear hidden units. By 2019, GPUs, frequently with AI-specific modifications, have displaced CPUs as the most popular method for training large-scale commercial cloud AI. OpenAI examined the amount of hardware needed in the largest deep learning projects from AlexNet (2012) to AlphaZero (2017) and found a 300,000-fold increase in the amount of compute needed, with a doubling-time trendline of 3.4 months. There are physical neural networks and embedded machine learning.

9.9.1   A PHYSICAL NEURAL NETWORK

Also known as a neuromorphic computer, is a sort of artificial neural network in which the function of a neural synapse is emulated by an electrically changeable substance. The term "physical" neural network refers to the use of physical hardware to simulate neurons rather than software-based techniques. Other artificial neural networks that use a memristor or other electrically adjustable resistance material to imitate a neural synapse are also known as memristor networks (Zaidan et al., 2017).

9.9.2   EMBEDDED MACHINE LEARNING

It is a sub-field of machine learning that uses embedded systems with low computing capabilities, such as wearable computers, edge devices, and microcontrollers, to run machine learning models (Fafoutis et al., 2018, Ajani et al., 2021). Running machine learning models in embedded devices eliminates the need to transport and store data on cloud servers for further processing, resulting in fewer data breaches and privacy leaks, as well as less theft of intellectual property, personal data, and company secrets. Embedded machine learning can be implemented using a variety of methods, including hardware acceleration, approximation computation, machine learning model optimization, and more (Agrawal et al., 2019, Branco et al., 2019).

Figure 9.4 summarizes the algorithms used by unsupervised, supervised, and reinforcement with detailed description of the learning types, categories, and types of algorithms under each.

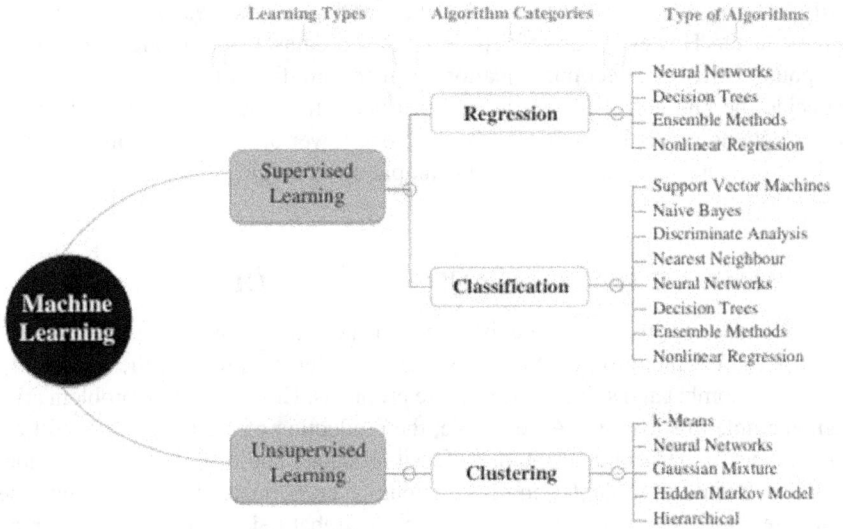

FIGURE 9.4 Algorithms of machine learning.

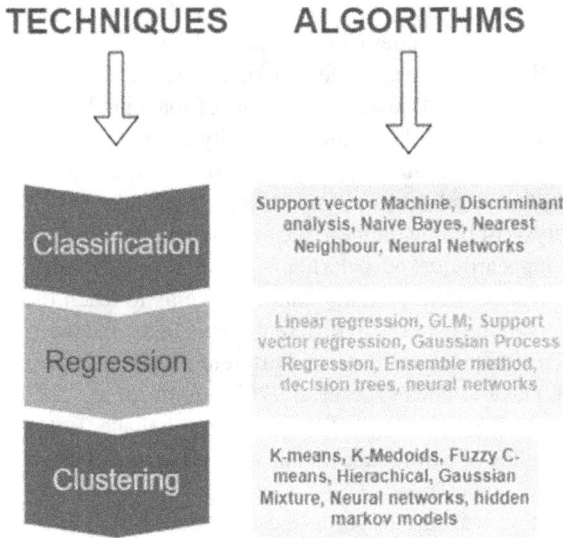

FIGURE 9.5 Schematic of techniques and algorithms in machine learning.

9.10 TECHNIQUES AND ALGORITHMS IN MACHINE LEARNING

Figure 9.5 shows the techniques and algorithms used in machine learning. Most forms of unsupervised learning are form of cluster analysis.

Clustering analysis separates data into two groups based on the nature of the data contained and shared characteristics in the data set. Clustering is further divided into hard clustering and soft clustering. Hard clustering is when each dataset belongs to one cluster. However, for soft clustering, each dataset can belong to more than one point cluster. A telecommunication engineer building a telephone tower needs to decide the number of towers to be constructed to achieve best signal reception as a telephone can only communicate with one tower at a time. In soft clustering, analysis of genes in cancer is a good example as no single gene can belong to one group alone.

9.11 MACHINE LEARNING AND HUMAN PROBLEM

A lot of challenges threaten the existence of planet Earth. The increasing growth of humanity and exploration continues to create problems. However, there have been attempts to combat and solve some of these problems. However, some problem arises from attempts to solve one. An example, the production of food, heating, and lightening gave rise to use of coal and other fossil products. These products caused global warming. Resulting in depletion of the ozone layer. Climate change has continued to gain research with aim to solve and keep the global rising temperature down and reduce greenhouse gases.

Machine learning can solve and assist humans in combating some of these. It can be employed in a predictable and crime-prevention manner. For example, prediction technology can be used to analyze millions of files and assaults to figure out what makes them tick. Companies can prevent future attacks by understanding mathematical DNA. ML can also accelerate experiential learning and knowledge sharing for humans. One of the most basic applications of machine learning is spam detection. Machine learning can solve spam or identity scam. It can also handle product recommendation and image and video recognition to avoid fakes, and it can detect fraudulent transactions and demand forecasting. It can also act as virtual personal assistant and analysis of public opinion. It can anticipate four different sorts of natural disasters, including earthquakes (Martínez–Álvarez and Morales–Esteban, 2019). Researchers can feed seismic imaging data to systems to teach them. The ML uses the data to learn about the patterns of different earthquakes and then predicts where an earthquake and its aftershocks will occur. Different researchers have applied machine learning to earthquake studies (Xie et al., 2020, Asim et al., 2017). These studies attempted to use machine learning to forecast the earthquake (Beroza et al., 2021), to estimate the magnitude of earthquake (Mousavi and Beroza, 2020), and to classify the earthquake (Mangalathu et al., 2020).

However, machine learning is yet to master the ability to reason. The ability to reason, which is a fundamentally human trait, is one area where ML has failed miserably. Another area is contextual limitations. Scalability, regulatory restrictions for data in machine learning, and the internal workings of deep learning are part of the areas that humans surpass machine learning.

9.12 IMPLEMENTING MACHINE LEARNING

As discussed earlier, machine learning involves three approaches, viz. unsupervised, supervised, and reinforcement learning. Supervised and reinforcement involve training a model using previous dataset to implement a solution. Different tools can be used in the training and execution of the model. MATLAB and Python are some of the software that has been used.

9.12.1 COMPARATIVE PERFORMANCE PARAMETERS

They are needed to obtain a superior and more accurate result from set of machine learning models. The major common comparative performance parameters include the root mean square error (RMSE) and correlation coefficient (r) used for determining the error and precision of models. They are represented in equations 9.1 and 9.2, respectively.

$$RMSE = \sqrt{\frac{1}{n}\sum_{i=1}^{N}\left(x_{ti} - x_{pi}\right)^2} \tag{9.1}$$

$$r = \left(1 - \left(\frac{\sum_{i=1}^{N}\left(x_{ti} - x_{pi}\right)^2}{\sum_{i=1}^{n}\left(x_{ti}\right)^2}\right)\right)^{1/2} \tag{9.2}$$

where n is the data point, x_{ti} is the target, and x_{pi} is the predicted value.

9.13 MATLAB FOR MACHINE LEARNING

MATLAB is one of the tools used for training models and implementing solution in machine learning. It has a strong environment for interactive exploration. It supports parallel computing, is easy to evaluate and iterate, and can choose the best algorithm. The trained models can integrate into other big data-related solution. MATLAB machine learning is classified as shown in Figure 9.6.

The workflow of a supervised machine learning is shown in Figure 9.7.

The data are collected and imported into MATLAB using different approaches. It can be imported from spreadsheet, notepad, or manually entered in MATLAB as shown in Figure 9.8. Thereafter, the data is cleaned prior to being explored. Several software products including MATLAB offer the ease of finetuning and generating the script and codes used for importing the data.

The data needs to match in terms of the array and column. Categorical data are converted into categorical arrays. It is therefore prepared for model selection. The data are aggregated into predictors and responses. The model is trained using previous dataset. The model is produced using input and output. The model is produced from known data and known response. The predicted response is obtained from the model and new data. The common apps available in MATLAB for machine learning are shown in Figure 9.9.

Classification helps in predicting the best group that a data belongs to. Classification helps in grouping a data you have not seen before. The approach involves training the classifier using different models. Thereafter, measure the accuracy of the model and

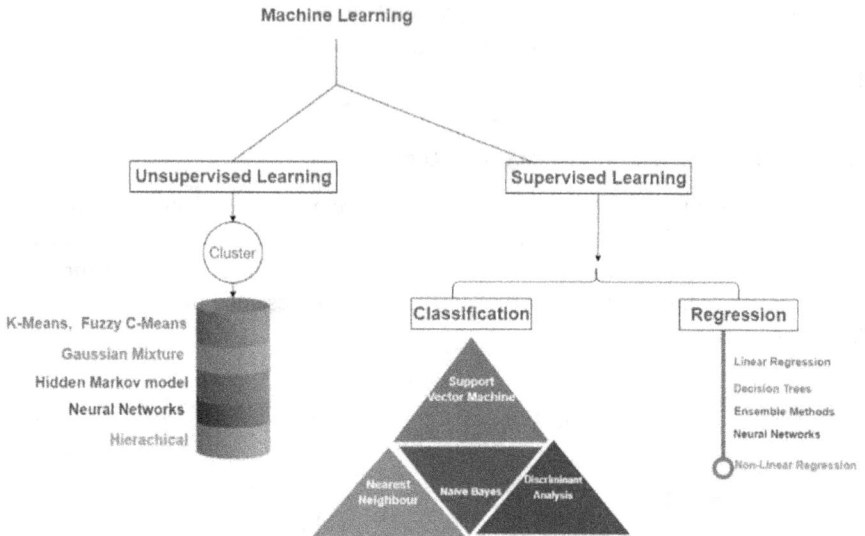

FIGURE 9.6   Classification of machine learning based on MATLAB tools.

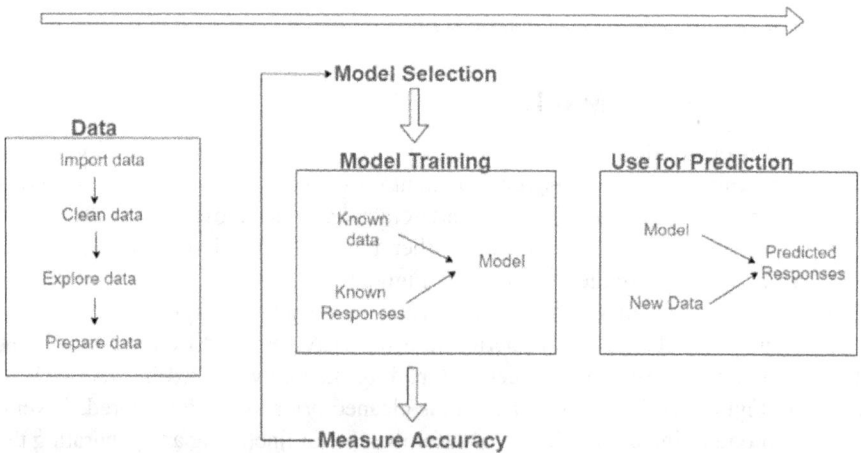

FIGURE 9.7   Supervised machine learning workflow.

compare models. The model complexity is reduced. The classifier is then used for prediction.

Neural pattern recognition app shown in Figure 9.10 is used for classification of input to a set of target categories.

A click on the neural pattern recognition takes the user to Figure 9.11 to start the neural network and input the data.

The input and the target are entered into the panel shown in Figure 9.12 to enable the model to get the training.

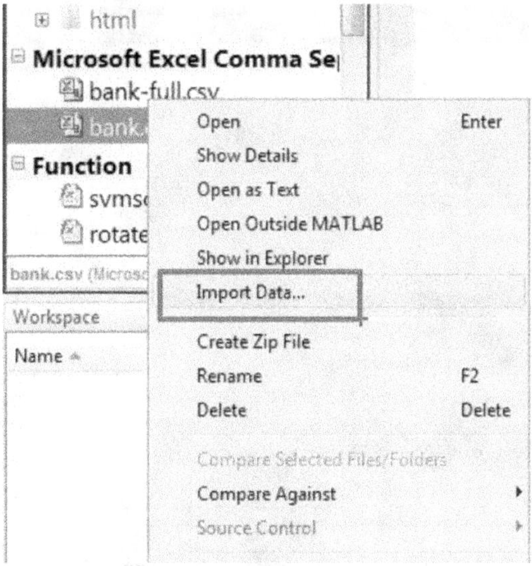

FIGURE 9.8   A screenshot of the MATLAB interface for importing data.

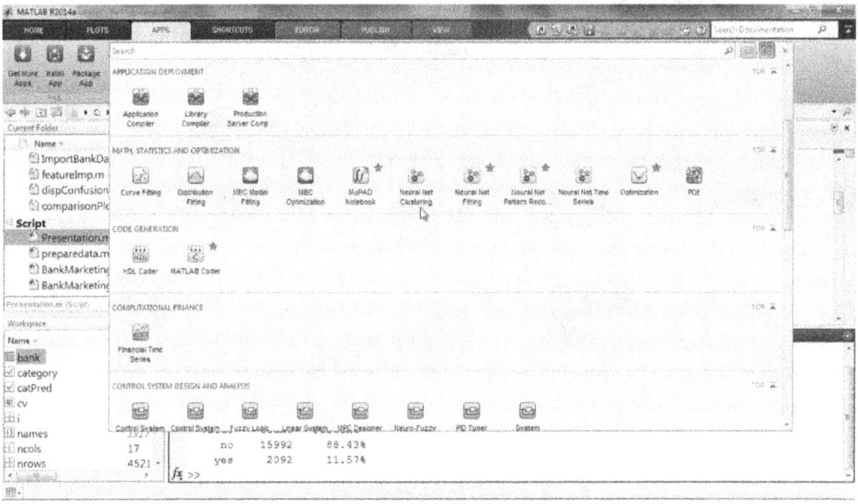

FIGURE 9.9   A screenshot of some apps in MATLAB used for supervised learning.

The input and target are entered into the panel and next is clicked. This will lead the user to Figure 9.13 that contains the panel for training the model.

The training is presented to the network during training. The network is adjusted based on the error. Validation is used to measure network generalization. To halt training when generalization stops improving. Testing panel is used to perform independent measure of the network performance during and after the training. They are in percentage and adjusted based on the desired outcome.

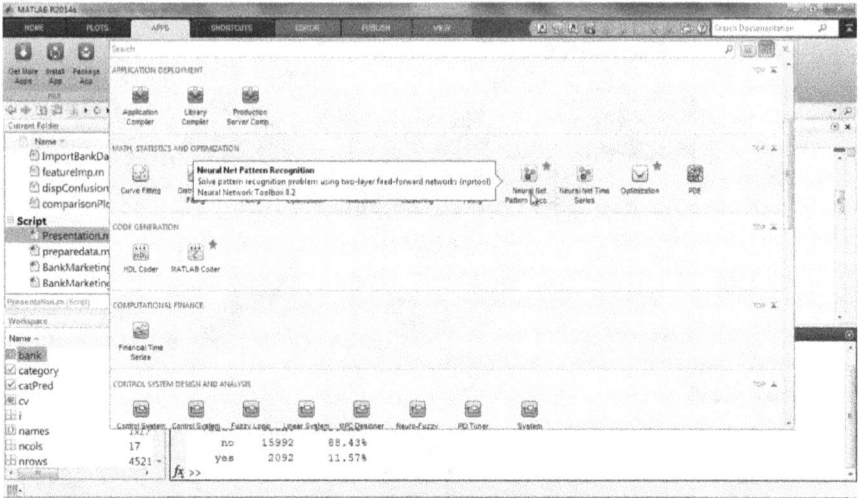

FIGURE 9.10 A screenshot of neural pattern recognition app in MATLAB.

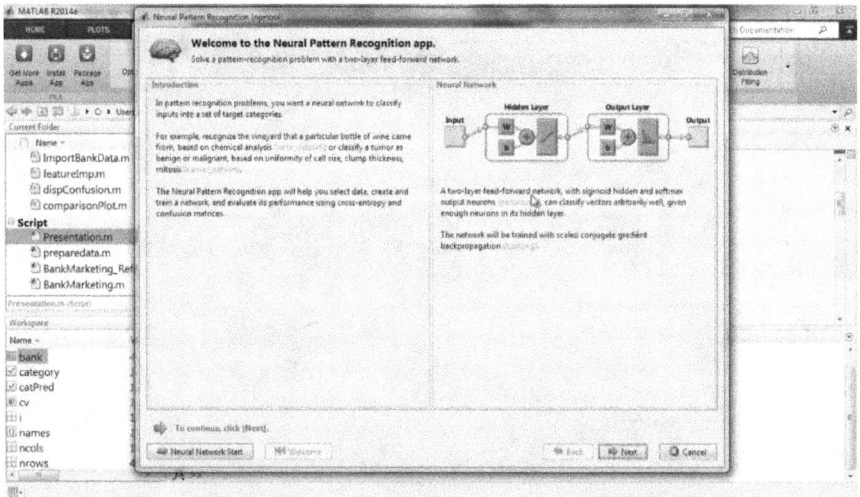

FIGURE 9.11 A screenshot of the input panel for neural pattern recognition.

One hidden layer 10 neurons neural network is shown in Figure 9.14.

The selection is trained, and the performance is analyzed by clicking on performance button shown in Figure 9.15.

The training can be stopped or allowed to complete before deciding to retrain or deploy the model as shown in Figure 9.16a and b, respectively.

A successful evaluation of the network can then lead to creation of function as shown in Figure 9.17. The result can be sent to Simulink for further processing and optimization.

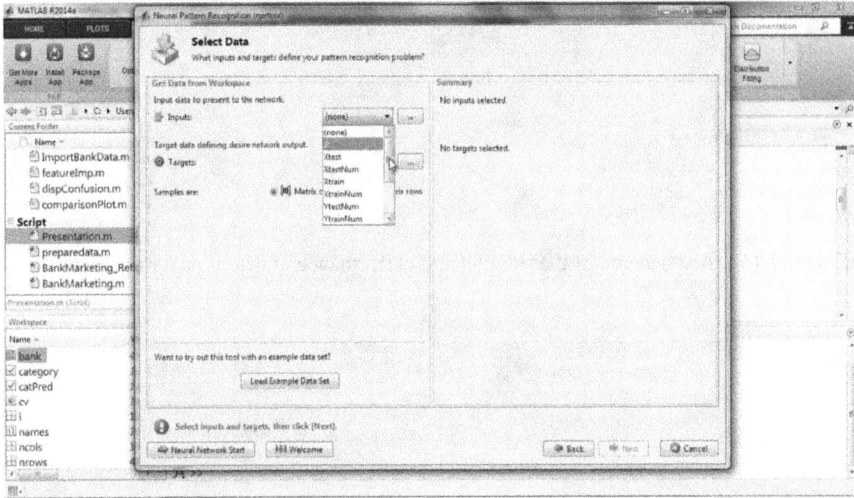

FIGURE 9.12 A screenshot of the data input panel of the neural pattern recognition.

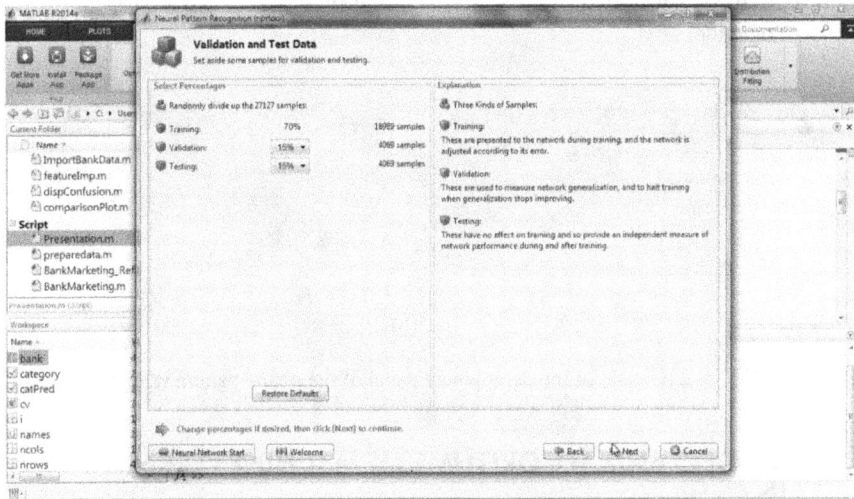

FIGURE 9.13 A screenshot of the validation and test data panel for neural pattern recognition.

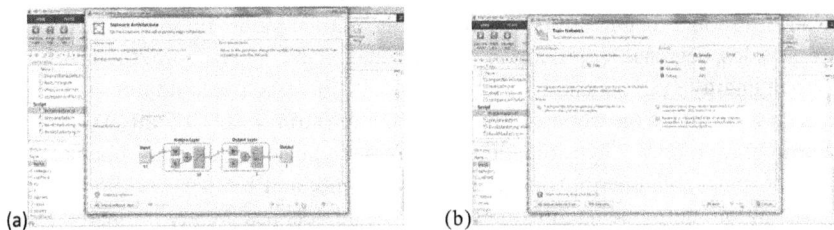

(a)   (b)

FIGURE 9.14 A schematic of one hidden layer 10 neurons neural network.

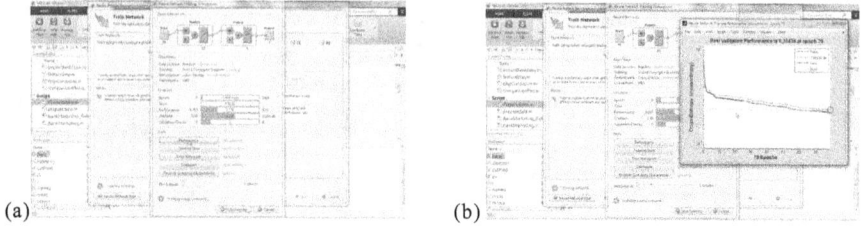

FIGURE 9.15   A schematic of the training and performance of neural pattern recognition.

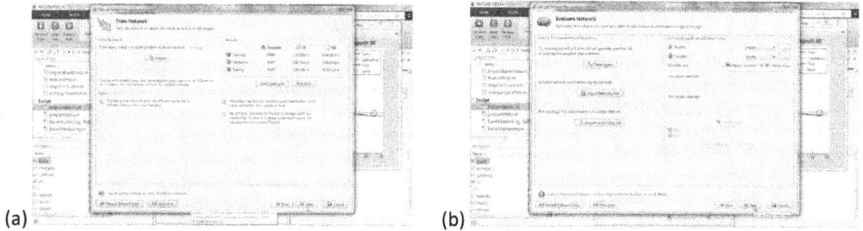

FIGURE 9.16   A schematic of (a) the retraining and (b) evaluate network panel of neural pattern recognition.

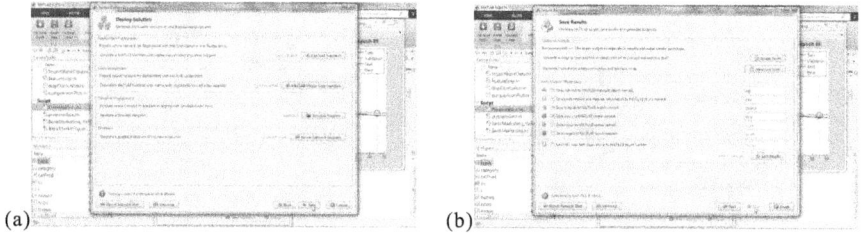

FIGURE 9.17   A schematic of the deployment panel of the neural pattern recognition.

9.14   MODELS AND ALGORITHMS OF MACHINE LEARNING IN ENERGY

Machine learning is used in the energy sector for the load, trading, and price forecasting. Electricity demand in kWh from the grid can be solved using regression (supervised learning). Produce better offers, predict client lifetime value, optimize retail prices, increase sales, decrease churn, forecast price and demand, and maximize consumption.

When energy systems use hybrid ML models, there is a notable improvement in the accuracy, precision, robustness, and generalization ability of the ML models. According to reports, hybridization can help enhance prediction models, especially for renewable energy sources like solar, wind, and biofuels. Additionally, the forecasting of energy demand using hybrid ML models has greatly improved energy governance and sustainability as well as efficiency.

About ten different machine learning models have been reported in literature for energy systems (Mosavi et al., 2019). These models include support vector machine (SVM), artificial neural network (ANN), MLP, wavelength neural network (WNN), embedded Machine learning, ANFIS, ensembles, deep learning, and decision trees.

Artificial neural network is used for curve fitting, forecasting, and regression. It is used for noisy data in energy system, to learn pattern, and fault-tolerant (a process that enables an operating system to respond to a failure in hardware or software), among others. It has been used for generation capacities, forecasting of wind speed, solar radiation, industrial building measurement and verification of energy saving, prediction of energy cost, and reducing wind power fluctuation, among others.

ANN was used by Anwar et al. (2016) to develop an innovative way of balancing the power smoothing and scheduling generation of a hybrid energy comprising of wind turbine and marine current. Genetic algorithms were utilized by Abbas et al. (2018) to maximize the generation capabilities of renewable energy systems coupled with storage systems. This study assessed whether adding energy storage systems to the electric grid would be economically feasible. The anticipated load model was validated using an artificial neural network. A chance-constrained model was used to address the uncertainties surrounding the renewable energy systems. Genetic algorithms were then used to resolve the issue. The proposed model's application to a scenario in Western China proved its resilience.

There was a nearly twofold increase in the usage of clean energy, which resulted in a drop in CO_2 emissions from 109 million tons to 38 million tons. As a result, when compared to the base example, the proposed scenario plays a more significant role.

9.14.1 ELM

It has a better speed of training and better generalization ability compared to ANN. It is useful for geothermal heat pump design. It is used for optimization of the geothermal heat pump. MLP is used for forecasting solar photovoltaic power generation. It can also be used for monthly average global solar radiation prediction. It is used for predicting the electrical load and consumption in a building.

9.14.2 MLP

It is an improved version of artificial neural network used for modeling and prediction of energy processes. It is used for training purposes for backpropagation and supervised learning. It is a feed-forward neural network and a control model. It has been used to forecast hourly solar irradiation using the autoregressive recurrent neural network. MLP has been used to classify power plants; potentials and estimate the power output of solar energy. It is also used for forecasting electricity load in a grid.

9.14.3 Support Vector Machine

It is used for reducing the structural risk and was built using statistical learning theory. It is the best model for analysis of regression, pattern recognition, and classification.

It is versatile for energy load forecasting due to the generalization ability. It is used for obtaining the optimum steam-oxygen ratio for ascertaining the heat requirement for heat deployment in coal gasification. SVM outperforms Bayes classifier, ANN, and future vector in power quality disturbance for real and synthetic data comparison. SVM can be used for estimating the electricity market in a fast execution time. It is also used for predicting the irradiance levels of solar photovoltaic electrical characteristics. It can predict the power quality of an energy system.

9.14.4   ANFIS

This model uses five main layers. It uses the Takagi-Sugeno fuzzy inference into the artificial neural network for modeling. It is a hybrid machine learning as it combines fuzzy logic and neural network. ANFIS is used as a protection device for a reverse power protection system. It is used for estimating the power demand of a plant to aid in energy flow optimization. It can be used for ascertaining the temperature of a module in a solar photovoltaic system. It is also used for biodiesel synthesis. ANFIS has been used in various aspects of Biochar.

9.14.5   WNN

This also combines two technologies, viz. neural network and theory of wavelength. A series of datasets is used to train the function and generate a desired output from a given input data. It is used for estimating the trend, computing, and process function. It has one hidden layer with an FFNN. The merit of WNN is that it has fast convergence and requires smaller training dataset compared to MLP model. WNN has been applied to estimate time series of renewable energy sources. It can also be used for estimating the heat load, simultaneously improving the fuel economy and charge state of a battery. It can reduce the uncertainty involved in estimating wind power. WNN is used for reduction of noise in raw data series in wind speed forecasting.

9.14.6   Decision Trees

It is versatile inductive inference model used in several energy applications. It is used for approximate discrete-valued target functions depicted in a decision tree form. It is used for energy storage control and planning. It is used for estimating the risk associated with power failure. It is useful for optimizing the electricity usage in a railway.

9.14.7   Ensemble Model

It is a statistical and machine learning tool that deploys multiple learning algorithms to achieve optimum performance. It has been used to estimate the electricity needs of buildings and compute and estimate the electricity billing. It can also be used to estimate the water level in an underground dam. It is used for obtaining nonlinear fault detection.

9.14.8 HYBRID MODELS

ANFIS and WNN are the first-generation hybrid machine learning models. There has been improvement and new hybrid models used in the energy sector. These help in forecasting short-term loads in microgrids, electricity price estimation, electric load estimation, and general energy issues. They are deployed in hydro, wind, solar, and microgrids.

9.15 EMERGING TRENDS IN MACHINE LEARNING

Machine learning is evolving in response to demands for better and more accurate results. The deployment of machine learning in the energy sector has seen tremendous improvement in service delivery and infrastructural growth. Load shedding, cost, speedy electricity supply, and other related issues are better delivered to targeted grid and end-users.

9.16 CONCLUSION

This chapter discussed machine learning with a view of making it easy to understand and implement. It also examines the application of machine learning to energy sector. It discussed most of the models used in the energy sector and specific areas of application of such models. ANN, ANFIS, WNN, and SVM are among the over ten models used in the energy sector. The applications of these models have helped in better service delivery and improving energy infrastructure, among others.

REFERENCES

Abbas, F., Habib, S., Feng, D., & Yan, Z. (2018). Optimizing generation capacities incorporating renewable energy with storage systems using genetic algorithms. *Electronics, 7*, 100.

Agrawal, A., Modi, A., Passos, A., Lavoie, A., Agarwal, A., Shankar, A., Ganichev, I., Levenberg, J., Hong, M., & Monga, R. (2019). TensorFlow Eager: A multi-stage, Python-embedded DSL for machine learning. *Proceedings of Machine Learning and Systems, 1*, 178–189.

Ajani, T. S., Imoize, A. L., & Atayero, A. A. (2021). An overview of machine learning within embedded and mobile devices–optimizations and applications. *Sensors, 21*, 4412.

Anderson, M., & Anderson, S. L. (2007a). Machine ethics: Creating an ethical intelligent agent. *AI Magazine, 28*, 15–15.

Anderson, M., & Anderson, S. L. (2007b). The status of machine ethics: a report from the AAAI Symposium. *Minds and Machines, 17*, 1–10.

Anderson, M., & Anderson, S. L. (2020). Machine ethics: Creating an ethical intelligent agent. In Association for the Advancement of Artificial Intelligence (Ed.), *Machine Ethics and Robot Ethics*. Routledge.

Anderson, M., Anderson, S. L., & Armen, C. (2004). *Towards machine ethics*. AAAI-04 Workshop on Agent Organizations: Theory and Practice, San Jose, CA.

Anwar, M. B., El Moursi, M. S., Xiao, W. (2016). Novel power smoothing and generation scheduling strategies for a hybrid wind and marine current turbine system. *IEEE Transactions on Power Systems, 32*(2), 1315–1326.

Asim, K., Martínez-Álvarez, F., Basit, A., & Iqbal, T. (2017). Earthquake magnitude prediction in Hindukush region using machine learning techniques. *Natural Hazards, 85,* 471–486.

Beroza, G. C., Segou, M., & Mostafa Mousavi, S. (2021). Machine learning and earthquake forecasting—next steps. *Nature Communications, 12,* 1–3.

Boyles, R. J. M. (2018). A case for machine ethics in modeling human-level intelligent agents. *Kritike, 12*(1), 182–200.

Bozinovski, S. (2020). Reminder of the first paper on transfer learning in neural networks, 1976. *Informatica, 44*(3), 291–302.

Branco, S., Ferreira, A. G., & Cabral, J. (2019). Machine learning in resource-scarce embedded systems, FPGAs, and end-devices: A survey. *Electronics, 8,* 1289.

Cassimatis, N., Luke, S., Levy, S. D., Gayler, R., Kanerva, P., Eliasmith, C., Bickmore, T., Schultz, A. C., Davis, R., & Landay, J. (2005). Reports on the 2004 AAAI Fall Symposia. *AI Magazine, 26,* 98.

Duda, R. O., & Hart, P. E. (1973). *Pattern classification and scene analysis.* Wiley New York.

Fafoutis, X., Marchegiani, L., Elsts, A., Pope, J., Piechocki, R., & Craddock, I. (2018). Extending the battery lifetime of wearable sensors with embedded machine learning. *2018 IEEE 4th World Forum on Internet of Things (WF-IoT).* IEEE, Singapore, 269–274.

Kohli, P., & Chadha, A. (2019). Enabling pedestrian safety using computer vision techniques: A case study of the 2018 uber inc. self-driving car crash. *Future of Information and Communication Conference, 2019.* Springer, Singapore, 261–279.

Mangalathu, S., Sun, H., Nweke, C. C., Yi, Z., & Burton, H. V. (2020). Classifying earthquake damage to buildings using machine learning. *Earthquake Spectra, 36,* 183–208.

Martínez–Álvarez, F., & Morales–Esteban, A. (2019). *Big data and natural disasters: New approaches for spatial and temporal massive data analysis.* Elsevier.

Mitchell, T. M. (1982). Generalization as search. *Artificial Intelligence, 18,* 203–226.

Mitchell, T. M. (1997). *Machine learning.* McGraw-Hill.

Moor, J. H. (2006). The nature, importance, and difficulty of machine ethics. *IEEE Intelligent Systems, 21,* 18–21.

Mosavi, A., Salimi, M., Faizollahzadeh Ardabili, S., Rabczuk, T., Shamshirband, S., & Varkonyi-Koczy, A. R. (2019). State of the art of machine learning models in energy systems, a systematic review. *Energies, 12,* 1301.

Mostajabi, M., Yadollahpour, P., & Shakhnarovich, G. (2015). Feedforward semantic segmentation with zoom-out features. *Proceedings of the IEEE Conference on Computer Vision and Pattern Recognition.* 3376–3385.

Mousavi, S. M., & Beroza, G. C. (2020). A machine-learning approach for earthquake magnitude estimation. *Geophysical Research Letters, 47,* e2019GL085976.

Samuel, A. L. (1959). Machine learning. *The Technology Review, 62,* 42–45.

Siler, C. (2015). Review of Anderson and Anderson's Machine Ethics. *Artificial Intelligence, 229,* 200–201.

Turing, A. M. (1956). Can a machine think. *The World of Mathematics, 4,* 2099–2123.

Waldrop, M. M. (1987). A question of responsibility. *AI Magazine, 8,* 28–28.

Wooldridge, M., & Jennings, N. R. (1995). Intelligent agents: Theory and practice. *The Knowledge Engineering Review, 10,* 115–152.

Xie, Y., Ebad Sichani, M., Padgett, J. E., & Desroches, R. (2020). The promise of implementing machine learning in earthquake engineering: A state-of-the-art review. *Earthquake Spectra, 36*, 1769–1801.

Zaidan, M. A., Canova, F. F., Laurson, L., & Foster, A. S. (2017). Mixture of clustered Bayesian neural networks for modeling friction processes at the nanoscale. *Journal of Chemical Theory and Computation, 13*, 3–8.

10 Demystifying 4IR and Emerging Modern Technology

Water Desalination, Smart Coatings, and Related Technologies

10.1 SMART COATINGS

Smart coatings are specialized films with predetermined characteristics that enable them to sense and react to external stimuli such as the surroundings. Innovative coatings, known as "smart coatings," react instantly to alterations in the microenvironment, such as heat and light irradiation.

Due to the intrinsic features and chemistry that smart coatings possess on modification, they have earned recognition in the material sciences, colloidal chemistry, biomedical sciences, and polymer chemistry. They are designed to have both passive and active components so that, when used, the advantages of their quick response depending on the necessary stimuli can be realized. Smart coatings are inventively manufactured for a variety of applications, and it is desirable to be able to respond to various cycles and span over a long period of time. Smart coatings have outperformed conventional coatings, thanks to the use of formulations with micro- to nanoparticles and mixtures of organic and inorganic phases. When compared to micro- and macroparticles, the performance of smart inhibitory materials has greatly improved thanks to the usage of nanoscale materials. The metal and metal oxide nanoparticles found in many smart coatings have a variety of functional features and improved performance. The multifunctional coatings are anticipated to be created by their hybrid nanoparticles. The special qualities of nanoparticles, such as their high surface area, surface activity, magnetic resonance relaxation, electronic sensitivity, etc., rely on their size and shape. Smart coatings are special because of remarkable advancements, and they are developing at an accelerated rate.

Large structures are a common sight in modern cities. The majority of these structures are totally glazed, which poses a maintenance issue. To address this issue and ease maintenance, a self-cleaning coating has been created. Similar to glazing, buildings' walls also need a lot of maintenance to prevent deterioration of the facades and the growth of dampness. Because of this, self-cleaning walls have been created, greatly reducing the care required for facades.

The use of self-cleaning surfaces makes certain items, like solar panels, much easier to maintain and, in certain situations, even more effective.

DOI: 10.1201/9781003364481-10

10.2   ENERGY STORAGE (HYDROGEN AND BATTERY)

Energy storage and conversion systems, including batteries, supercapacitors, fuel cells, solar cells, and photoelectrochemical water splitting, have played vital roles in the reduction of fossil fuel usage, addressing environmental issues, and the development of electric vehicles. The fabrication and surface/interface engineering of electrode materials with refined structures are indispensable for achieving optimal performances for the different energy-related devices. Atomic layer deposition (ALD) and molecular layer deposition (MLD) techniques, the gas-phase thin film deposition processes with self-limiting and saturated surface reactions, have emerged as powerful techniques for surface and interface engineering in energy-related devices due to their exceptional capability of precise thickness control, excellent uniformity and conformity, tunable composition, and relatively low deposition temperature. In the past few decades, ALD and MLD have been intensively studied for energy storage and conversion applications with remarkable progress.

Energy supplies that are environmentally friendly, sustainable, and safe for human consumption are critical and these provisions are heavily dependent on social, political, environmental, and economic challenges and fossil fuels. However, no single energy source can dominate and govern the global energy market. As a result, an energy-mix model depending on usable energy resources has gained worldwide attention. In the endeavor to replace fossil fuels as a source of energy, renewable energy resources emerge as a viable option. Hydrogen is the cleanest renewable energy source in the 21st century and it might pave a new way for energy supply in the future.

10.2.1   HYDROGEN STORAGE

Hydrogen storage is a key enabling technology for the advancement of hydrogen and fuel cell technologies in applications including stationary power, portable power, and transportation. Hydrogen has the highest energy per mass of any fuel. However, its low ambient temperature density results in a low energy per unit volume, therefore, requiring the development of advanced storage methods that have potential for higher energy density.

An alternative to conventional compressed and liquified hydrogen storage is materials-based storage. This technique uses materials solids or liquids that can absorb or react with hydrogen to bind it, due to their chemical attributes. Continuous and strong efforts on the improvement of hydrogen storage performances are necessary to meet the fast-growing demand for hydrogen storage. Meanwhile, the hydrogen storage gravimetric density should meet the requirement of 5.5 wt% from the US Department of Energy (DOE) by the year 2020.

Tremendous efforts are made to find novel materials for hydrogen storage. Material-based storage is broadly defined as hydrogen bound to solid materials, with its binding strength varying from physisorption to porous materials, such as zeolites and metal organic frameworks, to chemisorption in (complex) metal hydrides. Reversible adsorption energy and different improved electrochemical reactions are at the forefront of developments in this regard. In this regard, the acceleration of the development of new materials such as nanoporous materials, 2D nanostructures, and liquid organic hydrogen carriers (LOHC) incorporated with metal dopants as an enhancer or catalyzer for hydrogen storage is expected.

10.2.2 HYDROGEN PRODUCTION

From a variety of household sources, including biomass, fossil fuels, and water electrolysis with electricity, hydrogen can be created. The production method determines its energy efficiency and environmental impact. There is research into production methods that will lower the price of hydrogen production.

Gasification, electrolysis, liquid reforming, and fermentation are the current production method as shown in Figure 10.1. Natural gas, oil, coal, and electrolysis are the four main methods used to produce hydrogen for commercial purposes. These four methods account for 48%, 30%, 18%, and 4% of the hydrogen produced worldwide, respectively.

The primary source of industrial hydrogen is fossil fuels. Water splitting is the emerging production method of hydrogen. Water splitting using electricity and reforming are the two most common hydrogen production methods.

Electrolysis of water using electrolyzers is a common green hydrogen production method. These cells require a great deal of energy and involve the hydrogen evolution reaction (HER) on the cathode-side and the anode-side oxygen evolution reaction (OER). The hydrogen produced can later be oxidized to release energy and water. An alkaline electrolyzer is the most common type of water-hydrolysis cell, but polymer electrolyte membrane (PEM) and solid oxide electrolyzers (SOE) are also important. Optimizing these electrolyzers is the focus of much ongoing research.

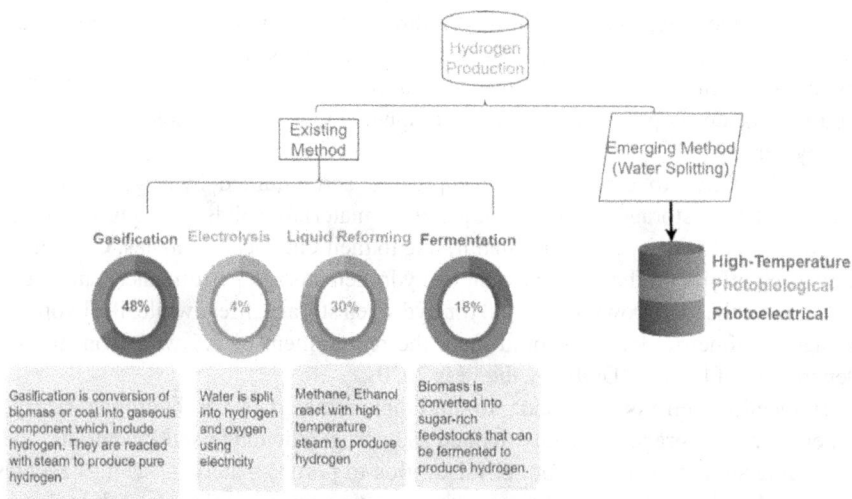

FIGURE 10.1   Different methods of hydrogen production.

Photocatalysts and electrocatalysts are currently in development to further improve electrolyzer efficiency.

Water splitting is divided into three, viz. high-temperature water splitting, photobiological water splitting, and photoelectrochemical water splitting. In high-temperature water splitting, chemical reactions break down water to produce hydrogen, with source of heat being nuclear reactor or solar. Semiconductors and solar energy produce hydrogen by breaking down water in a process called photoelectrochemical water splitting. Lastly, photobiological water splitting uses microbes and sunlight to break down water to produce hydrogen.

The shift to an energy system that is based on hydrogen will not happen overnight. The utilization of hydrogen is expected to be implemented in several applications that have been specifically targeted. The need for a new supply infrastructure could limit hydrogen use in countries adopting this strategy, especially in developing countries. Within the next 20 years, the development of new technologies relating to hydrogen, which can be viewed as a potential source of energy, without ignoring its application in the chemical industry, whose demand is expected to grow, will be of strategic importance and will present genuine technological challenges. The fact that hydrogen cannot be obtained for free from nature and must be generated in a way that requires energy expenditure means that the process of acquiring hydrogen needs to be as energy efficient as possible. Furthermore, due to its physicochemical characteristics, hydrogen presents difficulties in its storage and transport.

10.2.3 HYDROGEN DISTRIBUTION

Hydrogen is a delicate chemical that needs careful handling. Liquefied hydrogen tank and high-pressure tube trailers and pipelines are the major forms of distributing hydrogen to end users.

10.2.4 HYDROGEN USAGE

It is used in hydrocracking to produce lighter hydrogen fractions from heavier petroleum fractions. About 47% of global food is produced from fertilizers produced using hydrogen and synthetic nitrogen. It is also utilized in other processes including the aromatization, hydrodesulfurization, and Haber process, which produces ammonia. Fuel cells can generate power locally using hydrogen. It may also be utilized as a fuel for vehicles in the future. Electrolysis is used in industry to make chlorine, and as a byproduct, hydrogen is created. Hydrogen can be cooled, compressed, and refined for use in other operations on site or delivered to a customer through pipelines, cylinders, or trucks despite needing sophisticated technology.

10.2.5 BACKGROUND ON BATTERY

Battery plays crucial role in the durability and performance of equipment. Battery comes in different sizes and types depending on the equipment. Batteries are often designed for specific category of equipment. A battery is a chemical device that

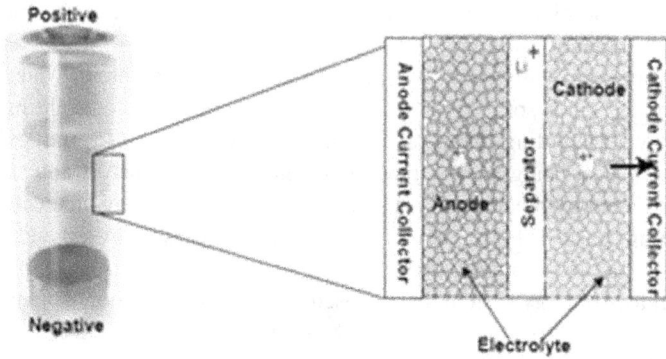

FIGURE 10.2 Schematic of battery.

stores electrical energy in the form of chemicals and uses electrochemical reactions as shown in Figure 10.2.

The anode equation is represented by equations 10.1 and 10.2 and the cathode equation is shown in equations 10.3 and 10.4.

Anode discharge:

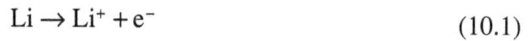

$$Li \rightarrow Li^+ + e^- \tag{10.1}$$

Anode charging:

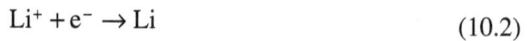

$$Li^+ + e^- \rightarrow Li \tag{10.2}$$

Cathode discharge:

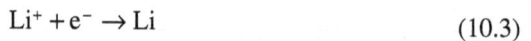

$$Li^+ + e^- \rightarrow Li \tag{10.3}$$

Cathode charging:

$$Li \rightarrow Li^+ + e^- \tag{10.4}$$

It converts the stored chemical energy into direct current (DC) electric energy. Alessandro Volta, an Italian Physicist, invented the first battery in 1800. The electrochemical reaction in a battery involves the transfer of electrons from one material to another (called electrodes) through an electric current.

Batteries are used in consumer electronics requiring Wh scales. It is used in electric vehicles requiring kWh scales of measurement. Also, it is used in storing energy generated by renewable energy sources such as solar and wind turbines. Different types of batteries exist. This is determined by the anode and cathode materials. The

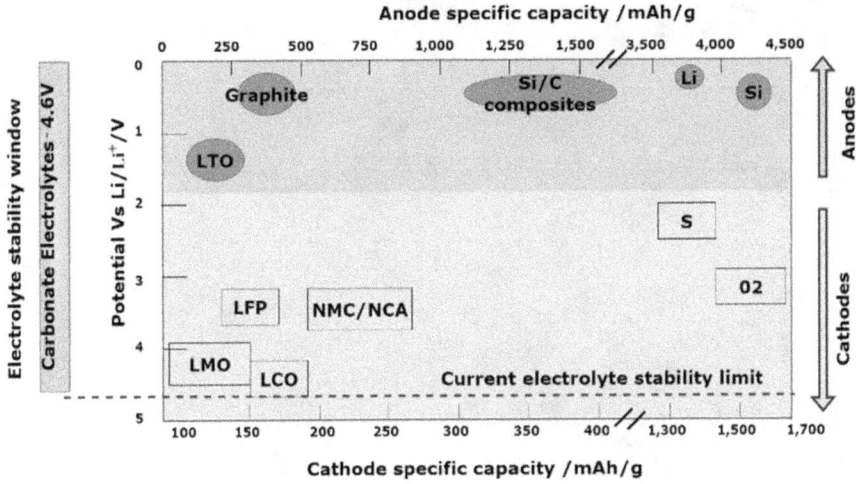

FIGURE 10.3 Chemistry of the battery.

anode is characterized by low voltage and high capacity. The cathode has high voltage and high capacity. They include nickel-cadmium, lead acid, and nickel-metal hydride. Others are lithium-ion, lithium-sulfur, lithium metal, and metal-air.

Figure 10.3 gives the chemistry of the battery with the current electrolyte stability limit and electrolyte stability window shown.

There are various techniques for manufacturing batteries depending on the type and shape of the battery. For example, lead-acid batteries are manufactured using several stages like oxide and grid production, pasting and curing, assembly, formation, filling, charge-discharge, final assembly, inspection, and dispatch. Solid-state batteries require different strategies for fabricating the solid electrolyte, which can be based on powder processing, thin-film deposition, or electrospinning. There are different anode and cathode materials. However, the most common anode material is graphite. There is current shift toward silicon and lithium-metal battery. However, lithium-ion battery is gaining interest. The most common cathode material is nickel manganese cobalt (NMC), but there is shift toward nickel-rich cathodes. The key performance metrics for battery include C-rate, capacity measured in Ah, and energy measured in Wh. Figure 10.4 represent the established and emerging battery technologies that currently exist.

10.2.6 BATTERY INDUSTRY STRUCTURE

The battery industry structure is an evolving one with a lot of intricacies. The structure comprises the raw materials, cells, applications, precursors, electrodes and their components.

The major raw materials of battery include aluminum, copper, manganese, nickel, cobalt and lithium, and graphite. The raw materials are influenced by supply chain, purity, and ethics. Graphite is the most common material used as anode material but gradually moving toward lithium-metal and silicon.

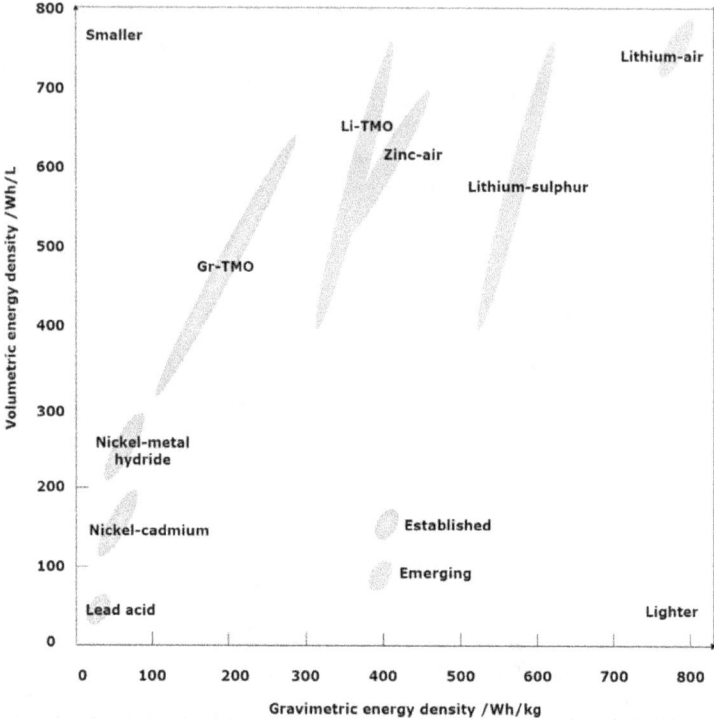

FIGURE 10.4 Schematic of established and emerging battery technologies.

The major precursors use in the battery industry include lithium carbonate and hydroxide, others are nickel and manganese sulfates, cobalt. The key determinant is the processing of the precursor and the crystal state (single and polycrystalline).

The cells of a battery are differentiated by the form factors, viz. cylindrical, prismatic, and pouch. Battery cells are classified based on the size and capacity. The size and capacity determine the applications and durability. Other pertinent issues related to battery cells include the electrolyte filling, formation, and useful life.

10.2.7 Battery Form Factor

Batteries have different form factors, namely cylindrical, prismatic, and pouch. Pouch cell uses low mechanical stability and high packing density. They are used in electronics such as drones. Prismatic cells have high mechanical stability, high packing density, and lower energy density than pouch. They are used in phones. Lastly, cylindrical cell has high mechanical stability with reduced cost. They are commonly available and used in power tools. They are usually around 18 mm in diameter and 65 mm long. The dimension of prismatic depends on the manufacturer and item to power.

Battery improvement is achieved by optimizing the chemistry of the battery, the microstructure, cell design and pack design, and the aesthetic in some applications. Figure 10.5 gives the graphical illustrations of the battery improvement factors.

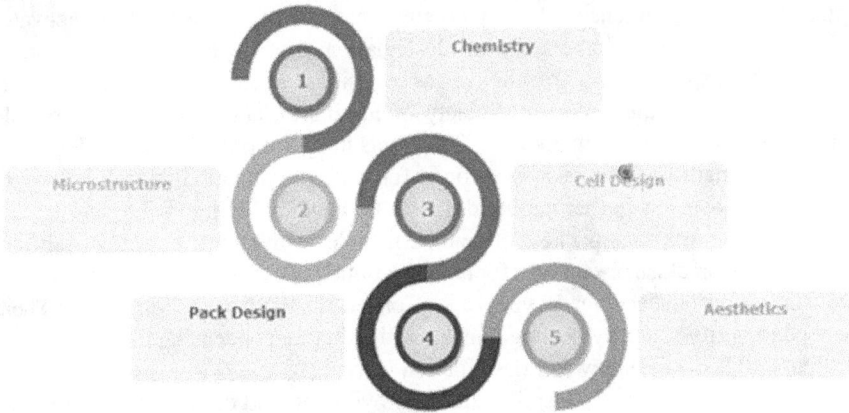

FIGURE 10.5 Factors to consider for improving battery.

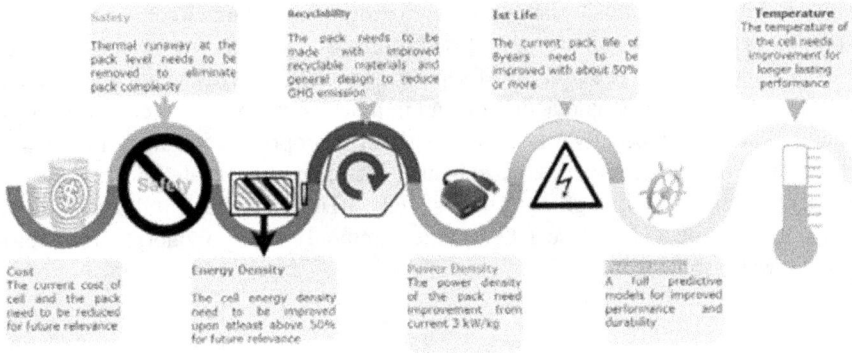

FIGURE 10.6 Gives the areas of possible improvement for future battery.

10.2.8 ANODE AND ELECTROLYTE MATERIAL

These include carbon-based material, alloy battery anode material, transition metal oxide, transition metal chalcogenides, transition metal oxalates, and transition metal carbides and nitrides (MXenes).

Figure 10.6 shows the key areas that need improvement for better performance of existing battery.

10.3 WATER DESALINATION

Water desalination is the process of making salt or ocean water clean and safe for drinking by removing the mineral and salt components. Desalination is also called desalting.

Desalination is an effective and efficient way of purifying water for man and animal consumption. However, the cost of installing water desalination plant is currently high and it is energy intensive. It also has environmental impact as hypersaline water is returned to the environment during desalination. The hypersaline water causes pollution thereby increasing greenhouse gas emissions and increasing fossil fuel reliance. According to the American Society of Engineers, about 300 million people depend on desalination, with about 20,000 plants in operation in 150 countries.

Different methods have been developed for water desalination, but all belong to either reverse osmosis or thermal desalination as shown in Figure 10.7.

Reverse osmosis uses packets of semipermeable membranes to remove salt and minerals from brackish or seawater for human, animal, and crop consumption. Reverse osmosis was introduced over 50 years ago. Conversely, thermal desalination and heat are used to purify water by heating to evaporation before condensing the water.

Figure 10.8 shows how water desalination works.

The use of solar energy to split water into hydrogen and oxygen in the presence of semiconductor photocatalysts has long been investigated as a viable method for producing clean, abundant hydrogen. High-temperature heat (500–2,000°C) is used in thermochemical water-splitting systems to power a series of chemical reactions that yield hydrogen. In this, solar is used as the source of heating. Each cycle's chemical input is recycled, resulting in a closed loop that uses only water and generates hydrogen and oxygen. The process of separating water into hydrogen and oxygen is known as electrolysis. This reaction occurs in a device known as an electrolyzer. In general, when a photocatalyst is altered with an appropriate cocatalyst, total water splitting can be accomplished. Consequently, it is crucial to create both cocatalysts and photocatalysts. Although amorphous TiO_2 ($aTiO_2$) has been utilized in some experiments, rutile and anatase TiO_2 are the commonly used polymorphs for photocatalytic water splitting.

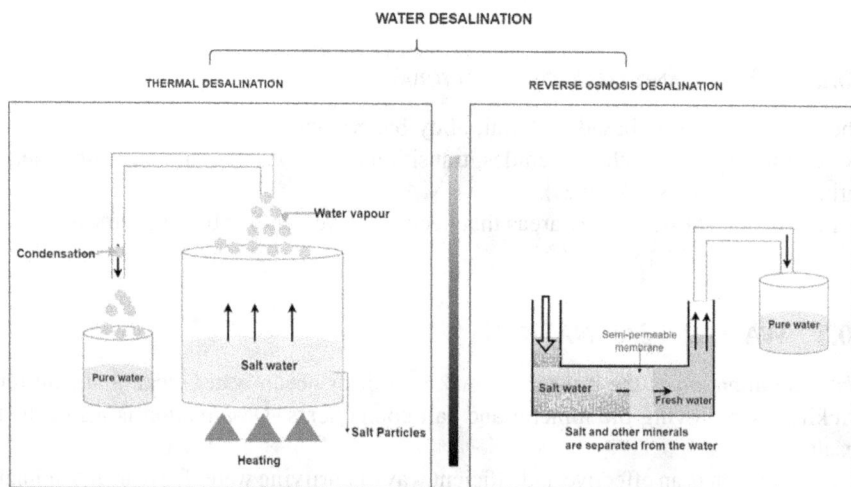

FIGURE 10.7 Schematic representation of water desalination methods.

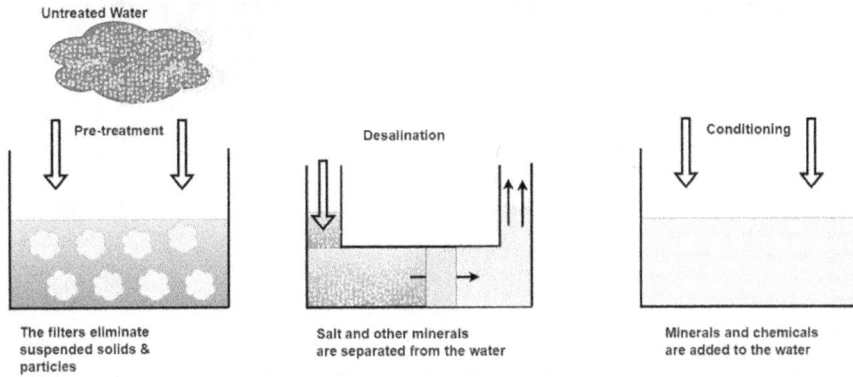

FIGURE 10.8 Schematic of water desalination process.

10.4 CONCLUSION

This chapter discussed water desalination, smart coatings, and energy storage. These are key emerging technologies that are shaping humanity and delivering the fourth industrial revolution. Water desalination is needed to reduce health and other hazards associated with untreated water. It is a process of purifying water for animal and human consumption. Energy storage using battery and hydrogen are key components used in various applications. Most renewable energy are not reliable, requiring a means to store the generated energy. Battery plays crucial role in the durability and performance of equipment. Battery comes in different sizes and types depending on the equipment. Batteries are often designed for specific category of equipment. The structure of battery comprises the raw materials, cells, applications, precursors, electrodes and their components. Natural gas, oil, coal, and electrolysis are the four main methods used to produce hydrogen for commercial purposes. These four methods account for 48%, 30%, 18%, and 4%, respectively, of the hydrogen produced worldwide. Water splitting using electricity and reforming are the two most common hydrogen production methods.

Index

Note: Page numbers in **bold** refer to tables, those in *italics* refer to figures.

For Product Safety Concerns and Information please contact our EU
representative GPSR@taylorandfrancis.com
Taylor & Francis Verlag GmbH, Kaufingerstraße 24, 80331 München, Germany

www.ingramcontent.com/pod-product-compliance
Lightning Source LLC
Chambersburg PA
CBHW060448240326
41598CB00088B/3955